# Linear Algebra

## An Interactive Laboratory Approach with *Mathematica*®

# Linear Algebra

## An Interactive Laboratory Approach with *Mathematica*®

## John R. Wicks

*North Park College*

**ADDISON-WESLEY PUBLISHING COMPANY**
Reading, Massachusetts • Menlo Park, California • New York
Don Mills, Ontario • Wokingham, England • Amsterdam • Bonn
Sydney • Singapore • Tokyo • Madrid • San Juan • Milan • Paris

| | |
|---|---|
| *Sponsoring Editor* | Marianne Lepp |
| *Managing Editor* | Karen Guardino |
| *Senior Production Coordinator* | Kim Ellwood |
| *Text Design* | John R. Wicks |
| *Manufacturing Supervisor* | Roy Logan |
| *Cover Design* | Meredith Nightingale |

Cover images © 1995 PhotoDisk, Inc.

Mathematica is a registered trademark of Wolfram Research, Inc.

Reproduced by Addison-Wesley from camera-ready copy supplied by the author.

ISBN    0-201-82642-9

1 2 3 4 5 6 7 8 9 10 VG 99989796

# Contents

To the Student     vii

To the Instructor     xiii

Acknowledgments     xxiii


## Part 1    Introductory Material     1

1   Introduction to *Mathematica*     3

2   Introduction to Linear Systems and Row Operations     15

3   Linear Systems and Applications     25


## Part 2    Linear Systems and Gaussian Elimination     33

4   Gaussian Elimination by Example     35

5   A Summary of Gaussian Elimination     41

6   Backsubstitution and Backaddition     45

7   Gaussian Elimination with *Mathematica*     49

8   The General Solution to Linear Systems     59

# Part 3    Matrix Algebra                                             63

9   Matrix Multiplication from a Geometric Viewpoint                   65

10  Matrix Arithmetic                                                  77

11  Matrix Arithmetic with *Mathematica*                              87

12  Properties of Matrix Algebra                                       93

13  Matrix Algebra and Block Matrices                                 105

14  Matrix Inverses: Definitions and Basic Properties                113

15  Computing Matrix Inverses                                         121

16  Matrix Inverses and Systems of Equations                         131

# Part 4    Systems from Two Advanced Viewpoints                     139

17  Elementary Matrices and the $P^T LU$ Decomposition              141

18  Applications of the $P^T LU$ Decomposition                      155

19  Discovering Determinants                                          165

20  Systems of Equations from a Geometric Viewpoint                  183

# Part 5    Abstract Vector Spaces and Bases                         197

21  Vectors and Vector Spaces                                         199

22  Coordinate Systems and Bases                                      209

23  Independent and Spanning Sets                                     227

24  Constructing Bases                                                245

25  The Theory of Bases                                    255

26  Subspaces and Linear Transformations                   271


**Part 6    Inner Product Spaces**                          **287**

27  Inner Products                                          289

28  Orthonormal Bases and Projections                       297

29  Gram-Schmidt Orthonormalization                         311


**Part 7    Change of Basis and Eigensystems**              **327**

30  Linear Transformations and Matrices                     329

31  The Effects of Changing Coordinates                     339

32  Discrete Dynamical Systems and Eigensystems             351

33  Eigenvectors and Eigenvalues                            357

Appendix of *Mathematica* Commands                          A-1

Index                                                       I-1

# To the Student

This text is probably quite different from any you have used before. Its format, content, and goals are unique, so we will take a minute for a quick overview of the text, to prepare you for what to expect. Although it may seem unusual at first, the entire text revolves around putting *you* in the "driver's seat." By using the computer, you will be able to take charge of your own learning process.

## Format of the Text

This text is unusual in two main respects. First, it is divided into a sequence of lessons, instead of chapters, which are designed to serve as interactive lectures on a particular topic of linear algebra. Thus they tend to be shorter than standard chapters, but longer than traditional sections of a text. They are designed to cover a single concept or a small number of closely related concepts in linear algebra, which are highlighted at the beginning and end of each lesson. Moreover, each lesson is intended to be read in an interactive manner, in that there are questions, experiments, and exercises interspersed throughout the text, which should be addressed *while one is reading the text*.

This leads to the second main distinction, in that various parts of the text actually "live" on a computer in what are called Notebooks, which are read by a program called *Mathematica*. For the most part, when you are reading a Notebook, it will seem that you are simply reading text off a computer screen. However, interspersed in the text are computer instructions, because *Mathematica* is actually a powerful computer program that can perform any routine algebraic manipulation as well as produce nice two- and three-dimensional graphics. This means that as you read about a certain concept, you will be able to hit a button and see a picture of the situation or an example worked out for you automatically! Moreover, you will be able to type in your *own* commands to work through examples independently.

We will use *Mathematica*, because Notebooks will allow us to exploit the power of this program without learning a lot of complicated commands beforehand. In fact, we will try to keep the number of commands that we use to a minimum and will only introduce new commands *as we need to use them*. Because most of the commands we will need are either very similar to standard mathematical notation or understandable English phrases, you will find them relatively easy to remember. Thus, even if you are not very experienced with a computer, you will quickly and painlessly be able to reap the rewards of the computer age.

Our goal is not to use the computer simply because it makes studying linear algebra more fun (although it does). Fundamentally, we want to become *independent* learners, because the vast majority of our education will take place after we leave school! This means that we need to develop the skills it takes to continue our education on our own. The computer can aid us in this process by helping us to:

- work through examples while we are learning a new concept,

- visualize new ideas through computer-generated two- and three-dimensional pictures and "movies",

- perform routine calculations, so we can focus on the main ideas, and

- monitor our own progress by allowing us to check our work.

Because our primary goal is to learn linear algebra and become independent learners in the process, much of this text is *not* on the computer. There are many ideas that can be explored simply with pencil and paper, and the computer would simply be an unnecessary distraction. Thus, some lessons in the text only appear in traditional printed form, while others are duplicated as *Mathematica* Notebooks. As a third option, some lessons are *accompanied* by a Notebook, to allow you to perform "scratch-work" easily as you follow along in the text. Although having three different formats for a text is unusual, this approach is designed to achieve the best possible learning environment.

Every lesson appears in printed form in this text, but because a good deal of the material also comes on computer disk, we have employed different devices to tell you about the ways the text and the computer portions work together. If a lesson comes in Notebook form, it will say "*Mathematica* Notebook" just below the title of the lesson and at the top of each subsequent page. The corresponding Notebook will be called "Less#.ma." For example, Lesson 1 comes with a Notebook called "Less1.ma." If a lesson only has an accompanying Notebook, it will say "Accompanied by Notebook" instead. The corresponding Notebook will be called "Less#a.ma." For example, Lesson 14.1 comes with a Notebook called "Less14.1a.ma." In any case, at each point at which you should employ the computer, the instructions in the text will contain a "bullet" (•) in the margin to remind you to use the computer.

No matter what format the text takes, the common denominator throughout the text is its dependence on *your* input. The text is full of questions, experiments, and exercises that you should respond to as you read. Each question will lead you further along into the subject, each experiment will explore some linear algebra concept, and each exercise will test your understanding while increasing your ability and confidence in your own understanding. The more you put into learning, the more you will get out.

You will notice that there are three types of labeled exercises. Besides the standard exercises, some exercises are marked as either "Drill Exercise" or "Advanced Exercise," and are indented slightly from the left margin. This is to indicate that these exercises are not as crucial to the main exposition and may be skipped on a first reading. Drill exercises are designed to provide extra practice in the standard algorithms of linear algebra, while advanced exercises are more challenging.

This text is designed to be read by small groups of two or more students, who have a computer easily available. It is designed as a conversation between you and its author that will expand to create discussion with your group members and your instructor.

Your group members will help you clarify your understanding, as you try to explain to each other what you are learning. By putting your heads together, you will be able to discover the truths and techniques of linear algebra *on your own*, as you are guided by the text. In a standard, Socratic lecture format, where the instructor asks leading questions, at most one student is required to think and respond at a time. In contrast, you will *always* need to be observing, thinking, and responding to what is being presented. Although this will take more work than in a traditional approach to linear algebra, you will find that you will learn a great deal more and have a lot more fun in the process!

## Content of the Text

This is a text on introductory linear algebra. Along with calculus, linear algebra is one of the most widely used subjects in mathematics. Much of mathematics revolves around solving equations, and when those equations are "linear," the language, concepts, and techniques of linear algebra can be applied. In this text, we will start by examining this problem and then use it as a theme to guide our discussion. This is only an introductory text, so we will limit our discussions to the basic concepts of linear algebra. These include:

- the two algorithms for solving linear systems, known as Gaussian elimination and Gram-Schmidt orthonormalization,

- the notions of vectors and vector spaces, matrices, and linear transformations,

- matrix algebra, rank, inverses and determinants, and

- subspaces, linear combinations, and bases.

We will be able to motivate all of these ideas by looking at the problem of solving systems of simultaneous, linear equations. Near the end of the text, we will also look at the second major application of linear algebra known as the "eigenvalue problem."

There are a number of themes that pervade the subject of linear algebra and help to unify the various topics that we will cover. For example, although matrix multiplication can be taught in high school algebra classes, we will learn to view multiplication from a number of different perspectives, as well as learn why it is defined in such an unusual way. For example, we will look at matrix multiplication geometrically as stretching, rotating, or collapsing space. We will also treat it from a purely algebraic viewpoint, so that we can solve systems of equations. Ultimately, we will see that it is best understood as representing composition of linear transformations.

Another important theme that has its roots in solving equations is that of the "existence" and "uniqueness" of solutions. A given problem may or may not have any solution; if there is one, there may be *only one* solution or many solutions. The thread of uniqueness will tie together the concepts of linear independence, left-invertibility, and full column rank. Similarly, the existence of solutions will relate the notions of a spanning set, right-invertibility, and full row rank. Although you do not understand these terms yet, you should be watching for this theme as it is developed in the text.

The most important theme in linear algebra is that of "changing our point of view." Just as matrix multiplication can be viewed in a number of different ways, almost every concept in linear algebra can be viewed in many different ways. We will discuss the concept of a basis that is crucial to converting a given problem into a *solvable* one. In general, we will learn a rich language for describing the same problem from a variety of perspectives. You will need to be very flexible, focus on learning the language, and practice translating a given problem between the various possible descriptions. Often the solution to a particular problem comes easily, once you have found the appropriate description of the problem.

In fact, it would be a great mistake to confuse any particular application with the subject of linear algebra as a whole. The profound usefulness of linear algebra comes from the fact that a great diversity of applications can be translated into a common language of linear algebra. Instead of focusing on any particular aspect or application, you should consider linear algebra as a particular *type of mathematical language* and *way of thinking*.

If it were *only* a language, linear algebra would not be so valuable. It is particularly useful for at least three more reasons. First, once you have framed a problem in the language of linear algebra, in most cases there are clear, easy-to-follow procedures that will definitely lead to a solution. Many times those procedures have already been programmed into a computer, so that you can reach an answer very quickly. Moreover, often the computations involved look and can be treated very much like high school algebra (even though the computations involve quantities that are more than simple numbers). *No matter what the application*, the *same* language and techniques can be applied, so you will receive a great return on your investment in studying linear algebra.

Finally, linear algebra is a beautiful blend of algebra and geometry. Most concepts in linear algebra can be understood from a geometric point of view. The truths of linear algebra will often seem obvious when viewed geometrically, because they will correspond to our own geometric intuition. Throughout the text, we will use the computer to produce pictures and "movies" to provide you with a concrete basis for understanding the various concepts.

## Goals of the Text

Although the subject of this text is linear algebra, our goals go far beyond simply learning how to get answers to any particular collection of problems. Because linear algebra can be used in such a wide variety of applications, it is not worthwhile to spend too much time on any particular one. An old proverb says: "Give a man a fish and he eats for a day; teach a man to fish and he eats for a lifetime." This text is designed to "teach you to fish" by teaching you to use the language of linear algebra, as well as the fundamental concepts and techniques that can be applied to many different settings.

As we said in the first section, this text is designed to encourage you to learn independently. By generating a large number of your own examples, you will learn by experimentation and observation, on your own as much as possible. Instead of simply memorizing and drilling, you will discover the definitions and algorithms of linear algebra. Instead of depending on some expert to tell you what to think, you will learn to make and test your own conjectures about linear algebra to determine what is true.

Beyond simply helping you to learn linear algebra, this text is designed to teach you *how to learn* mathematics, in general. Although you may have been able to get along up to this point by only reading the examples in a math text and skipping all of those messy proofs, that approach will not work much longer. The ability to read, understand, and learn from "formal" presentations of mathematics is an important skill to develop for anyone involved in a technical discipline that uses mathematics. As you advance within your discipline, you will find it necessary to learn new notation and apply mathematical facts to various situations. Those facts will be expressed as mathematical definitions and theorems, using precise mathematical language. Thus, in order to become independent learners, you will need to learn to use this language to a certain degree.

This text is written for a one-semester, introductory course, designed for a general audience of second-year college students that may include students of physics, chemistry, engineering, economics, biology, and so on. Its goal is not to produce a class full of math majors, but simply to increase your confidence in your ability to read and write mathematics. This may not be easy, and not everyone will reach the same level of proficiency, but we will start slowly and everyone's ability to make sense of mathematical definitions, theorems, and proofs will improve. As you work your way through the text, you will continue to practice translating mathematical language into English and vice versa. As a reward for your patient effort, your ability to learn mathematics will increase significantly.

# To the Instructor

This "text" is probably quite different from any you have used before. It is a complete set of interactive lectures on linear algebra. Each lesson is a self-contained lecture that can be covered in one or two class periods. The "text" consistes of a bound, paper text and a collection of *Mathematica* Notebooks and packages on computer disk. The Notebooks and packages have been extensively tested using *Mathematica* Version 2.2, but should support with Versions 2.0 and 2.1, as well.

In response to class testing and students' suggestions, this text has a unique organization. It consists of a sequence of lessons each of which uses the computer to a greater or lesser degree. On the one hand, the format of a *Mathematica* Notebook is well-suited for creating guided discovery experiences, where students can focus on the linear algebra and be directed through longer, more complex calculations (such as the $P^T LU$ decomposition). On the other hand, long prose passages that require a good deal of typeset mathematics and visual formatting of text become difficult to read on the computer. Thus, each lesson may be completely replicated within an associated *Mathematica* Notebook, or it may simply have an accompanying Notebook to aid in performing certain calculations, or it may not come with a Notebook at all. This also serves to emphasize the fact that *Mathematica* is simply a tool to be used as it is appropriate and not our primary focus of study. This is a text about linear algebra that happens to find a variety of uses for *Mathematica* in the course of the educational process.

We believe that students should be allowed to use the full power of *Mathematica* symbolic computation and graphical capabilities without spending a lot of time learning *Mathematica* or inputting long, complex instructions, or a number of commands.

The text includes a number of *Mathematica* packages of preprogrammed commands. These are used to perform a variety of tasks. Some commands simply carry out algorithms that the student has already mastered; students can use them to either speed up their progress or to verify their own calculations. Other commands provide more complete diagnostic information. These can give students the means to find potential errors or give "independent" corroboration to their results (by performing the calculation in an unknown manner). Some commands enable students to dynamically construct complex figures that would be tedious or even impossible for students to visualize or construct on their own.

Since students will use *Mathematica* throughout their careers, they are notified which commands are built-in *Mathematica* commands and which are provided in an ancillary package. In this way, they will understand which commands will be available in their future work and which are only of pedagogical value for use with this text. Although they are instructed on how to load each package manually, any preprogrammed command provided for use with this text will be loaded *automatically*.

# Philosophy

### Constructing Knowledge
The primary goal of this text is to provide a concrete foundation for each concept of linear algebra through first-hand experimentation, observation, and discussion. We include hand calculation (either arithmetic or algebraic), machine computation, along with two- and three-dimensional visualization. To as great a degree as possible, the text tries to exploit the students' pre-existing intuition to create a context for new ideas. After a mental "hook" has been created on which to hang a new idea, we present the topic in a more traditional, formal manner.

### Applicability *vs.* Applications
We have chosen to emphasize "hands-on" activities and experiments to involve students directly in the process of learning key ideas of linear algebra, without disrupting the flow of the presentation with applications that themselves require a learning curve.

Although the material in the text is not motivated primarily by "realistic" applications, each definition and theorem *is* motivated by how it leads to greater understanding and to its place in the conceptual framework that is linear algebra. Thus, this text takes great pains to emphasize the *applicability* of linear algebra, without spending too much time on particular applications in the main text.

Instead, the text contains an independent group-study project as an integral part of the course. This project guides a group of students in researching an application of its choice in depth. In this way the instructor can address both the relevancy issue and the problems associated with presenting a complicated electrical engineering example to a class of chemistry and computer science majors. We further believe that a single, detailed application that one chooses and pursues on one's own is as valuable as a number of prepackaged applications from the text.

A selection of applications has also been provided in ancillary Notebooks for use as outside reading or as a starting point for independent study.

### How to Use the Computer
*As a Calculator*     The computer is employed in two distinct ways by this text. It is used in the traditional manner, as a rapid "calculator," freeing the student from routine calculation to focus on the deeper conceptual issues of any given situation and allowing the student to carry out numerical, algebraic, and geometric experiments that would be overwhelming if done by hand. This is surely one way our students will use technology in their future, professional careers. This text encourages such practice.

We believe students should understand how the computer arrives at its results. However, to require students to have a "complete" understanding of the computer's algorithms beforehand would be like requiring someone to complete a course in auto mechanics before driving a car; even though it is advisable to have a good working understanding of how one's tools are designed, such understanding often comes *after* hands-on experience with those tools.

*As a Master*      In contrast, students are also encouraged to learn by observing the computer's responses, much as a novice tennis player would learn from watching an expert player or much as a scientist forms conjectures after taking experimental measurements.

For example, in Lesson 16, students are encouraged to use the Det[] function to "measure" invertibility before they know how to compute determinants by hand. By watching the computer carry out its calculations, students are led to form conjectures about linear algebra before they understand how to carry out the calculations for themselves. The computer can, thus, provide an immediacy and specificity to what would otherwise be an abstract (and messy) concept. This prepares the intellectual ground for a more rigorous presentation.

## How to Use the Text

Think of this text as an antidote to the lecture system. We are all well aware that learning proceeds at its own pace, sometimes when students are passively absorbing, sometimes when they are actively assimilating. This text carefully intersperses presentation with activities and questions, so that students begin to assimilate one idea before moving on to the next.

The exercises are designed to engage the students in actively following and understanding the material by requiring them to form conjectures, carry out experiments, make observations, form generalizations, or perform calculations. Most important results are "discovered" by the students before being presented or summarized in the text. Students are also encouraged to verify results with a specific example.

You will notice that there are three types of labeled exercises. Besides the standard exercises, some exercises are marked as either "Drill Exercise" or "Advanced Exercise," and are indented slightly from the left margin. This is to indicate that these exercises are not as crucial to the main exposition and may be skipped on a first reading. Drill exercises are designed to provide extra practice in the standard algorithms of linear algebra. Depending on the capabilities of the students, an instructor may choose to either curtail or extend such exercises. Likewise, the advanced exercises are more challenging, usually requiring a greater facility with general proofs and abstract reasoning.

The text is designed for use in a *small group* setting, stimulating experimentation and observation, and discussion. It has been class-tested in a computerized classroom, where students work independently in groups of two to four.

These materials could also be used with an overhead projector and classroom computer to provide an interactive lecture/discussion. For example, one student could be given the responsibility of operating the computer, with the support and suggestions of others in the class. The instructor could then direct the students through the presentation as given in the text, so as to focus attention, solicit participation, and summarize results. The role of the professor becomes that of facilitator instead of the source of all knowledge.

In addition, the text could be used to guide your students (or yourself) through linear algebra material prerequisite for another course, such as differential equations.

Although every lesson appears in printed form, we will use a number of different devices to indicate in which of the three ways we will use the computer for that lesson. If a lesson comes entirely in Notebook form, it will say "*Mathematica* Notebook" just below the title of the lesson and at the top of each subsequent page. The corresponding Notebook will be called "Less#.ma." For example, Lesson 1 comes with a Notebook called "Less1.ma." If a lesson only has an accompanying Notebook, it will say "Accompanied by Notebook" instead. The corresponding Notebook will be called "Less#a.ma." For example, Lesson 14 comes with a Notebook called "Less14a.ma." In any case, at each point that the computer may be used the instructions in the text will contain a "bullet" (•) in the margin to indicate that fact.

### Where to Use the Text

This is a complete text that can be used to teach or learn linear algebra. It can also be used for self-instruction for a later course that assumes knowledge of linear algebra. The text will serve to help bridge the gap between a computational course, such as calculus or differential equations, and more formal mathematics courses. The text encourages students to learn and use the formal language of mathematics. Through the exercises, it also requires students to give informal reasoning for their answers, fill in gaps in formal proofs, or give proofs that are similar to others that have already been given.

No matter which style of teaching is employed, the text tries to instill a sense of mathematical rigor throughout the course. However, it attempts to motivate definitions and theorem, and emphasize an appropriate level of abstraction by providing various types of explanations, such as:

- verification by well-chosen example,

- verification by picture,

- proof by informal reasoning, and

- formal mathematical proof.

Moreover, with relation to formal mathematics, it does more than simply present a technically correct proof. It also helps the student to read proofs, by commenting on the logical structure of the arguments used.

## Annotated Outline

Here is an annotated outline of the text. For each lesson, we provide a brief description that may discuss the content, the teaching style employed, the connections with other lessons, or the specific pedagogical objectives for that lesson. The lesson title followed by [*Mathematica* Notebook] indicates that the lesson *resides entirely* in a Notebook, while the title followed by [Accompanied by Notebook] implies that there is a *parallel* Notebook for "scratch-work" or graphical demonstrations.

**Part One**:    Introductory Material

Lesson 1:    Introduction to *Mathematica* [*Mathematica* Notebook]
This lesson is intended to bring a complete novice "up to speed" in *Mathematica*. It is not intended to be a complete description of *Mathematica*, but a gentle introduction, leaving additional material to be presented in context on a need-to-know basis.

Lesson 2:    Introduction to Linear Systems and Row Operations [Accompanied by Notebook]
This lesson introduces the notion of a linear system of equations and demonstrates why linear systems are so much "nicer" than nonlinear systems.

Lesson 3:    Linear Systems and Applications
This lesson shows how linear systems can arise in a number of practical applications. Specifically, it uses atomic balance in chemistry, simple resistive circuits from physics, equilibrium temperature distribution from thermodynamics, and a simple Markov process from economics.

**Part Two**:    Linear Systems and Gaussian Elimination

Lesson 4:    Gaussian Elimination by Example
This lesson guides students to discover the process of Gaussian elimination by examining a well-chosen example. Even those students who are somewhat familiar with the process from a previous algebra course will be encouraged to think more deeply about how the algorithm is designed and why it is designed as it is.

Lesson 5:    A Summary of Gaussian Elimination
This lesson highlights and solidifies the observations that students should have made in the previous lesson. It also serves as a reference for students on the algorithm and associated definitions.

Lesson 6:    Backsubstitution and Backaddition
This lesson follows the spirit of the previous two lessons to introduce the notions of a *reduced* row echelon matrix and the backaddition algorithm.

Lesson 7:    Gaussian Elimination with *Mathematica* [*Mathematica* Notebook]
This Notebook directs students in using *Mathematica* to carry out the process of Gaussian elimination. To this point, students are free to focus on the *process* of elimination, without getting distracted by the mechanics of actual hand computations. It also teaches them to create and solve their own exercises.

Lesson 8:    The General Solution to Linear Systems
This lesson emphasizes the three qualitative types of solutions to a linear system that students have discovered geometrically in Lesson 2. It also introduces the formal terminology and notation associated with such solutions.

**Part Three:**  Matrix Algebra

Lesson 9:  Matrix Multiplication from a Geometric Viewpoint [*Mathematica* Notebook]
This lesson encourages students to view matrices as more than algebraic objects.
By having *Mathematica* perform matrix multiplication *before* students learn how
to do this by hand, students are free to focus on its geometric significance. This
lesson begins to introduce the correspondence between the matrix of a transformation and its action on a basis.

Lesson 10: Matrix Arithmetic
Because many students have some intuitive understanding of matrix arithmetic,
this lesson is designed to force students to think about this material more critically.
It also begins to introduce them to formal notation and definitions.

Lesson 11: Matrix Arithmetic with *Mathematica* [*Mathematica* Notebook]
This lesson is a very quick introduction to matrices in *Mathematica*. It also shows
how systems of linear equations can be viewed in terms of matrix multiplication,
as well as introducing the notion of rank.

Lesson 12: Properties of Matrix Algebra
This lesson directs students to discover and/or check many of the basic, formal
properties of matrix algebra. It also encourages them to begin reading and writing
simple proofs.

Lesson 13: Matrix Algebra and Block Matrices
This lesson introduces the notion of block matrices. It also introduces the concept
of linear combination and relates this to matrix multiplication.

Lesson 14: Matrix Inverses: Definitions and Basic Properties [Accompanied by Notebook]
This lesson introduces the notions of left-, right-, and two-sided matrix inverses.
It also provides more practice in reading and creating simple proofs.

Lesson 15: Computing Matrix Inverses [Accompanied by Notebook]
This lesson motivates the standard algorithms for computing inverses, by using
block matrices. It also encourages students to begin thinking about necessary and
sufficient conditions for the existence and uniqueness of inverses.

Lesson 16: Matrix Inverses and Systems of Equations [Accompanied by Notebook]
This lesson continues along the vein of the previous one, relating matrix inverses
to rank. It provides students with more practice in reading and creating simple
proofs. Many of the proofs employ block matrices to emphasize this important
technique. It also introduces the notion of the determinant function.

**Part Four:**   Systems from Two Advanced Viewpoints

Lesson 17: Elementary Matrices and the $P^T LU$ Decomposition [*Mathematica* Notebook]
This lesson leads students to discover the $P^T LU$ decomposition of a matrix for
themselves. It also illustrates how one can produce a general $P^T LU$ decomposition by hand.

Lesson 18: Applications of the $P^T LU$ Decomposition
This lesson reinforces the previous one by giving a number of applications of the decomposition. These applications are more theoretical in nature but still serve to impress upon students the importance of this result. In the process, students are introduced to the concept of an eigenvalue.

Lesson 19: Discovering Determinants [*Mathematica* Notebook]
This lesson serves as an exercise in informal mathematics, as students are led to discover the important properties of the determinant. It also gives students more practice in simple proofs and mathematical notation.

Lesson 20: Systems of Equations from a Geometric Viewpoint [Accompanied by Notebook]
This lesson furthers the goals of Lessons 15 and 16 by helping students to think about the process of solving systems of equations geometrically. It also introduces the concepts of linear transformation, nullspace, and "image" of a transformation.

**Part Five**: Abstract Vector Spaces and Bases

Lesson 21: Vectors and Vector Spaces
This lesson introduces the formal notion of vector space in the context of three concrete examples: geometric vectors, column vectors, and polynomials. Students are directed to check the vector space axioms in these examples and are encouraged to practice using formal mathematical terminology.

Lesson 22: Coordinate Systems and Bases [*Mathematica* Notebook]
This lesson introduces the concepts of linear independence and spanning set in a concrete geometric setting. It then translates the students' intuitive notions into formal, mathematical definitions, and teaches students how to use the associated terminology.

Lesson 23: Independent and Spanning Sets [Accompanied by Notebook]
This lesson shows students how to perform concrete calculations in *Mathematica*, so that the abstract concepts from the previous lesson can "take root" in their minds. It also instructs students in creating and verifying their own exercises.

Lesson 24: Constructing Bases [*Mathematica* Notebook]
This lesson carries on from the previous one, showing students how to construct bases in a variety of settings.

Lesson 25: The Theory of Bases [Accompanied by Notebook]
This lesson builds on the ideas from Lesson 22 to show how a basis defines coordinates. It also introduces the idea of isomorphism between vector spaces and proves a number of standard results relating to bases and dimension.

Lesson 26: Subspaces and Linear Transformations
This lesson introduces the notions of subspace and linear transformation from a formal point of view, while providing concrete examples. It also serves to provide more practice in reading and producing simple proofs.

**Part Six:**    Inner Product Spaces

Lesson 27:  Inner Products [Accompanied by Notebook]
This lesson introduces the concept of an inner product and its basic properties. Although it focuses primarily on the standard, Euclidean inner product, students are encouraged to move beyond this to more general examples.

Lesson 28:  Orthonormal Bases and Projections [Accompanied by Notebook]
This lesson introduces the idea of orthonormal basis and shows how it can be used to solve overdetermined systems of equations.

Lesson 29:  Gram-Schmidt Orthonormalization [Accompanied by Notebook]
This lesson instructs students in the hand computation of the $QR$ decomposition of a matrix, which corresponds to Gram-Schmidt Orthonormalization. This lesson presents a novel approach that allows a majority of students to successfully master this difficult algorithm.

**Part Seven:**  Change of Basis and Eigensystems

Lesson 30:  Linear Transformations and Matrices
This lesson formalizes the observations that students have made in Lesson 9 on the matrix of a transformation. It also demonstrates the relationship between matrix multiplication and composition of transformations.

Lesson 31:  The Effects of Changing Coordinates [Accompanied by Notebook]
This lesson helps students to think through the effect of a change of bases on the coordinates of a vector or on the matrix of a transformation. It suggests a pictorial way of deriving the appropriate change-of-basis matrices and provides an intuitive framework for this process.

Lesson 32:  Discrete Dynamical Systems and Eigensystems [*Mathematica* Notebook]
This lesson directs students to discover the significance of eigenvectors and eigenvalues in the context of dynamical systems. It puts off the theoretical foundation of eigensystems until the following lesson.

Lesson 33:  Eigenvectors and Eigenvalues [Accompanied by Notebook]
This lesson gives formal definitions of eigenvectors and eigenvalues, relates eigensystems to a matrix decomposition, discusses the traditional approach to computing eigensystems, and provides the theoretical underpinnings for the students' observations from the previous lesson.

*Note*: We also provide an appendix describing all *Mathematica* commands used in this text.

## Outline by Teaching Style

While always emphasizing active student involvement, this text employs a wide variety of teaching styles. Some lessons emphasize visualization, others form a guided sequence of "proofs by experiment," several focus on particular algorithms, while many follow a more traditional, formal presentation. Only a few lessons are simply training in how to use *Mathematica*. In the following outline, we try to classify each lesson by the dominant style used in that lesson.

**Employing Visualization:**
Lessons 2, 9, 20, and 22 all employ *Mathematica* to dynamically create pictures, thus using students' geometric intuition as a starting point for each lesson. Lessons 13, 18, 21, 25, 26, 27, 28, and 29 also use visualization to a lesser degree.

**Guiding Investigations:**
Although every lesson tries to actively involve the student in the development of the material, Lessons 4, 6, 17, 19, 29, and 32 all depend quite heavily on student input. They are designed so that students will discover the concepts, theorems, algorithms, and proofs of linear algebra as independently as possible, by following instructive examples or other types of guided investigations.

**Learning Algorithms:**
Lessons 4, 5, 6, 10, 15, 17, 23, 24, and 29 are all primarily concerned with training students to perform certain standard algorithms. Lessons 19, 23, 24, 28, 31, and 33 also discuss algorithms in a more incidental fashion.

**Presenting Definitions, Theorems, and Proofs:**
Lessons 8, 12, 13, 14, 16, 18, 19, 21, 25, 26, 27, 28, 30, 31, and 33 rely heavily on a formal "definition-theorem-proof" style of presentation. Although it motivates the concepts geometrically, Lessons 22 also focuses on the formal definitions.

**Learning *Mathematica*:**
Lessons 1, 7, and 11 primarily teach necessary *Mathematica* commands. Lessons 17, 29, and 33 also present a number of useful *Mathematica* commands.

# Ancillaries

In order to keep the length down, a number of specific applications have been left out of the printed text. They are included solely as *Mathematica* Notebooks in the "Extras" folder. Some topics include:

- least-squares fitting of data,

- singular values and pseudo-inverses,

- linear differential equations, and

- the $QR$ algorithm.

Additional topics, including student-generated Notebooks, will be available via the internet from Addison-Wesley. In the instructor's edition, there are also complete, written solutions to all of the exercises, in the form of *Mathematica* Notebooks. A sample syllabus and other material are also provided, in PostScript format.

# Acknowledgments

I would like to thank all of the people who made this project possible. To my good friend Jon Peterson for inspiring me to write this text. To the organizers and participants of the 1991 NSF Workshop on Linear Algebra and Differential Equations at the University of New Hampshire for helping me to start on this project. Thanks also to David Lay and the other participants at the 1993 Rocky Mountain Mathematics Consortium for their encouragement to pursue publication of this work. I am indebted to Wolfram Research, Inc. for their fine software product, as well as for all of their technical support. Thanks go to my colleague, Alice Iverson, and the Dean of North Park College for their support through the years of development of this text. I also wish to thank the many reviewers, especially Alan Shuchat and Lynne Bauer for their helpful comments. I cannot thank Walter Carlip enough for all of his TeX-pertise. To all of the students from my Fall 1995 Linear Algebra class, thanks for your patience, your positive attitude, and for the invaluable suggestions to make this text as clear as possible. Thanks to the editorial and production staff at Addison-Wesley, especially Laurie Rosatone, Kim Ellwood, and Marianne Lepp, for their hard work and professionalism in shepherding this novice author through the winding road from proposal to finished product. Finally, I am grateful for the support and understanding of my family, friends, colleagues, and students during the past two years, and the grace that has made this text possible.

John R. Wicks

*Chicago, Illinois*
*December, 1995*

# Linear Algebra

**An Interactive Laboratory Approach with *Mathematica*®**

# PART ONE

# Introductory Material

# Introduction to *Mathematica*

## *Mathematica* Notebook

*Mathematica* is a powerful computer program that can perform any routine algebraic manipulation as well as create graphics that will help us see the geometry behind linear algebra. As one reads through this text, we will use *Mathematica* to:

- verify any hand calculations,

- perform routine calculations, so that we may avoid the frustration of arithmetic mistakes and focus on the patterns of thought in linear algebra,

- aid us in *discovering* the definitions and algorithms of linear algebra, as much as possible, and

- allow us to form and verify our *own* conjectures about linear algebra.

You should not be concerned if you are not a computer "whiz," because *Mathematica's* Notebook interface is very easy to learn to use. This entire text, and specifically this introduction, is designed for those readers who have never used *Mathematica* before. However, experienced *Mathematica* users will enjoy using these materials as well, because the Notebook interface allows the advanced features of *Mathematica* to be preprogrammed into the text, so that novices and advanced readers can use them with equal skill.

The first few sections of this lesson provide novices with a gentle introduction to this Notebook interface. However, this lesson also contains a number of important points about how this text is organized and some advanced tips on *Mathematica* that even experienced *Mathematica* users may not know. Thus, everyone should read this lesson very carefully.

This lesson, as well as many others in this text, are reproduced as *Mathematica* Notebooks. Although one may read them in printed form, they are designed to be read on the computer. If you are reading the printed text, you should open up the Notebook "Less1.ma" and begin following along on the computer.

- `Open Less1.ma now and begin reading at this point.`

*Important*: If the entire window of the Notebook is not visible on your screen, you should use the Tile Windows Wide command in the Window menu to resize the window to your screen.

## 1.1  *Mathematica* Notebooks

A Notebook consists of a sequence of "cells" that contain:

- text,

- graphics, or

- *Mathematica* commands.

Everything in a Notebook is contained in a cell, which is indicated by the innermost brackets along the right scroll-bar. A number of consecutive cells may be grouped together, indicated by another enclosing bracket. These groups, in turn, can be grouped still further with another bracket, etc. This feature allows us to open and close groups of cells to organize our work, similar to paragraphs, sections, and chapters in a standard textbook.

To begin we must learn to "open" and "close" groups of cells. Below is a sample group of cells with which you may experiment. The little box on the right indicates that the group is now closed. When a group is closed, the top cell will be visible, the bracket of the group will have a little hook on the bottom, and you will see a little box in the right-hand margin like that on the right. The size of the box indicates the amount of information contained in the group. Double-click now on the little box to open this group of cells. *Note*: When your cursor is on the white rectangle, it will change its appearance to a narrower arrow pointing to the left with a vertical line at the tip.

- This is the outermost group. It contains three cells. Open this group.
- Cell number one.
- Cell number two. This contains two more cells. Open this group.
- This is one of the innermost cells.
- This is the other.
- Cell number three.

Notice what happens when you open a cell. New cells appear, all of which are highlighted along the right-hand side. When cell brackets are highlighted, we say that those cells are "selected." When a group is selected, you may then close that group of cells by hitting Command-' (that is, hold down the Command key and then hit the right quote key ').

*Note*: On a Macintosh the Command key is the one with the little apple on it; on other computers use the Control key.

Hit Command-' now to close this group of cells again, and then reopen it. Experiment now with opening and closing the inner groups as well and observe what happens. Pay special attention to the way the brackets are arranged. When you are done, open the next group to learn how to select cells and groups of cells.

## 1.1.1   Selecting Cells

Take the mouse and place the cursor over the vertical brackets on the right. We will call this the "pointer," to distinguish this from the blinking vertical bar, which is usually referred to as the "cursor." Notice that the pointer changes to look like an arrow pointing to the left with a little vertical bar at the point. Click down once on the mouse. Notice that one of the brackets is then selected. By pointing precisely, you may select the bracket of any particular cell or group you wish.

Here is a copy of our sample group for practice. Open all the groups and subgroups.

- This is the outermost group. It contains three cells. Open this group.
- Cell number one.
- Cell number two. This contains two more cells. Open this group.
- This is one of the innermost cells.
- This is the other.
- Cell number three.

Try selecting each of the brackets in the sample. Try opening and closing each group and compare the results. Remember, to close a group of cells, you must select it by clicking on its bracket, and then hitting Command-'. *Warning*: You cannot open or close an individual cell with the Command-' command; this only works with groups of cells. *Hint*: As a short cut, you may also close a group of cells by simply double-clicking on the selected bracket. When you are done, open the next group to learn how to create your own cells.

## 1.1.2   Creating Cells

Take the mouse, place the pointer at the bottom in the middle of the screen, and slowly move the pointer up the screen. *Warning*: Avoid clicking on the bottom scrollbar, or the Notebook will scroll to the right and much of the text will disappear. Notice that it changes periodically from a vertical to a horizontal line. These changes occur at cell boundaries.

Position the mouse at a cell boundary and click down once on the mouse. Notice that a line appears. This indicates where the new cell will be inserted. Now if you begin to type, *Mathematica* will create a new cell containing the text that you type. Create a cell above this cell by typing in your name. You should see the blinking vertical bar (i.e., the cursor) appear to indicate where your typing will appear. Notice that another cell bracket appears to the right of your new cell. The key labeled "return" acts just as the carriage return of a typewriter. Open the next group of cells to learn how to edit cells.

## 1.1.3   Editing Cells

You may edit a Notebook in one of two ways. You may cut, copy, or paste either:

- a collection of cells or

- part of the contents of a single cell.

The commands to cut, copy, and paste are the usual ones: Command-x to cut, Command-c to copy and Command-v to paste. To cut or copy an *entire* cell or collection of cells, simply select them and then hit Command-x. To paste them back into the Notebook, place the cursor at a cell boundary as you did before to create a cell (i.e., position the mouse and click down) and paste using Command-v. Try this now by scrolling up to the cell that you created in the previous section, cut it out, and paste it just above this cell.

Your editing of the *contents* of a cell in *Mathematica* is the same as in your favorite word processor: simply highlight the text (by clicking and dragging with the pointer) and hit Command-x or Command-c. In this case, we must position the *blinking* cursor. Use the mouse to position the pointer within this cell and click down once on the mouse. Notice that a blinking cursor appears where the pointer was. To insert material within a cell, simply position the blinking cursor in the desired location and either paste or type. Practice cutting and pasting in the cell that you created earlier.

*Aside*: More advanced readers may wish to explore the commands in the Style menu. These can be used to format the cells in a Notebook to make it more readable. The Special styles are expressly provided for your own customization.

## 1.1.4 Evaluating Cells

In the remainder of this text, you will often be asked to carry out experiments in the course of your reading. These will often involve executing *Mathematica* commands that appear in accompanying Notebooks. To execute a *Mathematica* command, one "evaluates" the cell which contains the command. For this text we have instructed *Mathematica* to put a bullet (i.e., a small black-filled circle) in front of each such cell to remind you to evaluate such cells. To evaluate a cell, you simply position the blinking cursor anywhere in the cell and hit the return key, *while holding down the "shift" key.*

*Note*: On some computers, such as the Macintosh, you may be able to hit the *"enter"* key instead, usually located on the far right of your keyboard in the numeric keypad.

For example, to see a plot of $y = \sin(x)$, you would evaluate the following cell. However, you should should wait to do so until you have read completely through the following exercise.

●     `Plot[Sin[x],{x,0,5}]`

## Exercise 1.1

●     Before evaluating the previous cell, examine it closely. Then evaluate it, and write a few sentences describing all that occurs. For example:

- Did any new cells appear?

- Did any new text or graphics appear?

- What happens to the top of the window?

- What happens to the bracket of the cell?

– What happens if you evaluate the cell *again*?

*Note*: The previous cell was in red but appears in the printed text as a numbered exercise. Each exercise is designed to provoke a thoughtful written response. Err on the side of saying too much rather than too little! Make sure to check with your instructor as to how your answers should be written up.

*Warning*: The exercises will be numbered in your text, but these numbers will not appear in the Notebook. That is because the program that printed the text can automatically number the exercises, but *Mathematica* cannot.

### Exercise 1.2

- What do you think the $\{x, 0, 5\}$ does? Edit the cell to change the 0 and the 5 to some other numbers, re-evaluate the cell, and compare the results with the original. *Hint*: You may want to edit a copy of the cell. Experiment in this way until you determine what these numbers are telling the computer to do, and write your conclusions in a few sentences.

To this point, we have explicitly indicated when to open a group of cells. From now on, you should simply work your way down through this and every other Notebook, opening groups when necessary. We will use groups of cells to indicate logical groupings of text into sections, subsections, and exercises. You should be careful not to skip over any portion of material, unless specifically directed otherwise, since it may contain information that is crucial to understand future material or important *Mathematica* commands.

## 1.2 More Advanced Topics

Congratulations! You have now mastered 60 percent of the skills necessary to explore the *Mathematica* Notebooks that will be used in the remainder of this text. Although we want to start slowly and not overburden you with too many *Mathematica* commands, there are a number of more advanced topics with which you should be familiar.

## 1.2.1 Variables and Assignments

As its name implies, *Mathematica* is designed to do mathematics. This means, for example, you may create variables and give them specific values, like: (evaluate this)

- x = 2

or: (evaluate this)

- y = 3

You may also do algebra and save the results in a new variable, such as: (evaluate this)

- `z = x^3 - y^2`

  or: (evaluate this)

- `w := x^3 - y^2`

    Notice that the first three assignments returned the assigned value, while the last one did not. That is because we used ":=" in place of "=". This is type of assignment is used to establish a functional relationship between variables, instead of actually performing a specific assigment.

    To determine the value *Mathematica* thinks a variable possesses, simply evaluate a cell containing only that variable. For example, to determine *Mathematica*'s values for $x$, $y$, $z$, and $w$, evaluate each of the following cells:

- `x`

- `y`

- `z`

- `w`

## Exercise 1.3

-     Record the results of these four evaluations. Now evaluate the cell:

- `x = 1`

    and recheck the values of $x$, $y$, $z$, and $w$ by re-evaluating the four previous cells. How does this compare with the first time? Explain the similarities or differences.

    Notice that *Mathematica* only remembers the *most recent* definition that was *actually evaluated*. Thus, it's important to be careful to evaluate cells in the proper order. For commands given in this text, this will usually be the order in which they appear.

    If you want *Mathematica* to forget a definition for a variable, say $x$, simply evaluate a cell containing the Clear[] command like: (evaluate this)

- `Clear[x]`

## Exercise 1.4

-     Evaluate this cell and then check what *Mathematica* thinks about $x$. Record your results. Does *Mathematica*'s behavior surprise you? Explain.

*Important*: All of the cells in this Notebook that you were to evaluate have had a bullet in front. These are called Input cells, and in this text we have instructed *Mathematica* to put a bullet in front of every Input cell to remind you to evaluate that cell when you come to it. Eventually we will no longer explicitly tell you to evaluate each cell, but if you see a bullet in front, you should should make sure to do so. In general, a bullet anywhere in the text will indicate that you should use *Mathematica*.

To illustrate *Mathematica*'s algebraic capabilities, we will evaluate the equation:

- `Clear[x,y]; x = 278; y = x^20 - x^10 + x^3`

To *Mathematica* the "^" symbol indicates the "power operation." *Aside*: Notice how we cleared the variables before we used them. This generally serves to avoid confusion, in case the variables had been assigned values earlier in this Mathematica session. Notice that *Mathematica* gives an *exact* answer with all 49 digits! If you wish to see your result in scientific notation to 12 places, you would use the N[] command:

- `N[y, 12]`

Here N[] stands for "numerical output." This is the best value you would have obtained without *Mathematica*, using a hand-held calculator.

## 1.2.2  Some Miscellaneous Pointers

We may save time by typing many commands in the same cell, but we must separate each command with a semicolon. Only the output of the last command will appear on the screen, but *all* the commands will be evaluated. For example, the first few commands of the last section may be combined into a single cell, such as:

- `x = 2; y = 3; z = x^3 - y^2`

Because *Mathematica* generally ignores spaces, tabs, and carriage returns, this may be entered as:

- `x=2; y =        3;`
  `z = x^3-y^2`

Spaces are only important when the input is ambiguous. For example, it is not clear whether:

- `z = xy`

means "give *z* the value of the variable called *xy*" or "give *z* the value of the product of the two variables *x* and *y*." You may mean the latter, but *Mathematica* will assume the former, because it has few preconceived notions of what a variable name should be. Thus, if you want to multiply *x* and *y*, you would enter:

- `z = x y`

Although spaces, tabs, and carriage returns are ignored, they are helpful for making one's work more readable. While we may skimp a bit on the carriage returns in these Notebooks in order to eliminate "vertical white space" from the printed text, you should be sure to make your own work as readable as possible.

We conclude this section with two more useful facts. *Mathematica* is case-sensitive. This means that to *Mathematica* "x" and "X" are two *different* variables. For example, consider the results obtained by evaluating the following:

- `x = 3; X = 5; Print["x = ", x,"  while X = ",X];`

Also, it is possible to print out on paper any collection of cells in a Notebook, instead of the entire Notebook, by selecting the cells you want and choosing the Print Selection option (in the File menu). Although you can print an entire Notebook with the standard Print command from the File menu, this will usually waste a lot of paper and time. Thus, it is better to be selective in what you choose to print.

### Exercise 1.5

- Summarize at least three important ideas from this section.

## 1.2.3   In/Out Labels and Memory Management

By this point, you should have noticed that each time you evaluate a cell, that cell is labeled with "*In[] :=* " and consecutive numbers appear in the brackets. If the command produces output, this will usually appear in a new cell that is similarly labeled with "*Out[] =* ", where the number in the brackets is the same as the corresponding In[] number. These labels can be used in a variety of ways. To see this, evaluate the following cell:

- `test = 2`

These labels are written as equations because they actually represent variable assignments. *Mathematica* remembers every command that you evaluate and every value that it outputs by defining the sequence of variables In[1], In[2], ... and Out[1], Out[2], ....

### Exercise 1.6

- Determine the In[] number assigned to the "test = 2" command. Edit the following cell to place that number in the brackets.

- `In[]`

  Now evaluate this cell. Did it return the value that you expected? Explain. Experiment with changing the In[] number and reevaluating. Describe what you expect to happen in each case and report the results. Try a similar experiment with the Out[] command.

*Note*: As soon as you edit a cell, its label will disappear, since the data in the cell may no longer agree with the stored value.

The In[] and Out[] labels can help you in two ways. By paying attention to the In[] labels, you may track your progress through a Notebook. You should always be careful to evaluate the cells of a Notebook in the order requested, or you may get surprising results! Also, you may avoid retyping a previous command or restore old values to a variable by determining the corresponding In[] number. To determine the In[] number of a command that you have deleted, you can simply use trial-and-error to search for the command. More on this later.

However, because *Mathematica* saves everything, it may run out of memory (especially after creating many three-dimensional graphs). The "Start.ma" Notebook defines the command ClearInOut[], which will clear the In[] and Out[] definitions, thus freeing up some memory. If *Mathematica* begins to "complain" that it is running low on memory, you should try evaluating the ClearInOut[] command (with no arguments). If you do not wish to lose *all* In[] and Out[] definitions, you may evaluate something like ClearInOut[20], which will only clear the first 20 definitions. If you want to be more precise, something like ClearInOut[{1, 3, 28}] will only clear the definitions for commands numbered 1, 3, and 28.

Because *Mathematica* is divided into two separate programs, known as the Front End and the Kernel, it actually uses memory in two different ways. The Kernel is the program that actually evaluates any *Mathematica* command from a Notebook. Thus, variable assignments use memory from the Kernel, so, for example, the ClearInOut[] command will only free memory for *that* program.

On the other hand, the Front End is the program that allows one to *view* a Notebook. In particular, it "renders" any plot produced by the Kernel so that you may actually view it. Thus, if you view a large number of graphics, the Front End may run out of memory, and we must free more memory for this program in a different manner.

The key to this is to remember two basic facts:

- The Front End only uses memory when a picture is actually viewed.

- The Front End will free memory when a Notebook is closed.

This suggests the following strategy, when one is running low on Front End memory (as indicated by a warning message, or by the memory usage indicator on the bottom-left of a Notebook). When using a Notebook that contains a number of graphics cells (i.e., cells containing pictures), one should:

- close all groups of cells that contain graphics,

- save the Notebook,

- close the Notebook, and

- re-open the Notebook.

Although all of the graphics are still in the Notebook, you should notice that the memory usage is much less. That is because the graphics are not rendered until you actually view them.

## 1.2.4   Loading and Using Packages

Our final advanced topic concerns loading new *Mathematica* commands. *Mathematica* comes with a large number of built-in commands, but it also allows one to write one's own commands and save them in a file for future use. We have provided a number of additional commands and designed a system so that all such commands for this text will be loaded automatically, so that you may use them very easily. The only requirement is that, from now on, one should always first open the file "Start.ma" before working through any other Notebooks or utilizing any specialized commands provided with this text.

In this section, we will explain the precise function of "Start.ma" and discuss how one would load packages by hand. Moreover, throughout the text, we will differentiate between commands and packages that are provided especially with this text and those that come with the *Mathematica* program. In this way, you will know what commands will be available to you in the future, when you use *Mathematica* in contexts other than while reading this text. A list of commands used with this text is given in the Appendix to the printed version of the text.

## Using Standard *Mathematica* Packages

We begin by showing how one would use the standard *Mathematica* packages that are distributed with the commercial program. For example, if one tries to use the command Polyhedron[] from the "Polyhedra.m" package:

- `Show[Polyhedron[Cube]]`

one sees that it does not seem to work. That is because we must first load its definition into *Mathematica's* memory, using the Needs[] command:

- `Needs["Graphics`Polyhedra`"]`

    - The name of the file is in double quotes.
    - Since the "Polyhedra.m" package is stored in the Graphics directory (called a "folder" on a Macintosh computer), we must include this directory name in the argument.
    - Each name in the argument ends with a back-quote (which is *not* the same as an ordinary single-quote — take the time to locate it now on your keyboard).
    - We have dropped the ".ma" extension from the filename.

Although this command has been loaded, it still does not work:

- `Show[Polyhedron[Cube]]`

This occurs because we attempted to use this command before loading the correct definition. In this case, *Mathematica* has created a "dummy" definition that overrides the correct definition. This phenomenon is called "shadowing." Compare this with the error messages we obtained from the Needs[] command. If we remove the dummy definitions for Polyhedron[] and Cube:

- `Remove[Polyhedron,Cube]`

the correct definition becomes "visible," so that it now will work:

- `Show[Polyhedron[Cube]];`

## Using Custom Packages for this Text

In the "Start.ma" file, we have included *Mathematica* instructions to load any packages used in the text *automatically*. In this way, as long as you start each session by opening "Start.ma" and evaluating, you will never need to load any *Mathematica* package explicitly. That is, you may simply *use* the commands given in the text without worrying about the difficulties previously discussed.

However, if you forget to start with "Start.ma" and use a command that is not built-in, you will encounter the same problem we had when we first tried to use the Polyhedron[] command. For example, to draw arrows in the plane we provide a *Mathematica* command called MyArrow[] located in the "GVects.m" package. If one tries to use the command without first loading it:

- `Show[MyArrow[{0,0},{1,2}]]`

one sees that it does not seem to work.

If one *now* remembers to open the "Start.ma" Notebook, it will automatically remove the shadowing problem and instruct *Mathematica* to automatically load any necessary packages. Use the Open command in the File menu to open the "Start.ma" file now. A dialog box will appear and ask "Do you want to evaluate all initialization cells in the Notebook 'Start.ma'?" Make sure to click on the Yes button.

Now MyArrow[] will work easily:

- `Show[MyArrow[{0,0},{1,2}]]`

To summarize, at the start of every *Mathematica* session one should begin by opening the "Start.ma" Notebook. If you do so, all commands in this text should work automatically. However, if you forget to do so, commands that are not "built-in" to *Mathematica* will not work, until you remember to open "Start.ma."

## Obtaining Additional Information

Once it has been loaded, you may obtain a brief description of any *Mathematica* command by using the Function Browser in the Help menu, or the ? operator. For example, to learn about the MyArrow[] command, you would evaluate:

- `?MyArrow`

## Exercise 1.7

- What does this tell you about the MyArrow[] command?

You may determine if a command has been loaded, indirectly, by using the ? operator.

## Exercise 1.8

- What happens if you use the ? operator on a command that has not been loaded, such as:

- `?UndefinedCommand`

*Note*: Once you have loaded a command, *Mathematica* will remember it until you Quit the program, so you only need to load any given file *once* per session.

## 1.3   Summary

In this lesson you should have learned how to:

- open and close groups of cells,

- create, select, and edit text within cells, as well as entire cells,

- evaluate cells;

- clear values from a variable, assign values to a variable, and learn the value assigned to a variable,

- use spaces, carriage returns, and semicolons to format input to *Mathematica*,

- print small sections of a Notebook,

- free up memory when it runs low, and

- load *Mathematica* packages.

You should have also written responses to eight exercises.

In the remainder of the text, we will use Notebooks to involve you more closely in the learning process. We will group cells to create sections and subsections with their own titles and exercises. By including cells to evaluate, the text will "come alive" and allow you to use the computer to experiment with the concepts while you read about them! As you have seen, you will not need to know how to program, because we will include all necessary instructions right in the text and provide packages for more complicated commands.

You will be asked repeatedly to observe, discuss, and write down your conclusions in short essay responses. Thus, this will probably be quite different from any other math text you have used. In fact, it will read a bit like a science lab manual, where you are expected to experiment and discuss your findings. Along the way, you will learn to think and communicate in a more mathematically mature manner.

# Introduction to Linear Systems and Row Operations

## Accompanied by Notebook

*Important*: This lesson is *accompanied* by a Notebook (as opposed to being simply text or being completely reproduced in a Notebook). This means that many of the calculations that one must perform may be accomplished easily in *Mathematica*. Many of the necessary *Mathematica* commands are even *pretyped*. Thus, while one reads this lesson, one should follow along in the corresponding Notebook and evaluate the corresponding *Mathematica* commands.

In this lesson, we will:

- define what we mean by a "system" of "linear" equations,

- investigate the nature of the corresponding solutions,

- discuss the process of solving a system of linear equations,

- describe how we can write down an infinite solution set, and

- investigate the type of allowable operations involved in the solution.

## 2.1 Equations, Solutions, and Graphing

Before we discuss systems of equations, we must first clarify what we mean by a "solution" and learn to use *Mathematica* to visualize solution sets. As usual in algebra, a "solution to a single equation" is an ordered list of values, which when substituted for the corresponding variables produces equality. The "solution set of an equation" in $n$ variables is the set of all solutions, expressed as $n$-tuples of numbers. The "graph of an equation" is simply the graph of all of the points in its solution set.

For example, the solution set for $x^2 + y^2 = 4$ contains an infinite number of points. You may easily check that $(0, 2)$ gives a point in the solution set. You may even use *Mathematica* to verify this, by evaluating the following command in the accompanying Notebook:

- ```
  Clear[x,y];
  x^2 + y^2 == 4 /.{x->0, y->2}
  ```

Notice that we use a "==" in place of a single "=" to instruct *Mathematica* that we do not wish to make an assignment, but rather to perform a *logical test* for equality. Also, the arrow symbol ("->") instructs *Mathematica* to perform the indicated replacement of values for the variables in the equation.

## Exercise 2.1

Give four more solutions for this equation. Also, determine two pairs of values that are *not* solutions.

We often visualize *all* of the solutions by graphing. This may be accomplished by using *Mathematica* in one of two ways. One way is to solve the equation for one of the variables and use the Plot[] command from Lesson 1. For example, to graph $x^2 + y^2 = 4$ you would solve for $y$ to get $y = \pm\sqrt{4 - x^2}$ and then evaluate the corresponding Plot[] command. Because this is not a single function, we must actually construct two plots simultaneously, using the the following command:

- ```
  Clear[x];
  Plot[{Sqrt[4 - x^2],-Sqrt[4 - x^2]},{x,-2,2}];
  ```

Notice that although these commands should produce a circle, the plot seems to be an ellipse. That is because the computer has chosen different scales on the $x$- and $y$-axes, so that the graph will fit conveniently on the computer screen. To force the scales on the axes to be indentical, one would use the AspectRatio option:

- ```
  Clear[x];
  Plot[{Sqrt[4 - x^2],-Sqrt[4 - x^2]},{x,-2,2},
        AspectRatio->Automatic];
  ```

Notice how we make sure to Clear[] the variable $x$ before executing a Plot[] command. A variable used in this manner in a Plot[] command is often referred to as a "dummy variable." Dummy variables must not have a preassigned value by the time a Plot[] command is evaluated.

Another way to plot an equation in two variables does not require one to solve for any of the variables, but rather uses the concept of a "contour plot." For example, one could think of the equation $x^2 + y^2 = 4$ as part of the graph of $z = x^2 + y^2$ at the value $z = 4$. To plot this equation, you would evaluate the following ContourPlot[] command:

- ```
  ContourPlot[x^2 + y^2,{x,-2,2},{y,-2,2},
          Contours->{4},
          ContourShading->False];
  ```

This method has a few disadvantages. It may not be as accurate as using the Plot[] command, and we need to include a range on the $y$-values. However, when the equation involved is complicated, this method may be the only feasible one, in that it may be impossible to solve the equation for a single variable.

### Exercise 2.2

What is the function of the Contours option? What would happen it we changed the 4 to a 1? *Hint*: Look carefully at the equation we are plotting and compare with the arguments to the ContourPlot[] command.

## 2.2   Systems of Equations

Now we proceed to consider systems of equations. In this case the corresponding *Mathematica* commands become more complex. By a "system of equations", we mean a collection of equations that we want to consider all at the same time, as a single unit. In this case, a solution to the system must satisfy *all* the equations at the same time. That is, the solution set for a system of equations is simply the intersection of the solution sets of each of its equations.

In principle, solution sets should be easy to visualize. In practice, coercing a program, such as *Mathematica*, to plot more than one solution set at a time is rather difficult. For example, to plot the solution set for:

$$2xy = 4 \quad \text{and} \quad 3x - y = 2$$

we would solve each for $y$ to get:

$$y = \frac{2}{x} \quad \text{and} \quad y = 3x - 2$$

and then use the Plot[] command:

- `Plot[{2/x, 3x - 2}, {x, -1, 1}];`

You should notice that it is difficult to see both graphs, because the scale on the $y$-axis is so large. We may set the scale on each axis with the PlotRange option, as follows:

- `Plot[{2/x, 3x - 2}, {x, -1, 1},`
                    `PlotRange->{{-1,1},{-5,5}}];`

*Mathematica* provides two nice features for manipulating graphs. First click on the picture. You should notice that the pointer now looks like a little pair of axes. You may now resize the picture by clicking and dragging any one of the corners of the graphic. Because *Mathematica* tries to automatically scale the graph and the axes, the picture may originally be too small or the axes too crowded. Resizing the picture often helps.

Moreover, if you hold down the Command key and place the pointer over the graph, you will notice that the pointer changes again (this time to little cross-hairs) and the coordinates of the point on the graph appear in the lower left of the window. *Note*: The Windows version of *Mathematica* does not fully support this feature. This feature allows one to find the approximate coordinates of any point in a graph. If you click down with the mouse on any given point, *Mathematica* will temporarily show a small black dot at the point that you have selected. The Copy and Paste commands may then be used to print the coordinates of that point directly into the Notebook. *Note*: Windows users *may* Copy and Paste to obtain coordinates.

## Exercise 2.3

- Use the mouse to determine the approximate coordinates of the intersection point (i.e., solution) that is visible in this graph. Set the two equations equal and solve algebraically for the exact $x$-coordinate and the corresponding $y$-coordinate.

You should notice from the graph that there appears to be another point of intersection that is not pictured. In general, a computer can never show us the entire graph, since any graph is theoretically infinite. Thus, we must have some idea as to where the interesting parts of the graph are, in order to use the computer intelligently.

## Exercise 2.4

- Replot this graph, changing the PlotRange values so that the other point of intersection becomes visible, and use the mouse to find the coordinates of that point. Compare this with the exact values of this solution.

To repeat the previous exercise using ContourPlot[], you cannot simply type both functions into ContourPlot[], since it is designed to plot only a single function at a time. If you wish to see two contour plots together, you must generate the plots separately and then Show[] them together. For example, to carry out the previous exercise, we would use the following sequence of commands:

```
plot1 = ContourPlot[2x y,{x,-10,10},{y,-10,10},
           Contours -> {4},
           ContourShading->False,
           DisplayFunction->Identity];
plot2 = ContourPlot[3x - y,{x,-10,10},{y,-10,10},
           Contours -> {2},
           ContourShading->False,
           DisplayFunction->Identity];
Show[plot1,plot2,
      PlotRange->{{-10,10},{-10,10}},
      DisplayFunction->$DisplayFunction];
```

Here we generate the two plots separately and save the pictures as plot1 and plot2. The DisplayFunction->Identity option suppresses the drawing of each plot separately, while the DisplayFunction-> $DisplayFunction option in the Show[] command causes the plots to be shown together as usual.

When equations involve three variables, solutions must be plotted in three dimensions. For example, to visualize the solutions to:

$$x^2 + y^2 = z \quad \text{and} \quad 2x + 2y - 1 = z \qquad (2.1)$$

we would use following commands:

```
plot1 = Plot3D[{x^2 + y^2,Red},
         {x,-2,2},{y,-2,2},
           DisplayFunction->Identity];
plot2 = Plot3D[{2x + 2y - 1,Green},
```

```
        {x,-2,2},{y,-2,2},
          DisplayFunction->Identity];
Show[plot1,plot2,
     DisplayFunction->$DisplayFunction];
```

*Note*: We have plotted each graph in a different color to highlight the intersection. Notice that this method is not accurate enough to produce accurate *numerical* solutions, but it does provide a rough *qualitative* impression of the solution set.

### Exercise 2.5

Describe the shape of the intersection of these two graphs.

### Exercise 2.6

Determine which of the following points are solutions to Eq. 2.1 and which are not. Explain your answers.

(a) $(2, 2, 8)$          (b) $(1, 2, 5)$          (c) $(3, 2, 4)$

## 2.3    Linear *vs.* Nonlinear Systems

From one point of view, linear algebra is simply the study of systems of linear equations. Such a description requires us to define what we mean by "linear." Everyone would agree that something like:

$$2x + 4y = 6$$

should be called a linear equation, since its graph is a straight line. However, one might be surprised to learn that the equation:

$$2x - 4y + 5z = 6$$

is also considered linear. The graph of this equation is a plane, as you may verify by plotting $z = (6 - 2x - 4y)/5$ with the Plot3D[] command:

● 
```
Clear[x,y];
Plot3D[(6 - 2x - 4y)/5,{x,-2,2},{y,-2,2}];
```

Geometrically, one could argue that a plane consists of straight lines and so should also be considered to be linear. Algebraically, we observe that both of the previous examples were defined using only the operations of addition and multiplication by constants. For this reason, these two operations are referred to as a "linear operation." An equation, in any number of variables, defined using only these two operations is likewise referred to as a "linear equation." Thus, even an equation such as

$$2x - 4y + 5z - 6w + 3u = 6$$

that can only be graphed in five-dimensional space is still considered linear.

Conversely, an equation such as:

$$2x^2 - xy = 6$$

is called a "nonlinear equation." Intuitively, while a linear equation should have a "flat" solution set, that of a nonlinear equation should be "curved." We may see that this is true in this example, using the ContourPlot[] command:

- ```
Clear[x,y];
ContourPlot[2x^2 - x y,{x,-12,12},{y,-12,12},
        Contours -> {6},
        ContourShading->False];
```

For the remainder of this section, we will begin to discover why *linear* algebra is so important. By exploring various systems graphically, you should discover the following intuitive principle:

**Observation:**

Solutions to linear equations tend to be much more well-behaved than those to nonlinear equations.

For example, while linear equations have flat, connected solution sets, those of nonlinear equations may curve and come in many pieces. An extreme example of this is given by the equation $\sin(x)\cos(2y) = 0.5$, which you may plot as follows:

- ```
Clear[x,y];
ContourPlot[Sin[x]Cos[2y],{x,-5,5},{y,-5,5},
        Contours -> {.5},
        PlotPoints->30,
        ContourShading->False];
```

*Note*: The PlotPoints option is included to increase the number of points plotted (from the usual 15 points) so that the graph plots smoothly.

These qualitative features distinguishing linear from nonlinear equations carry over to *systems* of linear equations as well. Consider a *general* system of two linear equations in two variables, $x$ and $y$. Its solution set corresponds to the intersection of two lines in the $x$-$y$ plane.

## Exercise 2.7

What are the *qualitatively* different types of solution sets that could arise? *Hint*: There are three different cases to consider. Describe each possible case.

## Exercise 2.8

Repeat Ex. 2.7 for three linear equations in three unknowns. How are the possibilities similar? How are they different?

Unlike linear systems whose solutions sets are relatively easy to describe, no matter how many equations or variables one is using, nonlinear systems will behave very badly and encompass a much wider range of possibilities.

### Exercise 2.9

• Consider the following system:

$$|x| - y = 2$$
$$y - mx = b$$

for various values of $m$ and $b$. Repeat Ex. 2.7 with this system. That is, describe the four possible types of solution sets that may arise, as one varies $m$ and $b$.

### Exercise 2.10

• Repeat Ex. 2.9, but with the system:

$$x^3 - x - y = 2$$
$$y - mx = b$$

*Note*: In this case, there are only three possibilities.

## 2.4   Complex *vs.* Simple Systems

At this point, you should be convinced that the solution sets of linear systems are much simpler than those of nonlinear systems. The linear system solutions are all "flat," "connected" sets (in other words, points, lines, and planes) which, qualitatively, are completely described by their dimension. Moreover, once you know the number of equations and variables in a linear system, you may narrow the possible types of solutions down to a very small list. On the other hand, nonlinear systems can have practically any type of solution set (limited only by one's imagination); even with only two equations and two unknowns, one may create a wide variety of solution sets.

Thus, linear systems should be much easier to study than nonlinear ones. In general, one is much more likely to obtain accurate results from a linear model of any given situation than a nonlinear one. This is why linear algebra is so widely used. Now it remains to learn a general method, called Gaussian elimination, for computing specific numerical values for the solutions to a linear system. We will do so in Lessons 4–8. In the remainder of this lesson, we discuss the general philosophy behind this method.

### A First Introduction to Gaussian Elimination

Think about solving the following system of equations:

$$x + y = 2$$
$$x - y = 0$$

This system has a unique solution, namely:

$$x = 1$$
$$y = 1$$

*Mathematica* can find this solution very easily with the Solve[] command. However, we want to *understand* how *Mathematica* arrived at that solution.

To do so, we must examine this solution from a fresh perspective. Notice that this solution is a system of equations, as well. It happens to be in such a simple form that we may read off the solution directly. From this point of view, solving such a system of equations involves not simply "finding the $x$- and $y$-values," but *transforming* the original system into a *simpler* system that has the *same solution set*. This idea of transforming a hard problem into a simpler one is one of the *most fundamental ideas, not just in linear algebra, but in all of mathematics!*

In linear algebra, this point of view is particularly important, because the solution sets may consist of more than a single point. For example, consider the system:

$$\begin{aligned} x &+ y &+ 2z &= 3 \\ x &+ 2y &+ 5z &= 5 \end{aligned} \tag{2.2}$$

Geometrically, this solution is the intersection of two planes. Thus, there are an infinite number of solutions. We must describe all these solutions more simply. We will learn how to transform this system into the following equivalent system (i.e., having the same solutions):

$$\begin{aligned} x & & - z &= 1 \\ & y &+ 3z &= 2 \end{aligned} \tag{2.3}$$

or

$$\begin{aligned} x &= 1 &+ z \\ y &= 2 &- 3z \end{aligned} \tag{2.4}$$

Equation 2.4 is visibly simpler than Eq. 2.2, since it contains fewer variables. Moreover, by choosing various values for $z$, Eq. 2.4 automatically produces the corresponding values of the other variables in the solution. An engineer would say that these equations are "decoupled," in that we may legitimately compute values from each equation in Eq. 2.4 without considering the other equations. In this way, we may generate all possible solutions. For example, if we take $z = 1$, then we get the solution $(x, y, z) = (2, -1, 1)$. Likewise, if $z = 2$, $(x, y, z) = (3, -4, 2)$.

Notice that Eq. 2.3 had fewer variables in each equation than Eq. 2.2. In other words, in transforming the equations we *eliminated* some of the variables, in a very specific pattern. The most widely used, systematic method for eliminating variables from a linear system is named after its inventor Karl Friedrich Gauss, and is called *Gaussian elimination*. Although there are many *possible* methods, Gauss' method is the most widely used method because of its simplicity, generality, and ability to be done quickly and easily by a computer.

You may have already learned other *ad hoc* methods for solving systems of two or three variables, but such methods tend to break down when the number of equations and variables increases. Before discussing Gaussian elimination in detail, we take the remainder of this lesson to experiment. We will try to determine the kinds of operations with which we may transform a system of equations *without changing its solution set*. In this way, we will motivate the three basic operations used in Gaussian elimination.

## 2.5   Legal Operations on a System

Consider the following system of equations:

$$\begin{aligned} x + 2y &= 4 \\ x - y &= 1 \end{aligned}$$

(2.5)

### Exercise 2.11

For each part, we have applied the indicated operation to Eq. 2.5 to obtain the given system. Determine which of the following operations are "legal", that is, those which do not change the solution set of the system. Feel free to use *Mathematica*.

(a) Multiply one equation by a constant.
$$\begin{aligned} 2x + 4y &= 8 \\ x - y &= 1 \end{aligned}$$

(b) Erase a variable.
$$\begin{aligned} x + 2y &= 4 \\ - y &= 1 \end{aligned}$$

(c) Switch order.
$$\begin{aligned} x - y &= 1 \\ x + 2y &= 4 \end{aligned}$$

(d) Square both sides of an equation.
$$\begin{aligned} x + 2y &= 4 \\ (x - y)^2 &= 1 \end{aligned}$$

(e) Add two equations and save all.
$$\begin{aligned} x + 2y &= 4 \\ x - y &= 1 \\ 2x + y &= 5 \end{aligned}$$

(f) Add one equation into another.
$$\begin{aligned} x + 2y &= 4 \\ 2x + y &= 5 \end{aligned}$$

(g) Add two equations.     $2x + y = 5$

(h) Solve one equation and substitute it into the other.
$$x + 2(x - 1) = 4$$

(i) Solve one equation, substitute it into the other, and save the original equation.
$$\begin{aligned} x + 2(x - 1) &= 4 \\ x - y &= 1 \end{aligned}$$

(j) Multiply one equation by 0.
$$\begin{aligned} 0x + 0y &= 0 \\ x - y &= 1 \end{aligned}$$

## 2.6    Summary

Linear algebra is a wonderfully flexible and applicable subject. It may be naively described as the study of systems of linear equations. Such systems arise in a wide variety of applications in mathematics, physics, chemistry, business, and economics. In fact, almost every academic discipline that employs mathematics will use linear algebra, because *linear* systems are the simplest and most well-behaved type of system.

In Lessons 4–8 we will learn how to employ Gaussian elimination to obtain the solution set of any linear system. The remainder of the text will then attempt to refocus this process to the more abstract point of view of "linear transformations" and "vector spaces." Beyond providing greater clarity and coherence to the subject, these abstractions ultimately enable us to broaden the applicability of linear algebra to include geometry, computer graphics, differential equations, and statistics. Along with calculus, linear algebra is one of the single most useful fields of mathematics!

In this lesson, you should have learned:

- the difference between a linear equation and a nonlinear one,

- what we mean by a system of equations and a solution to such a system,

- the different possible types of solutions that arise for linear systems and how these are much more well-behaved than nonlinear ones, and

- what types of operations may legally be applied to a system of equations.

As you might expect, for the most part, the only allowable operations (i.e., those that do not change the solution set) that one may perform on a system of equations are the linear ones of addition and multiplication by a constant. In Lesson 4, we will examine in detail how to apply these linear operations in an organized manner to systematically eliminate as many variables as possible from any given linear system of equations. But first, we dedicate the next lesson to convincing you of the wide applicability of linear algebra.

# Linear Systems and Applications

## Accompanied by Notebook

In this lesson, we will discuss how linear systems of equations arise naturally in a variety of applications. Specifically, we will examine:

- chemical reactions,

- heat distribution,

- circuit analysis, and

- economic models.

We will see how the concepts of equilibrium and balance naturally lead to systems of equations. We will also discuss why, in most applied settings, the equations involved tend to be linear.

## 3.1  Linear Equations in Chemistry

Chemical reactions lie at the heart of a wide variety of household and industrial processes. For example, many of us have used drain cleaner to unclog a sink. This product contains sodium hydroxide (NaOH) and aluminum (Al) which when mixed with water produce the reaction:

$$NaOH + H_2O + Al \longrightarrow NaAl(OH)_4 + H_2$$

This generates hydrogen gas ($H_2$) and a harmless precipitate of $NaAl(OH)_4$, which bubbles inside the plumbing to unclog the drain. You may observe that this equation does not make chemical sense as written, because we have only written the reactants and products, without indicating their proper proportions. For example, we cannot have three atoms of hydrogen entering a reaction and only two atoms coming out.

The actual reaction may be written:

$$x NaOH + y H_2O + z Al \longrightarrow w NaAl(OH)_4 + u H_2$$

for some choice of $x$, $y$, $z$, $w$, and $u$. For this equation to balance, we must choose values for the variables that lead to equal numbers of each type of atom on either side. Each of the four types of atoms leads to an equation in $x, y, z, w$, and $u$:

$$
\begin{aligned}
x &= w & \text{to balance Na} \\
x + y &= 4w & \text{to balance O} \\
x + 2y &= 4w + 2u & \text{to balance H} \\
z &= w & \text{to balance H}
\end{aligned}
$$

At this point, you may argue that these equations may be solved by hand, without the theory of linear algebra, and you would be right. In more realistic industrial problems, there will be a large number of separate interlocking equations. These will result in many more equations in many more variables, which must be solved quickly, over and over again with small changes in the control parameters. Thus, in practice, this problem requires an efficient, systematic, programmable procedure for solving linear systems. Although this simplified example is not completely realistic, it serves to illustrate, in principle, how linear equations could arise in industry.

### Exercise 3.1

•     Find one particular solution for these equations that looks reasonable (i.e., contains all positive, integer values), either by hand or by using *Mathematica*. Verify that the resulting chemical reaction balances.

### Exercise 3.2

•     When a coroner wishes to test for arsenic (As) poisoning, she would combine some zinc (Zn) and hydrochloric acid (HCl) with a sample of hair from the victim. If arsenic is present, the following reaction will occur:

$$x\,\mathrm{Zn} + y\,\mathrm{H_3AsO_4} + z\,\mathrm{HCl} \longrightarrow w\,\mathrm{ZnCl_2} + u\,\mathrm{H_2O} + v\,\mathrm{AsH_3}.$$

One can determine if arsene gas ($\mathrm{AsH_3}$) is one of the gaseous products of the reaction by heating the discharge in a long tube and measuring the distance to the point at which it leaves a residue on the tube. Record the system of five equations that you must solve to balance this reaction, compute a particular solution, and verify that the resulting equation balances.

### Exercise 3.3

Explain why systems of equations arising in this way from chemical reactions will never have a single, unique solution.

## 3.2   Linear Equations in Electrical Engineering

By the middle of the twentieth century, electricity and electrical devices had reshaped Western society. In particular, the availability of small, inexpensive, digital computers has revolutionized almost all aspects of our lives. To build any electrical device, one must understand the physical principles behind electrical circuits. In this section, we will examine how linear algebra enables us to analyze simple, electrical resistance circuits.

As with chemical reactions, in practice, most circuits are not as simple as the ones we will examine. They may contain many more types of components, such as capacitors and inductors, whose behavior may be time-dependent or may depend on changeable parameters. Also, working engineers have developed a number of short cuts (such as Thévenin equivalence) for analyzing electrical circuits. However, such simplified examples should convince you that, in principle, linear equations are useful in electrical engineering, as well.

To make things more interesting, we will create a circuit in the context of a moderately realistic situation. Imagine that it is a hot summer night and Linda is going to a movie with her friends. She plugs in her 40-watt (W) curling iron and takes a shower. As she is dressing, she decides to press her pants, so she plugs in a 1200-W iron. Since they will not eat until after the movie, she decides to make a tuna melt sandwich in her 1350-W toaster oven. She lives in a college dorm, so all of these appliances are plugged into a single outlet via a system of extension cords. She is already running a 1350-W air conditioner and one might wonder if she can can run her 1600-W hair dryer without blowing a fuse! *Note*: These wattages were taken from actual appliances. This situation suggests the following circuit:

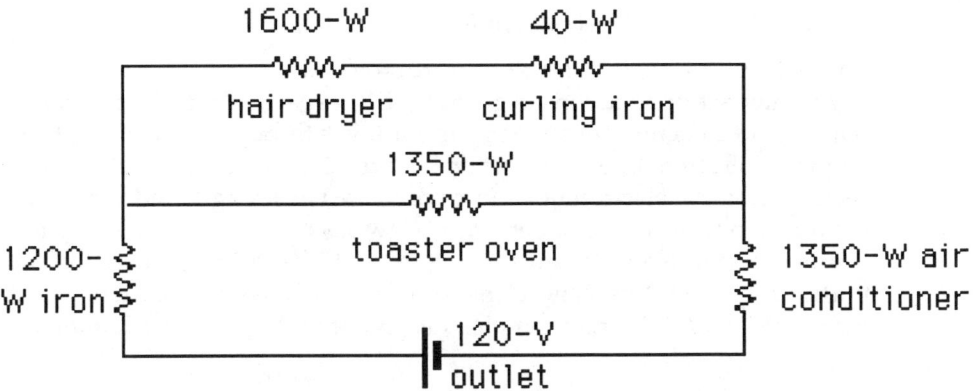

*Note*: For you home wiring experts, we stop to indicate where this circuit has been simplified. Here we are treating the outlet as a DC voltage source, when in practice it provides AC current. Moveover, in actuality, any circuit created by standard appliances will be connected in parallel (so as to provide a consistent voltage for each appliance), so this diagram is not technically accurate. However, this example illustrates the basic principles of circuit analysis better than the actual circuit would.

Although we know that a standard wall outlet provides 120 volts of electricity, there are still a number of unknowns in this circuit. Specifically, we do not know the amount of current, $I_j$, flowing through each part of the circuit and the electrical potential (or "electomotive force"), $V_i$, at each point in the circuit. To analyze this circuit, we must first determine the resistance of each appliance. We may determine the average, internal resistance $R$ (in ohms) of each appliance from the wattage $W$ by the formula $R = \frac{V^2}{W}$, where $V = 120$. If we relabel each appliance by its resistance and label the remaining unknowns, we obtain the following circuit diagram:

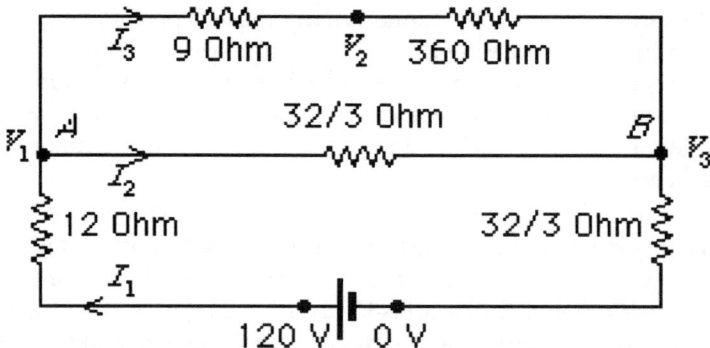

At this point, we may deduce two sets of linear equations. At each point where the circuit branches or joins, we have a "balancing equation" of currents. Because current is just movement of electrons, the total current entering any given point must equal the total current flowing out. This leads to two (redundant) linear equations:

| | | | |
|---|---|---|---|
| At point $A$: Total current in | $I_1 = I_2 + I_3$ | Total current out |
| At point $B$: Total current in | $I_2 + I_3 = I_1$ | Total current out |

This balancing principle is known as Kirkoff's law.

Another set of equations is given by Ohm's law, which describes how the electric potential in a circuit drops when current flows through a resistor. Intuitively, the current is at full strength of 120 V when it leaves the outlet on the left. As it flows through each resistor, the electromotive force decreases in strength and is totally exhausted by the time it returns to the outlet. Ohm's law says that the voltage drop across a resistor is proportional to the current flowing through the resistor, where the constant of proportionality is the "resistance," measured in ohms. As an equation, Ohm's law may be written $\Delta V = IR$. In our example, this gives the five voltage equations:

$$
\begin{aligned}
120 - V_1 &= 12I_1 \\
V_1 - V_2 &= 9I_3 \\
V_2 - V_3 &= 360I_3 \\
V_1 - V_3 &= \frac{32}{3}I_2 \\
V_3 - 0 &= \frac{32}{3}I_1
\end{aligned}
$$

In total, we have seven linear equations in six unknowns. For this system to have a unique solution, exactly one of these equations must be redundant. We have already seen that one of Kirkoff's equations is redundant, so we should be left with a unique solution. Using *Mathematica* we may easily solve these equations.

## Exercise 3.4

• Fuses are rated in amperage, that is, by the maximum current that they will allow to pass through without breaking. Assuming that Linda's outlet is on a 3 Amp fuse, will the fuse blow? Explain your answer. *Hint*: Which part of the circuit should have current below 3 Amp?

### Exercise 3.5

•   Record the equations expressing Ohm's and Kirkoff's laws in the following circuit. Solve for the indicated currents and voltages, either by hand or using *Mathematica*. Verify that your solution satisfies all necessary equations.

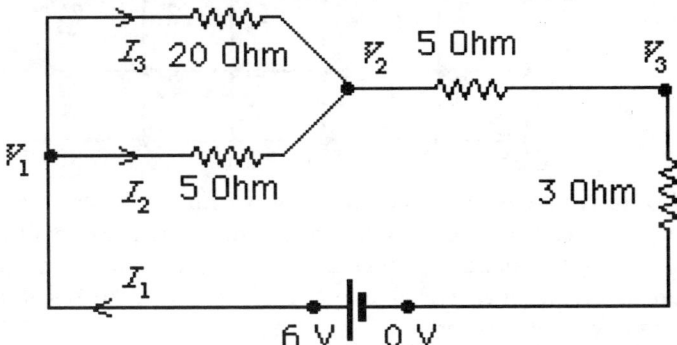

## 3.3   Linear Equations in Thermodynamics

Linear equations also arise in the analysis of heat flow. For example, consider a studio apartment with one radiator, two south-facing windows, and one small window facing east:

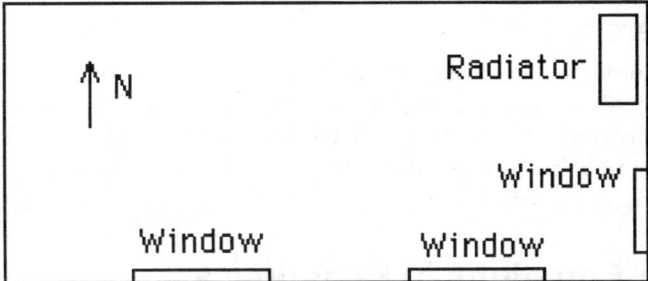

As a very practical matter, one would be concerned with the temperature distribution in the room during the winter. As usual when there are a large number of variables in a completely realistic problem, some reasonable, simplifying assumptions will allow us to describe the problem by a system of linear equations.

For example, we will assume that on a typical winter day, the southern windows are heated by the sun and maintain a temperature of 50°. On the other hand, the eastern window is covered with a thin layer of frost because it receives no direct sunlight and is only 30°. The external walls to the south and west are a well-insulated 60°. The remaining walls border the common hallway and remain about 65°. The only source of heat is the radiator, which remains at 150°. When the radiator is functioning, the heat will flow out from its corner to the rest of the room. We may trace the flow of heat by measuring the heat at a grid of points around the room, as follows:

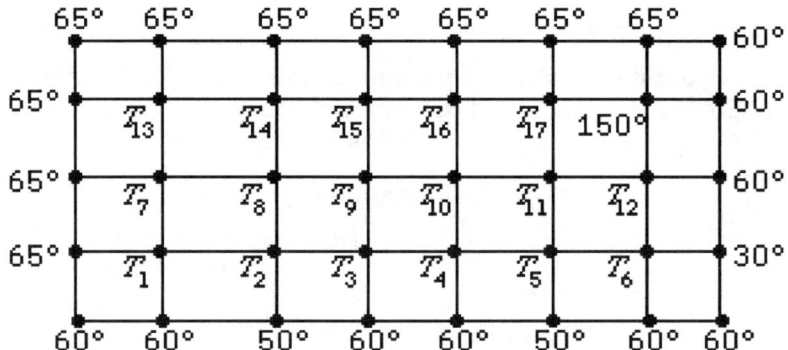

To derive a system of equations governing this situation, we will assume that the room *eventually* reaches an equilibrium temperature distribution. This means that temperatures will balance in a certain way. For example, we might assume that the temperature at any point will be an average of the temperatures of its neighbors. For example, in the lower-left corner, we would have:

$$T_1 = \frac{60 + T_2 + T_7 + 65}{4}$$

*Note*: Although one might choose to include the diagonal "neighbors" in this average as well, possibly with smaller weights, we will stay with the simpler model. This type of reasoning leads to 17 linear equations in 17 unknowns. We have already entered 15 of the equations into *Mathematica*.

### Exercise 3.6

•     Enter the remaining two equations into the Solve[] command in the accompanying Notebook and solve for the equilibrium temperature distribution. From the resulting plot and table of values, determine the point(s) $T_i$ that have the highest and lowest equilibrium temperature, respectively. Do your answers correspond to your intuition? Explain.

## 3.4   Linear Equations in Economics

A final example arises in economics, when one analyzes the distribution of market share among a number of competing industries. For example, in the automotive industry we might focus on three main sectors of the industry: American-, European-, and Far Eastern-made cars. Every year each sector commands some fraction of the market; that is, a certain percentage of cars owned in the United States in any given year are American, European (German, Swedish, and so on) , and Far Eastern (Japanese, Korean, and so on). We may represent these fractions by variables, as follows:

$$x_n = \text{fraction of American-made cars owned in year } n,$$
$$y_n = \text{fraction of European-made cars owned in year } n, \text{ and}$$

$$z_n = \text{fraction of Far Eastern-made cars owned in year } n$$

One might expect market share to vary from one year to the next, depending on such factors as advertising, customer satisfaction, and reliability. In general, these changes could depend on the level of market share over many previous years in a rather complex manner. However, in order to begin to analyze this model, we begin by assuming that these changes depend in a *linear* fashion on only the *previous* year's levels. You may argue that this is overly simplistic, but we must begin with a model that we may hope to analyze successfully, before we employ more complex assumptions.

For example, the change in market share for American-made cars from the $n$th year to the $(n+1)$st year might be given by the following equation:

$$x_{n+1} - x_n = -0.1x_n + 0.07y_n + 0.05z_n$$

Intuitively, this says that 10% of former U.S. car owners switched to foreign models, but U.S. car manufacturers picked up sales from 7% of former European car owners and from 5% of Far Eastern car owners.

As a number of European models, such as Volvo and Mercedes-Benz, tend to last a long time, their manufacturers may lose a smaller fraction of their market each year. On the other hand, they also may not attract the loyal customers of the other sectors, leading to a "transition" equation, such as:

$$y_{n+1} - y_n = 0.03x_n - 0.08y_n + 0.02z_n$$

## Exercise 3.7

Explain why these two equations require the third transition equation to be:

$$z_{n+1} - z_n = 0.07x_n + 0.01y_n - 0.07z_n$$

*Hint*: One sector's loss is another sector's gain so that there is a necessary balance that must be achieved here.

Over time, assuming no changes in these transition percentages from year to year, one might expect the level of market share for each sector to reach an equilibrium value, say $x_\infty$, $y_\infty$, and $z_\infty$, respectively. Thus, the market shares should satisfy the equilibrium equations:

$$
\begin{aligned}
x_\infty - x_\infty &= -0.1x_\infty + 0.07y_\infty + 0.05z_\infty \\
y_\infty - y_\infty &= 0.03x_\infty - 0.08y_\infty + 0.02z_\infty \\
z_\infty - z_\infty &= 0.07x_\infty + 0.01y_\infty - 0.07z_\infty
\end{aligned}
$$

Because the third equation may be deduced from the first two, it is redundant, and this system of equations will not have a unique solution. However, being percentages, the market shares must also satisfy the equation:

$$1 = x_\infty + y_\infty + z_\infty$$

## Exercise 3.8

Explain why this additional equation should be true.

### Exercise 3.9

- Solve this system of equations, either with *Mathematica* or by hand. Which sector ends up with the smallest market share? Is this intuitively plausible, based on the transition equations? Explain.

*Note*: One might argue that it is unrealistic to assume that the transition percentages will remain constant long enough for the market to reach equilibrium. Even if that is the case, this equilibrium analysis would still interest automakers, since it predicts the long-term consequences of their present-day actions. If the equilibrium result is unfavorable to their sector, they must then pursue a more aggressive advertising campaign, or some other strategy to gain market share.

## 3.5   Summary

"Balancing relations" are common in a variety of disciplines. In economics, all ledgers must balance. In the sciences, balance is usually expressed in terms of conservation laws. In chemical equations, we have conservation of atoms. In electrical circuits, we have conservation of charge and current. In thermodynamics, we have conservation of thermal energy. Even in situations where there is no conservation law in the short term, some sort of balance may be assumed in the long run if we eventually reach an equilibrium. Thus, we are natually led to systems of equations in a wide variety of situations.

The most common type of relationship between different quantities is a direct proportion, such as that given by Ohm's law. Even when the actual relationship is more complicated, we will often *assume* it is a proportionality (as in Hooke's law for springs), just to make careful analysis possible. Thus, along with calculus, linear algebra is one of the most widely used tools in applied mathematics, and it is well worth your time spent in careful study.

To summarize, in this lesson, one should have learned:

- that linear algebra is a ubiquitous tool of applied mathematics, arising in a variety of disciplines,

- how the concepts of proportionality, balance, conservation, and equilibrium lead to linear systems of equations, and

- some specific situations where linear algebra may be applied.

These are certainly not the only practical examples of linear algebra, but they will serve us well to illustrate a number of different concepts. Thus, we will return to them periodically throughout the text.

So far, we have used *Mathematica* to solve the linear systems that we have encountered, without understanding *how* it accomplishes this. In Lesson 2 we discussed some operations that may applied to simplify a system of equations by eliminating variables. In Lesson 4, we will see how Gauss used some of these operations in a systematic fashion to simplify any system of linear equations. This is one of the most commonly used techniques for solving linear systems and is, in fact, the way *Mathematica* has been solving our equations all along.

# PART TWO

# Linear Systems and Gaussian Elimination

# Gaussian Elimination by Example

In this lesson, we begin to see what *Mathematica* is *really* doing when we ask it to solve a system of linear equations. The computer uses a technique called Gaussian elimination, credited to C. F. Gauss (although it was later found to be used originally by the Chinese about one thousand years earlier). Thus, we will:

- discuss the allowable operations in Gaussian elimination,

- discuss the use of matrix notation,

- investigate the order in which the operations are applied, and

- investigate the goals of Gaussian elimination.

## 4.1 Discovering the Patterns of Gaussian Elimination

In Lesson 2, we discovered a number of allowable operations that may legitimately be applied to simplify a system of equations. For example, we saw that we could multiply one of the equations by a nonzero number, or add one of the equations to another. Because any combination of allowable operations will also be allowable, there are a large number of possible operations. However, for Gaussian elimination, only three operations are used. One is allowed to do the following:

(A) Divide an equation by a nonzero number.

(B) Interchange two equations.

(C) Subtract a multiple of one equation from another.

Notice that the first operation may be thought of as multiplication, while the last operation is a combination of multiplication and addition. For reference, we establish notation for each type of operation:

| Verbal Description | Shorthand Notation |
|---|---|
| Divide the first equation by $-3$. | $E_1 \div (-3)$ |
| Interchange the first and second equations. | $E_1 \leftrightarrow E_2$ |
| Subtract 3 times the first equation from the second. | $E_2 - (3) \times E_1$ |

In our shorthand notation, $E_i$ refers to the $i$th equation. As with any notation, this one is somewhat arbitrary. You may choose to employ another notation, as long as it is clear and unambiguous.

To understand Gaussian elimination, we must answer a number of questions:

- How do we choose the equations on which to operate and when do we apply each type of operation?

- If we use operations A or C, how do we choose the divisors and multipliers?

- When do we stop?

In the next exercise, you will deduce the answers to these questions by carefully analyzing a well-chosen example.

## Exercise 4.1

Examine the following example. We have labeled each step with an arrow and the letters $a$ through $i$. At each step:

- Determine which equation(s) were changed. *Note*: Operation C changes two equations, while A and B only change one.

- Decide which type of operation from A to C was used. *Note*: Once you eliminate C as a possibility, it is easy to tell whether or not A was used.

- If operation A was used, determine the divisor.

- If operation C was used, determine the multiplier and the other equation whose multiple was subtracted.

- Determine a shorthand description of the complete operation, as suggested above.

Record your observations in a chart such as this:

| Step | Type of Operation Used | Row(s) Changed | Shorthand Description |
|------|------------------------|----------------|-----------------------|
| $a$ | A | 1 | $E_1 \div 2$ |
| $\vdots$ | $\vdots$ | $\vdots$ | $\vdots$ |

Armed with this chart, we will go back and search for patterns in the process to determine the general rules for Gaussian elimination.

$$
\begin{array}{rcl}
2x + 4y - 2z + 4w &=& 8 \\
-x - 2y + 4z + 10w &=& 11 \\
3x + 5y - 6z + 5w &=& 7 \\
2x + 5y - 3z - 11w &=& -7
\end{array}
\quad \xrightarrow{a} \quad
\begin{array}{rcl}
x + 2y - z + 2w &=& 4 \\
-x - 2y + 4z + 10w &=& 11 \\
3x + 5y - 6z + 5w &=& 7 \\
2x + 5y - 3z - 11w &=& -7
\end{array}
\quad \xrightarrow{b}
$$

$$
\begin{array}{rcl}
x + 2y - z + 2w &=& 4 \\
3z + 12w &=& 15 \\
3x + 5y - 6z + 5w &=& 7 \\
2x + 5y - 3z - 11w &=& -7
\end{array}
\quad \xrightarrow{c} \quad
\begin{array}{rcl}
x + 2y - z + 2w &=& 4 \\
3z + 12w &=& 15 \\
-y - 3z - w &=& -5 \\
2x + 5y - 3z - 11w &=& -7
\end{array}
\quad \xrightarrow{d}
$$

$$
\begin{array}{rcl}
x + 2y - z + 2w &=& 4 \\
3z + 12w &=& 15 \\
-y - 3z - w &=& -5 \\
y - z - 15w &=& -15
\end{array}
\quad \xrightarrow{e} \quad
\begin{array}{rcl}
x + 2y - z + 2w &=& 4 \\
-y - 3z - w &=& -5 \\
3z + 12w &=& 15 \\
y - z - 15w &=& -15
\end{array}
\quad \xrightarrow{f}
$$

$$
\begin{array}{rcl}
x + 2y - z + 2w &=& 4 \\
y + 3z + w &=& 5 \\
3z + 12w &=& 15 \\
y - z - 15w &=& -15
\end{array}
\quad \xrightarrow{g} \quad
\begin{array}{rcl}
x + 2y - z + 2w &=& 4 \\
y + 3z + w &=& 5 \\
3z + 12w &=& 15 \\
-4z - 16w &=& -20
\end{array}
\quad \xrightarrow{h}
$$

$$
\begin{array}{rcl}
x + 2y - z + 2w &=& 4 \\
y + 3z + w &=& 5 \\
z + 4w &=& 5 \\
-4z - 16w &=& -20
\end{array}
\quad \xrightarrow{i} \quad
\begin{array}{rcl}
x + 2y - z + 2w &=& 4 \\
y + 3z + w &=& 5 \\
z + 4w &=& 5 \\
0 &=& 0
\end{array}
\quad \text{STOP!}
$$

At this point, Gaussian elimination is complete. The final system has a very special form, called "row echelon" form, which is the ultimate goal of elimination.

## Exercise 4.2

Based on this example, briefly describe what you think a general system in row echelon form should look like. *Hint*: "Echelon" is the French word for "stairs."

## Exercise 4.3

Based on the previous example, identify the intermediate goals of Gaussian elimination by answering the following questions:

(a) How are steps $a$, $e$, $f$, and $h$ similar? What seems to be the desired effect of these steps? How is the system simpler as a result of these steps?

(b) How are steps $b$ through $d$, step $g$, and step $i$ similar? How are they different? How is the system simpler as a result of these steps? What seems to be the desired effect of these steps?

## 4.2   Using Matrix Notation

We may save ourselves a lot of writing by employing matrix notation. You may have heard of matrices before. From a purely formal point of view, a "matrix" is a rectangular array such as this:

$$\begin{bmatrix} 1 & 0 & 3 \\ 3 & -5 & 0 \\ -6 & 2 & 9 \end{bmatrix} \quad \text{or} \quad \begin{bmatrix} 1 & 2 & 0 & 3 \\ 4 & 5 & -1 & 6 \\ 7 & 0 & 2 & 9 \end{bmatrix} \quad \text{or} \quad \begin{bmatrix} 1 & 2 \\ 3 & 4 \\ 5 & 6 \\ 7 & 9 \end{bmatrix}$$

Notice that there is no restriction on the shape or size of a matrix, except that it is rectangular and all the entries are filled. Thus, something like

$$\begin{bmatrix} 1 & & 3 \\ 3 & -5 & 0 \\ -6 & 2 & \end{bmatrix}$$

is not a legal matrix. Like the word "index" or "appendix," "matrix" is the singular form and its plural is "matrices."

As an interesting side note, the word "matrix" comes from "mater," the Latin word for mother. Historically, this word was used because matrices were thought to give birth to determinants! Near the turn of the century, much of anaytic geometry revolved around determinants, although today they are used mainly in theoretical work. We will discuss determinants more fully in Lesson 19.

Using matrix notation, we may write the system of equations

$$\begin{array}{rcl} 2x + 4y - 2z + 4w &=& 8 \\ -x - 2y + 4z + 10w &=& 11 \\ 3x + 5y - 6z + 5w &=& 7 \\ 2x + 5y - 3z - 11w &=& -7 \end{array} \quad \text{as} \quad \left[\begin{array}{rrrr|r} 2 & 4 & -2 & 4 & 8 \\ -1 & -2 & 4 & 10 & 11 \\ 3 & 5 & -6 & 5 & 7 \\ 2 & 5 & -3 & -11 & -7 \end{array}\right]$$

Similarly, we may write

$$\begin{array}{rcl} x + 2y - z + 2w &=& 4 \\ y + 3z + w &=& 5 \\ z + 4w &=& 5 \\ 0 &=& 0 \end{array} \quad \text{as} \quad \left[\begin{array}{rrrr|r} 1 & 2 & -1 & 2 & 4 \\ 0 & 1 & 3 & 1 & 5 \\ 0 & 0 & 1 & 4 & 5 \\ 0 & 0 & 0 & 0 & 0 \end{array}\right]$$

These are examples of "partitioned" (or "augmented") matrices. The partition is the vertical bar separating the matrix into two submatrices (just like a partition in a room). In this case, the vertical bar indicates where the equal signs are to go.

Now each row corresponds to an equation. Our operations on a system of equations correspond to operations on the rows of the corresponding matrix. Thus, we modify our terminology slightly and refer to them as "row operations."

**Definition 4.1** We will use only three types of operations, which we refer to as *elementary row operations*:

(a)  Divide a row by a nonzero number.

(b)  Interchange two rows.

(c)  Subtract a multiple of one row from another.

These operations are "elementary" in that they are the most fundamental operations of Gaussian elimination, in the same way that elementary particles are fundamental to physics.

## Exercise 4.4

Briefly describe what a matrix in row echelon form should look like.

## *Drill Exercise 4.5*

Write out the example of Gaussian elimination, given in Section 4.1, as a sequence of matrices. Outline the general procedure for Gaussian elimination using the language of rows, columns, and matrices.

# A Summary of Gaussian Elimination

This lesson summarizes the important facts of Lesson 4. Namely, we give a precise description of row echelon form of a matrix and a step-by-step description of Gaussian elimination.

## 5.1 Defining Row Echelon Form

In matrix form, an arbitrary row echelon system looks something like this:

$$\begin{bmatrix} 1 & 2 & -1 & 2 & 4 \\ 0 & 1 & 3 & 1 & 5 \\ 0 & 0 & 0 & 1 & 5 \\ 0 & 0 & 0 & 0 & 0 \\ 0 & 0 & 0 & 0 & 0 \end{bmatrix}$$

*Note*: The dotted lines are not part of the matrix — they are included simply to highlight the pattern. This type of pattern is called "row echelon," because of the stairstep pattern of 1's and 0's. In fact, the word "echelon" literally means "stairs" or "ladder" in French.

**Definition 5.1** A *row echelon* matrix must satisfy the following two properties:

(a) Each row must have a 1 as its first nonzero entry, unless the row consists entirely of 0's. This entry is referred to as a "leading 1."

(b) The leading 1 in any given row should be further to the right than the leading 1's in any previous row.

*Important*: There are no other requirements or restrictions. In particular, a row echelon matrix may have 0's or 1's or anything else *above* the "stairs." These two rules force any rows of 0's to the bottom of the matrix and lead to this stairlike pattern.

### Exercise 5.1

Decide whether the following matrices are in row echelon from. If you say that one is not, you should also explain why.

$$
\text{(a)} \begin{bmatrix} 1 & -1 & 3 \\ 0 & -5 & 1 \\ 0 & 0 & 1 \end{bmatrix} \qquad
\text{(b)} \begin{bmatrix} 1 & -1 & 3 \\ 0 & 0 & 1 \\ 0 & 1 & 4 \end{bmatrix} \qquad
\text{(c)} \begin{bmatrix} 0 & 1 & 3 \\ 0 & 0 & 1 \\ 0 & 0 & 0 \end{bmatrix}
$$

## 5.2  A Precise Description of Gaussian Elimination

The steps of Gaussian elimination are largely determined by its goal of achieving row echelon form, while only using the three operations of Definition 4.1. It may be outlined as follows:

1. Start in the first row of the first column.

2. Attempt to obtain a leading 1 in this position by dividing the row by the entry in this position.

   (a) If this entry is 0, we interchange its row with a row *below* it that contains a nonzero entry in the same column.

   (b) If all lower entries in its column are 0, we continue to work on the same row, but skip to the next column and start again at step 2.

3. Obtain 0's below the leading 1 created in step 2 by subtracting multiples of this row from the rows *below*. *Note*: We sometimes refer to this leading 1 as the "pivot."

   (a) In each successive row, focus on the entry below the pivot and subtract that multiple of the pivot row from the current row.

   (b) If a row already contains a 0 below the pivot, skip to the next row.

4. When one has obtained all 0's below the pivot, move over one column, down one row, and start again at step 2, continuing in this way until one achieves row echelon form.

This detailed description may also be summarized by the following three rules:

   – work your way down and to the right,

   – use the operations described in Definitions 4.1a and 4.1b to obtain a leading 1, and

   – use Definition 4.1c to obtain 0's below.

### Exercise 5.2

Compare these instructions with the example in Lesson 4. Rewrite them in your own words.

### Exercise 5.3

Use Gaussian elimination to convert the following matrices into row echelon form. Make sure to indicate clearly the row operations you use and the order in which you apply them.

(a) $\begin{bmatrix} 1 & 1 & -8 & -14 \\ 3 & -4 & -3 & 0 \\ 2 & -1 & -7 & -10 \end{bmatrix}$

(b) $\begin{bmatrix} 1 & -2 & 1 & 4 \\ 2 & -3 & -1 & 2 \end{bmatrix}$

Answer: $\begin{bmatrix} 1 & 1 & -8 & -14 \\ 0 & 1 & -3 & -6 \\ 0 & 0 & 0 & 0 \end{bmatrix}$

Answer: $\begin{bmatrix} 1 & -2 & 1 & 4 \\ 0 & 1 & -3 & -6 \end{bmatrix}$

(c) $\begin{bmatrix} 1 & 2 & 3 & 4 & 7 \\ 0 & 5 & 7 & 6 & 8 \\ 0 & 5 & 2 & 4 & 7 \\ 0 & 0 & 2 & 1 & 0 \end{bmatrix}$

(d) $\begin{bmatrix} 4 & -1 & 2 & 6 \\ -1 & 5 & -1 & -3 \\ 3 & 4 & 1 & 3 \end{bmatrix}$

Answer: $\begin{bmatrix} 1 & 2 & 3 & 4 & 7 \\ 0 & 1 & 7/5 & 6/5 & 8/5 \\ 0 & 0 & 1 & 2/5 & 1/5 \\ 0 & 0 & 0 & 1 & -2 \end{bmatrix}$

Answer: $\begin{bmatrix} 1 & -1/4 & 1/2 & 3/2 \\ 0 & 1 & -2/19 & -6/19 \\ 0 & 0 & 0 & 0 \end{bmatrix}$

# Backsubstitution and Backaddition

In the last lesson, we learned to apply Gaussian elimination to a system of equations to obtain another system in row echelon form that we believe to have the same solutions. We have yet to give a formal *proof* of this fact. We will ultimately do so in Lesson 18, once we have developed the necessary theoretical tools. For now, we will assume that this is true, based on our experiments in Lesson 2, and proceed to discuss two different methods for solving the resulting row echelon system, referred to as "backsubstitution" and "backaddition," respectively.

## 6.1   Solving a Row Echelon System by Backsubstitution

Remember that an arbitrary row echelon matrix is of the form:

$$\begin{bmatrix} 1 & 2 & -1 & 2 & 4 \\ 0 & 1 & 3 & 1 & 5 \\ 0 & 0 & 0 & 1 & 5 \\ 0 & 0 & 0 & 0 & 0 \\ 0 & 0 & 0 & 0 & 0 \end{bmatrix}$$

If this corresponds to a system in $x$, $y$, $z$, and $w$, we may rewrite this matrix as the following equations:

$$\begin{aligned} x + 2y - z + 2w &= 4 \\ y + 3z + w &= 5 \\ w &= 5 \\ 0 &= 0 \\ 0 &= 0 \end{aligned}$$

Notice that we omitted the partition before the final column in the original matrix. It is common in linear algebra to insert or omit partition lines, depending on the context. For example, Gaussian elimination (and backaddition) proceeds irrespective of any partitioning, so any partitions may be omitted during elimination. Likewise, *Mathematica* will not show partition lines. When a matrix is intended to represent a system of linear equations, however, we must remember to insert a vertical partition before the last column to indicate where the $=$ sign belongs. Ultimately, one must learn to keep track of partitions on one's own, based on the context of the calculation.

In general, we translate between the matrix and equation forms of a system of equations by associating variables with each column but the last. From the row echelon form of the system, we classify each variable (and column) as either "free" or "basic." The variables corresponding to the columns with leading 1's are called "leading," "dependent," or "basic" variables, while the remaining variables are called "free" or "independent" variables.

### Exercise 6.1

Identify each of the four variables in this system ($x, y, z,$ and $w$) as either basic or free.

This terminology comes from the fact that the free variables may be freely assigned values, which then uniquely determine values for the basic variables. The term "basic" refers to the more advanced notion of a basis, which we will not discuss in detail until Lesson 24.

To see this practice, we will rewrite the equations from the previous exercise in a more convenient form, using the process of "backsubstitution." In a single word, this term accurately describes the entire process. This technique proceeds "bottom-up," using successive substitution and algebraic simplification. In this example, starting at the bottom, we observe that the equations $0 = 0$ are trivially true for any values of the variables and thus tell us nothing about the solution. We proceed by moving up to the first nontrivial equation, which tells us that $w = 5$. We then substitute this value into the next equation, moving upward, to obtain:

$$y + 3z + 5 = 5 \qquad \text{or} \qquad y = -3z$$

Continuing in this way, we also deduce that:

$$x + 2(-3z) - z + 2(5) = 4 \qquad \text{or} \qquad x = -6 + 7z$$

Summarizing, the solution set may be described by the following equations:

$$
\begin{aligned}
x &= -6 + 7z \\
y &= -3z \\
w &= 5
\end{aligned}
$$

Notice that we have solved for the basic variables in terms of the free variables. Thus, $z$ may be assigned any value, and the corresponding values for $x, y,$ and $w$ are then easily computed.

## 6.2   Solving a Row Echelon System by Backaddition

Another method for obtaining the same results resembles Gaussian elimination and is referred to as "backaddition" or "Gauss-Jordan" elimination. Its goal is to convert the system to "reduced row echelon" form. That is, to eliminate more entries by obtaining 0's *above* the leading 1's as well. As in Lesson 4, we will discover the process of backaddition by carefully analyzing a particular example.

## Exercise 6.2

Describe the row operations used at each step in the following example:

$$
\begin{bmatrix} 1 & 2 & -1 & 2 & 4 \\ 0 & 1 & 3 & 1 & 5 \\ 0 & 0 & 0 & 1 & 5 \\ 0 & 0 & 0 & 0 & 0 \\ 0 & 0 & 0 & 0 & 0 \end{bmatrix} \xrightarrow{a}
\begin{bmatrix} 1 & 2 & -1 & 2 & 4 \\ 0 & 1 & 3 & 0 & 0 \\ 0 & 0 & 0 & 1 & 5 \\ 0 & 0 & 0 & 0 & 0 \\ 0 & 0 & 0 & 0 & 0 \end{bmatrix} \xrightarrow{b}
$$

$$
\begin{bmatrix} 1 & 2 & -1 & 0 & -6 \\ 0 & 1 & 3 & 0 & 0 \\ 0 & 0 & 0 & 1 & 5 \\ 0 & 0 & 0 & 0 & 0 \\ 0 & 0 & 0 & 0 & 0 \end{bmatrix} \xrightarrow{c}
\begin{bmatrix} 1 & 0 & -7 & 0 & -6 \\ 0 & 1 & 3 & 0 & 0 \\ 0 & 0 & 0 & 1 & 5 \\ 0 & 0 & 0 & 0 & 0 \\ 0 & 0 & 0 & 0 & 0 \end{bmatrix} \text{STOP!}
$$

## Exercise 6.3

Compare and contrast the procedures of backaddition and Gaussian elimination.

Because we have already obtained leading 1's, backaddition is easier than Gaussian elimination. Moreover, as we work our way up and to the left, the arithmetic is particularly simple, because of all the 0's entries in a row echelon matrix. If we now convert this reduced row echelon matrix to its corresponding system of equations:

$$
\begin{aligned}
x - 7z &= -6 \\
y + 3z &= 0 \\
w &= 5
\end{aligned}
$$

and solve for the leading variables

$$
\begin{aligned}
x &= -6 + 7z \\
y &= -3z \\
w &= 5
\end{aligned}
$$

we obtain the same answer as before using backsubstitution.

Comparing the two methods, we may make two observations:

- Backsubstitution is convenient for those who prefer algebraic manipulation, and it emphasizes the connection with a system of equations.

- Once one has mastered Gaussian elimination, backaddition is the natural way to bring a system to its most simple form.

Although it may seem that backsubstitution is faster, when we count the number of required additions and multiplications, in general, both techniques employ the *same* number of arithmetic operations. However, in Lesson 7 we will see how backaddition is more convenient for computerized computation, and in Lesson 15 we will learn how backaddition may be applied more quickly to solve "parallel" systems of equations. Although you may begin by using whichever method you prefer, be aware that the remainder of the text primarily uses backaddition.

## Exercise 6.4

Assume that the matrix $\begin{bmatrix} 1 & 3 & 5 & -2 & 4 \\ 0 & 1 & -1 & 3 & 2 \\ 0 & 0 & 1 & 2 & 1 \\ 0 & 0 & 0 & 0 & 0 \end{bmatrix}$ represents a system of equations in $x, y, z,$ and $w$.

(a) Identify the basic and the free variables.

(b) Use backsubstitution to write out the general solution.

(c) Perform backaddition on the matrix, and compare the resulting equations with your answer from 6.4b.

(d) Pick a specific value (such as 2) for each of the free variables and solve for the corresponding basic variables. Verify that the resulting values give a valid solution to the original system.

## Exercise 6.5

Solve the following linear systems in the indicated variables, given in matrix form, either by backsubstitution or backaddition. Verify each of your answers.

(a) $\begin{bmatrix} 1 & -2 & 4 \\ 0 & 1 & -3 \end{bmatrix}$ as a system in $x$ and $y$.

(b) $\begin{bmatrix} 1 & 2 & 3 & 6 \\ 0 & 1 & 4 & 3 \\ 0 & 0 & 1 & 2 \end{bmatrix}$ as a system in $x, y,$ and $z$.

# Gaussian Elimination with *Mathematica*

*Mathematica* Notebook

## 7.1  Introduction

In this lesson, we discuss how to manipulate matrices in *Mathematica*. In particular, we will see how to perform Gaussian elimination and backaddition. We will discuss techniques for organizing one's work in *Mathematica*, as well as how to create, solve, and verify self-generated exercises. We conclude by discussing the rank of a matrix. Although we will not discuss its uses until after discuss matrix algebra and inverse, it is a simple concept that is naturally introduced at this point.

## 7.2  Matrices in *Mathematica*

*Mathematica* can perform many different operations, but it uses notation for matrices that differs from standard mathematical notation. For example, *Mathematica* will print the matrix:

$$\begin{bmatrix} 1 & 2 & 3 \\ 4 & 5 & 6 \\ 7 & 8 & 9 \end{bmatrix}$$

as:

```
1  2  3

4  5  6

7  8  9
```

Actually, *Mathematica* treats a matrix as a list of rows. Thus the above matrix would be entered as:

- `{{1, 2, 3},{4, 5, 6},{7, 8, 9}}`

To see a matrix in its rectangular form, we must use the MatrixForm[] command:

- `MatrixForm[{{1, 2, 3},{4, 5, 6},{7, 8, 9}}]`

*Warning*: As we said in Lesson 1, *Mathematica* is case-sensitive, so we must always remember to capitalize the letters M and the F in MatrixForm[].

We may assign a matrix to a variable just as before. We will usually reserve uppercase variables names for matrices, such as:

- `A = {{1, 2, 3},{4, 5, 6},{7, 8, 9}}`

It is a standard convention to refer to the entries of a matrix by using subscripts on variable name for the matrix. For example,

$$A_{2,3}$$

would refer to the entry in the second row, third column of the matrix $A$. *Mathematica* uses a similar convention.

## Exercise 7.1

- Verify that

- `A[[2,3]]`

    returns the correct value, when evaluated. Notice how we use *double* brackets and a comma between the subscripts. Experiment by changing the subscripts and re-evaluating. Describe what you expect to happen in each case and report the results.

In a similar manner, one may refer to an entire *row* of a matrix simply by dropping the second subscript. For example,

- `A[[1]]`

should return the first row of $A$.

## Exercise 7.2

- Verify that this is true. As before, experiment by editing the subscript. Report on your results.

One may also assign new values to any entry or row, in the obvious way.

## Exercise 7.3

- Describe what should happen if you evaluate:

- `A[[2,3]] = 12; A[[3]] = {9,8,7}; MatrixForm[A]`

    Test your conjecture by evaluating the cell. Experiment with these commands to change the entries of $A$. Report on your results.

## 7.3    Row Operations in *Mathematica*

*Mathematica* will perform row operations in a very natural manner. To illustrate, we will reproduce the steps in Lesson 4. We begin with:

- ```
  A = {{2,4,-2,4,8},
       {-1,-2,4,10,11},
       {3,5,-6,5,7},
       {2,5,-3,-11,-7}};
  MatrixForm[A]
  ```

Remember that, even though the spaces and carriage returns are unnecessary, they are helpful for readability.

Although we originally phrased it in terms of division, the first operation may be described as "multiply the first row by $1/2$." This suggests that the operation may be performed by the following command:

- ```
  A[[1]] = 1/2 A[[1]];
  MatrixForm[A]
  ```

Literally, this instructs *Mathematica* to multiply the first row by $1/2$, place the result back in the first row, and print the resulting matrix in a readable form. This illustrates how to perform the row operation described in Definition 4.1a.

Notice that this command destroys our original matrix $A$. To recover the original matrix, remember from Section 1.2.3 how *Mathematica* saves the output of each command in a sequence of Out[] variables.

### Exercise 7.4

- Determine the number of the command that originally defined the matrix $A$. If it were $9$, then the command $A =$Out[9] would reset $A$ to its original value. Edit and evaluate the following command:

- ```
  A = Out[];
  ```

  to restore $A$ to its original value, recording the exact command you use.

*Important*: In this way, if you ever make a mistake, you may correct your mistake without starting over from the beginning of the Notebook. You may instead simply backtrack to a point where you are confident in your results, restore the values of all necessary variables to their values at that point, and then perform the *correct* steps from that point on.

Because we have restored $A$ to its original value, before proceeding to the second step of Gaussian elimination, we must again perform the first operation:

- ```
  A[[1]] = 1/2 A[[1]];
  MatrixForm[A]
  ```

Now we must subtract $-1$ times the first row from the second, placing the result back in the second row. This is the row operation of Definition 4.1c, which may be performed with the following command:

- ```
  A[[2]] = A[[2]] - (-1) A[[1]];
  MatrixForm[A]
  ```

## Exercise 7.5

- Record the command which would perform the next row operation, namely to obtain a 0 in the third row, first column. Edit the previous cell to create that command. Verify that your command achieves the desired result. *Remember*: If your first attempt is incorrect, you may backtrack as in Ex. 7.4.

## Exercise 7.6

- Repeat the previous exercise again to perform the next row operation. Record the necessary *Mathematica* instruction.

At this point, we see that we must use the operation in Definition 4.1b. This is a bit more complex than the other two types of operations. That is why we provide the command SwitchRows[] in the package "Matrices.m". *Remember*: If you have not opened the Notebook "Start.ma" yet, you should do so before evaluating the following command. For example, to switch the second and third rows of $A$, we would use the command:

- ```
  A = SwitchRows[A,2,3]; MatrixForm[A]
  ```

Explicitly, this performs the following sequence of commands:

- ```
  tmp = A[[2]]; A[[2]] = A[[3]]; A[[3]] = tmp;
  MatrixForm[A]
  ```

We must store one of the rows temporarily; otherwise when we evaluate $A[[2]] = A[[3]]$, the original second row will be lost and and we will be unable to assign it to the third row. Although this sequence of commands is not too difficult, by using the command SwitchRows[] we may focus on the intent of this operation, instead of on its precise implementation.

Having seen how to perform all three elementary row operations, we can complete Gaussian elimination using the computer.

## Exercise 7.7

- Use *Mathematica* to perform each of the remaining steps of Gaussian elimination. You may either copy, edit, and evaluate the previous cells or type in the necessary commands directly. You may verify each step by comparing your results with those of Lesson 4. You should record the necessary commands in the order in which you use them.

We may also use *Mathematica* to verify our work. We provide the command Gauss-Eliminate[] in the package "Decomps.m" to be used in two ways. It will show "the" final result of Gaussian elimination on a matrix. Because there is some freedom of choice
  ```

in Gaussian elimination, specifically in choosing which rows to interchange, the resulting row echelon form is not unique. This command only switches when necessary (i.e., when there is a 0 when one would want a leading 1) and then only with the *first* row below with the necessary nonzero entry. As long as one follows these same conventions, one should obtain the same result. For example, the final matrix from Ex. 7.7 should have been:

- ```
  B = {{2,4,-2,4,8},{-1,-2,4,10,11},{3,5,-6,5,7},{2,5,-3,-11,-7}};
  MatrixForm[GaussEliminate[B]]
  ```

Notice that this command also prints the intermediate matrices that one should obtain after elimination in each column. Thus, if one's final answer seems to be incorrect, by comparing one's hand-calculations with the printout of this command, it is possible to narrow down the precise location of one's error. *Important*: Choosing to interchange rows in a different manner will lead to a different answer than that of the computer. To verify such a result, one can only plug one's answer back into the original equations by hand, without the aid of the computer.

We may, of course, use the same commands to continue on to achieve *reduced* row echelon form.

### Exercise 7.8

- Perform backaddition on the same problem. Again, record the necessary commands in their proper order.

Again, *Mathematica* is useful for verifying our answer. The appropriate command is called RowReduce[]. For example, one should have obtained the following reduced row echelon matrix:

- ```
  MatrixForm[RowReduce[B]]
  ```

*Note*: This is a built-in command. This means that we may use it at any time on any computer that has *Mathematica*, without loading any packages. Because it is a fundamental *Mathematica* command, in the remainder of the text, we will begin to rely more and more heavily on it. In this case, it does not matter how we choose to interchange rows, because the *reduced* row echelon form of a matrix *is* unique.

## 7.4  Homework Problems

With the commands described in the previous section, the computer can aid us in learning the process of Gaussian elimination by performing all of the associated arithmetic. Although one could obtain the answer with a single GaussEliminate[] command, you should use the computer to practice Gaussian elimination by performing *each* individual row operation, until you are completely familiar with the method. Ultimately, you should solve a few problems *completely by hand*, to make sure that you are prepared to do this when the computer is unavailable, such as on an exam.

To complete the exercises in the remainder of this lesson, you may wish to create a *new* Notebook (using the New command from the File menu) to contain all of one's scratch-work in a neat and orderly fashion, complete with comments describing your thought process. This Notebook will serve a number of purposes. It will enable you to show your work to someone else, say, to check your computations, in a systematic manner. When completely correct, such a Notebook also serves as a valuable study aid. Finally, if you must end a *Mathematica* session before completing a problem, you will be able to reconstruct your work to the point at which you quit, simply by re-evaluating all of the cells in that Notebook. For your convenience, we have summarized the commands that we have used so far in this Notebook, grouped according to their function. You may choose to use them only for reference, or you may copy and edit them into your new Notebook to save typing.

**To store your original matrix, with a spare copy for later:**

- 
```
B = A = {{2,4,-2,4,8},
    {-1,-2,4,10,11},
    {3,5,-6,5,7},
    {2,5,-3,-11,-7}};
MatrixForm[A]
```

**The three types of row operations:**

- 
```
A[[1]] = 1/2 A[[1]];
MatrixForm[A]
```

- 
```
A[[2]] = A[[2]] - (-1) A[[1]];
MatrixForm[A]
```

- 
```
A = SwitchRows[A,2,3]
```

**To check your work:**

- 
```
MatrixForm[GaussEliminate[B]]
```

- 
```
MatrixForm[RowReduce[B]]
```

## Exercise 7.9

- Use the computer to perform Gaussian elimination and backaddition on the following matrix:

```
0   -1   -2   1

2    6    2   8

1    6   11   9
```

Record all necessary row operations in their correct order. Make sure to check your work with the computer.

As independent learners, we also must learn to promote and evaluate our progress on our own by creating exercises and verifying our results. Throughout this text, one should be on the lookout for ways to check one's understanding and to create practice problems. For example, it is easy to create linear systems of equations:

– Start with any set of equations, leaving the right-hand side blank.

– Pick a solution.

– Plug in to determine the corresponding right-hand sides.

This method has the advantage of assuring us of at least one correct solution. However, its disadvantage is messy arithmetic (i.e., fractions), which may tend to appear in the resulting calculations.

Once we have discussed the algebra of matrices in detail (in Lesson 17), we will be able to design matrices for which Gaussian elimination will involve only integer arithmetic. For now, we will use the command MakeSquareSystem[]. This command is provided with this text and defined in the package "Decomps.m". It will create a matrix corresponding to a system of $n$ equations in $n$ unknowns. For example, we may generate three equations in three unknowns with the following command:

● 
```
A = MakeSquareSystem[3];
MatrixForm[A]
```

### Exercise 7.10

● Use the following MakeSquareSystem[] command to generate a practice problem.

● 
```
A = MakeSquareSystem[3];
MatrixForm[A]
```

Write this out in equation form in $x$, $y$, and $z$ and then solve. You may use the computer as you wish to solve this. You should make a record of all necessary row operations for Gaussian elimination, in their correct order. You may then choose to use backaddition or backsubstitution to obtain the actual solution. If you use backaddition, you should record the steps for this as well. If you use backsubstitution, you should show all necessary algebra.

### *Drill Exercise 7.11*

● For additional practice, generate and solve three more practice problems as in the previous exercise. Your problems should all have three or more variables and at least one should have four or more.

## 7.5  The Rank of a Matrix

Associated with Gaussian elimination is the very important property of "rank." We may define the rank of a matrix as follows:

**Definition 7.1** The *rank* of a matrix is the number of leading 1's in any row echelon form of the matrix.

One may well wonder if this is a good definition. Because the exact row echelon form obtained from Gaussian elimination depends on the row interchanges used, it is not clear that the number of leading 1's is the same in every row echelon form of a matrix. To understand why this definition makes sense, we must first develop some theoretical concepts of linear algebra.

With this example, we begin to emphasize the point that theory is neither irrelevant nor arbitrary. Throughout this text, we will introduce just enough terminology and prove just enough results to obtain clear answers to important questions. Almost the entire text will be concerned with establishing a comprehensive understanding of the problem of solving linear systems. One should notice that even the terminology we introduce is not arbitrary but is loosely connected with standard English usage of the words.

## Exercise 7.12

- Compute the rank of the following matrices.

$$
\text{(a)} \begin{bmatrix} 1 & 1 & -8 & -14 \\ 3 & -1 & -3 & 0 \\ 2 & -1 & -7 & -10 \end{bmatrix} \quad
\text{(b)} \begin{bmatrix} 1 & -2 & 1 & 4 \\ 2 & -3 & -1 & 2 \end{bmatrix} \quad
\text{(c)} \begin{bmatrix} 1 & 2 & 3 & 4 & 7 \\ 0 & 5 & 7 & 6 & 8 \\ 0 & 5 & 2 & 4 & 7 \\ 0 & 0 & 2 & 1 & 0 \end{bmatrix} \quad
\text{(d)} \begin{bmatrix} 4 & -1 & 2 & 6 \\ -1 & 5 & -1 & -3 \\ 3 & 4 & 1 & 3 \end{bmatrix}
$$

*Hint*: These matrices are from an earlier exercise. Try to explain why we use the word "rank" to describe this concept. *Hint*: Consider the various dictionary definitions of this word.

We provide the command MatrixRank[] in the package "Decomps.m" to allow us to verify our answers to the previous exercise. For example,

- 
```
A = Partition[{-4, 12, -8, 12, 12, 3, -9, 6, -9, -9, 1, -3, 2, -3, -2, 0,
-3, -12, -12, 9}, 5];
Print["The reduced row-echelon form of A is: ", MatrixForm[RowReduce[A]]];
Print["A has rank ", MatrixRank[A]];
```

## 7.6  Summary

In this lesson, one should have learned:

- how to create, modify, and display matrices in *Mathematica*,

- how to perform each of the three types of row operations,

- how to use Out[] variables to backtrack in a computation,

- how to create, solve, and check new exercises in Gaussian elimination and backaddition, and

– the definition of the rank of a matrix.

While most of the systems that we examined in this lesson had unique solutions, in practice this is often not the case. In the next lesson, we will examine the full range of possible solution sets for linear systems. We will discuss how to distinguish each case by examining the row echelon form of the associated matrix. We also discuss how to write each type of solution out in precise mathematical terms.

# The General Solution to Linear Systems

In the last lesson, we discussed how to compute the solution of a system of linear equations using *Mathematica*. In Lesson 2 we began to investigate the different types of solution sets that one might expect to see for a linear system. Qualitatively, each type of solution set could be classified by its dimension, starting from "no solution" to a single point (i.e., zero-dimensional), to a line, a plane, and so on. In this lesson, we want to reclassify these possibilities into three main groups and discuss how these three different possibilities will arise, from the point of view of matrices and Gaussian elimination. We will also learn to use another helpful *Mathematica* command, MakeSystem[].

## 8.1  Three Types of Reduced Systems

We know that any system of equations may be converted to an equivalent reduced row echelon system by using Gaussian elimination and backaddition. One of the simplest possible results would be the following:

$$\begin{bmatrix} 1 & 0 & 0 & 4 \\ 0 & 1 & 0 & 5 \\ 0 & 0 & 1 & 5 \\ 0 & 0 & 0 & 0 \end{bmatrix}$$

This result implies that the corresponding system of equations has a *unique* solution. For example, if this corresponds to a system of equations in $x$, $y$, and $z$, the only solution would be:

$$x = 4, y = 5, \text{ and } z = 5$$

We say that this system is "uniquely determined."

In a certain sense, this simple situation occurs quite rarely in practice. Usually a system of equations is either "overdetermined" or "underdetermined." Altogether systems of equations may be classified into three general categories.

**Definition 8.1** The following three terms are used to describe a system of linear equations, depending on the qualitative nature of its solution set:

(a) *Uniquely determined* means that there is one and only one possible solution to the system.

59

(b) *Overdetermined* means that there is no possible solution to the system. Intuitively, that is because there are too many conflicting conditions imposed on the variables. The term "inconsistent" is also used to describe this situation.

(c) *Underdetermined* means that the system has many possible solutions. Intuitively, there are not enough (independent) conditions imposed on the variables to narrow down the solution to a single possibility.

## Exercise 8.1

Assuming that each of the following matrices represents a system of equations in $x$, $y$, $z$, and $w$, decide which one is overdetermined and which is underdetermined. Give reasons for your answers. *Hint*: Hypothesize potential solutions for each system and verify that they work.

(a) $\begin{bmatrix} 1 & 0 & -7 & 0 & | & -6 \\ 0 & 1 & 3 & 0 & | & 0 \\ 0 & 0 & 0 & 1 & | & 5 \\ 0 & 0 & 0 & 0 & | & 0 \\ 0 & 0 & 0 & 0 & | & 0 \end{bmatrix}$
(b) $\begin{bmatrix} 1 & 0 & -7 & 0 & | & -6 \\ 0 & 1 & 3 & 0 & | & 0 \\ 0 & 0 & 0 & 0 & | & 1 \\ 0 & 0 & 0 & 0 & | & 0 \\ 0 & 0 & 0 & 0 & | & 0 \end{bmatrix}$

## Exercise 8.2

For each part of this exercise, formulate general rules to quickly determine to which of the three cases a given system belongs. In each part, your rules should only refer to the indicated information.

(a) Assume that you have the result of RowReduce[] applied to the system.

(b) Assume that you have the result of GaussEliminate[] applied to the system.

(c) Assume that you know the rank of the augmented matrix for the system, the rank of the matrix without the last column, and the number of variables.

Use your observations from the third part of this exercise to formulate a precise mathematical theorem.

## Exercise 8.3

Classify each of the following systems as either overdetermined, underdetermined, or uniquely determined. You should use *Mathematica*, where appropriate; however, you should explain your reasoning in each case.

(a)
$$\begin{aligned} 3x - y &= 7 \\ 6x + 2y &= 10 \\ -3x + 4y &= -10 \end{aligned}$$

(b)
$$\begin{aligned} 3x - y + z - 4w &= 2 \\ 6x + 3y - z - 4w &= 3 \\ 9x + 2y \phantom{{}+z} - 8w &= 6 \end{aligned}$$

$$\begin{aligned} 3x - y + z - 5w - u &= 0 \\ \text{(c)} \quad 6x - 2y + 2z - 9w + u &= 0 \\ -9x + 3y - 3z + 11w - u &= 0 \end{aligned} \qquad \begin{aligned} -x + 2y - z &= -4 \\ \text{(d)} \quad 3x + 4y + 2z &= 15 \\ -4x + 6y + z &= -7 \end{aligned}$$

### Advanced Exercise 8.4

- Consider the system $\begin{aligned} 3x + 6y &= 1 \\ 2x + Ky &= 5 \end{aligned}$, where $K$ is some unknown constant. For which value of $K$ does the system have no solution? Can this have infinitely many solutions? Explain your answers. *Hint*: You may use the RowReduce[] command in *Mathematica* even when the entries of the matrix are unknowns.

## 8.2  Using *Mathematica* to Construct Examples

At this point, you should be proficient at Gaussian elimination. Thus, we will begin to focus our energy on *interpreting* the results of elimination and not on the mechanics. However, if you are not yet confident with Gaussian elimination, you should continue to solve additional exercises *by hand*, above and beyond your usual assignments, until you become proficient. The *Mathematica* command MakeSystem[] is a generalization of MakeSquareSystem[], also defined in the package "Matrices.m", which you may find helpful. With this command, we may specify any number of variables and equations. We may also indicate the qualitative nature of the desired system. For example

- `MakeSystem[3, 5, SolutionType->UnderDetermined, Rank->2]`

will generate a system of three equations in five unknowns. The Rank option specifies the rank of the matrix of the system *without* the final column, that is, the number of basic variables of the system. Thus, this system will then have three free variables. Such a system may either be under- or overdetermined, so we must specify that we want an underdetermined system.

*Warning*: *Mathematica* will not evaluate if you ask for a system that is impossible.

### Exercise 8.5

Evaluate the command

- `MakeSystem[3, 5, SolutionType->UnderDetermined, Rank->6]`

in *Mathematica* and describe what happens. Explain why such a system is impossible.

### Exercise 8.6

Give two more examples of impossible situations, explaining why each example is self-contradictory. Try to illustrate a variety of contradictions.

*Warning*: If you do not specify any options, the computer will assume that you desire a system of equations with a single solution and maximum possible rank. This implies that a command like

- `MakeSystem[3,4]`

is equivalent to:

- `MakeSystem[3,4, SolutionType->Determined, Rank->Full]`

*Aside*: You may always determine the default options of any *Mathematica* command by using the Options[] command. For example, the command Options[MakeSystem] returns:

`{SolutionType->Determined, Rank->Full}`

## Exercise 8.7

- Evaluate the command MakeSystem[3,4] and describe exactly what *Mathematica* returns. Expain why this result occurs.

### *Advanced Exercise 8.8*

Appealing to your theorem from Ex. 8.2, explain *in general* (i.e., for any system of equations, with any number of equations and variables) why we cannot have any possibilities other than:

- – no solution,
- – a single solution, or
- – infinitely many solutions.

In particular, you should explain why a linear system, unlike a pair of quadratic equations, cannot have exactly two solutions.

# Matrix Algebra

# Matrix Multiplication from a Geometric Viewpoint

*Mathematica* Notebook

## 9.1 Introduction

To this point, we have treated matrices as simply a shorthand for systems of equations. However, they are much more. As with any other interesting type of object, matrices may be viewed from many different points of view. One of the most important is that of a "transformation."

Transformations arise in many practical geometric settings. For example, suppose you are using a computer graphics package and you wish to transform (i.e., change) the picture on the left to obtain the one on the right:

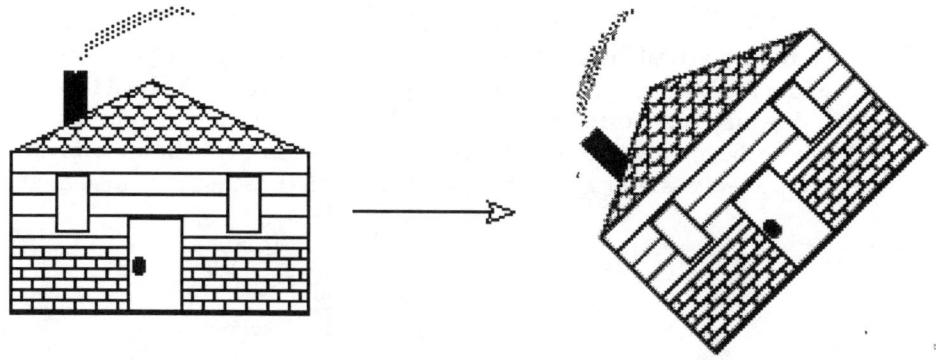

That is, you wish to perform a 45° rotation in a counterclockwise direction. The computer will carry out this operation by multiplication with a particular 2 × 2 matrix.

Likewise, as a physicist or chemist, to precisely characterize the reflective symmetry of water, you would observe geometrically that the water molecule on the left is essentially the same as that on the right.

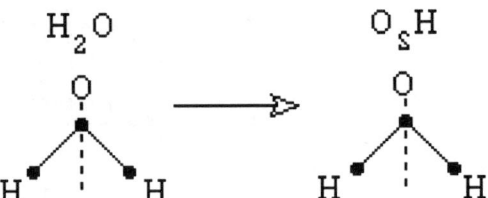

This picture illustrates the transformation of reflection through the dotted vertical line. Because this transformation does not substantially affect the molecule, we say that water has a reflective symmetry. This reflection may be represented by a matrix and is one element in the "symmetry group" of water. A complete understanding of its symmetry group is crucial for quantum mechanical analysis of any molecule.

Transformations may be used quite generally to transform a solution to a particular problem to that of a related problem. For example, in thermodynamics it is easy to analyze the heat distribution in a room where one wall is 60 degrees and the opposite wall is 20 degrees, assuming that the remaining walls are well-insulated. Heat will flow directly from one wall, as indicated with the following picture:

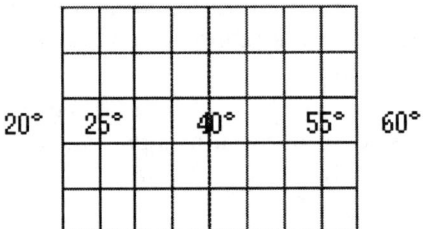

Here the vertical lines represent isotherms ("iso" = "same"; "therm" = "temperature"), while the horizontal lines show the path along which the heat flows.

We may transform this picture to obtain the solution to a similar heat-flow problem, where the temperature difference is now between two windows on the ends of the same wall, again assuming that the remaining walls are perfectly insulated:

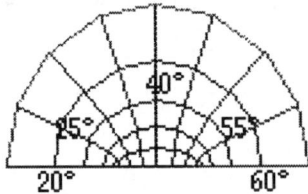

Electrostatic fields, equipotentials, and other similar physical quantities may be determined using a similar mapping technique. In this lesson, we will see that even such nonlinear transformations may be approximated by matrices via differentiation.

In general, we will observe how matrices act as transformations by examining a number of different, specific examples. We will visualize each transformation in the plane, using various new *Mathematica* commands. In general, matrices may represent transformation because they can be made to "act" on points through a special type of "multiplication." We will intentionally postpone the discussion of how to actually perform the requisite matrix multiplications until the next lesson, allowing the computer to perform them for us, so that we may focus on the *geometric* significance of matrix multiplication. Our primary objectives will be to:

- become familiar with the basic elements of linear algebra, the so-called elementary matrices,

- discover the direct connection between a transformation and its matrix, and

- begin to notice the importance of the standard basis.

These three issues are quite important and will recur throughout the text. In this way, *Mathematica* proves particularly helpful, in that it enables us to investigate important topics and develop our intuitive understanding, well before we have developed the theoretical basis for a careful mathematical analysis.

## 9.2   A Matrix as a Transformation

We begin by illustrating how a matrix acts on points to transform them into other points. In this section, we will concentrate on the matrix:

- `A = {{0,1},{1,0}}; MatrixForm[A]`

As a $2 \times 2$ matrix, this has a natural action on points in the plane, through a special type of multiplication, indicated in *Mathematica* by the "." symbol. For example, the point:

- `InPoint = {3,2}`

is transformed under multiplication by $A$ to the point:

- `OutPoint = A . InPoint`

In this way, multiplication sets up a correspondence between inputs and outputs.

Mathematically, such a correspondence is called a "function." We are all familiar with functions that take in a single value and output a single value, such as $\sin(x)$ or $\log(x)$. The function given by multiplication with $A$, in contrast, takes in *two* numbers and outputs two numbers, expressed as ordered pairs. In such a situation, when a function has unusual inputs or outputs, although we often use the terms "mapping," "operator," or "transformation" in place of the term "function," one should remember that all these words refer to the *same* underlying concept.

We may emphasize the functional nature of this correspondence by using functional notation:

- `f[{x_,y_}] := A.{x,y}`

It is common, when writing functions of several variables, to drop the inner pair of parentheses. Thus, we could also write:

- `f[x_,y_] := A.{x,y}`

We may now use the letter $f$ to refer to this mapping of inputs to outputs. Although the notation is different, one may verify that $f$ produces the same result as before:

- `f[InPoint]`

   In the "MPicts.m" package, we provide various *Mathematica* commands to enable us to visualize the effect of multiplication by a matrix by "plotting" its associated mapping. To plot a function of a single variable, such as $\sin(x)$, we would set the output of the function equal to $y$ and plot solutions to the equation $y = \sin(x)$ to obtain its graph in the $x$-$y$ plane. To plot a function such as $f[\{x, y\}]$, however, requires two additional variables, call them "$u$" and "$v$", to specify a single output via the equation $u, v = f[\{x, y\}]$. Thus, to picture f requires *two* coordinate planes, the $x$-$y$ plane for the inputs and the $u$-$v$ plane for the outputs. For example, we would picture the mapping of $\{x, y\} = \{3, 2\}$ to $\{u, v\} = \{2, 3\}$ with the MatrixPlot[] command:

- `MatrixPlot[A, {3,2}];`

*Note*: You may need to resize this picture by clicking on it and dragging on a corner, to make it more readable. If this command does not work, you have probably forgotten to open "Start.ma". If so, simply open that Notebook now and re-evaluate the previous command.

   Geometrically, $A$ reflects the $x$-$y$ plane around the line $y = x$ to give the $u$-$v$ plane. To see this more clearly, we may use the MatrixPlot[] command to plot more points:

- `MatrixPlot[A, {{3,2},{4,1},{2,2}}];`

Notice how we enter more than one point as a list (i.e., enclosed by an extra pair of brackets). Each point is numbered, so that one may easily indentify the output corresponding to each input.

   Observe the labels on the graph of inputs on the left, as well as the graph of outputs on the right. Because a transformation is simply a type of function, we use much of the same terminology as is used in basic math courses. In general, the set of possible input points for a transformation is called the "domain" of the transformation, while its set of possible output points is called its "range" (or "codomain"). For any particular point in the domain, we refer to the corresponding point in the range as its "image" via the transformation.

## Exercise 9.1

- What do you expect the image $\{u, v\}$ of $\{x, y\} = \{1, 2\}$ under $A$ will be? You may verify your conjecture algebraically by evaluating:

- `{u, v} = A . {1,2}`

   or geometrically with MatrixPlot[]:

- `MatrixPlot[A, {1,2}];`

   Reflection is a fundamental geometric transformation, in that we may perform a reflection through a given line entirely geometrically (say, with a ruler and compass) without imposing a coordinate system on the plane. We say that the matrix $A$ "represents" the geometric transformation "reflect around the $y = x$ line." This is because when we use coordinates to represent points in the plane, the effect of multiplication by

*A* on the coordinates of a point is the same as the effect of the geometric transformation of the point itself.

*Aside*: This distinction is rather subtle, and may seem somewhat technical at the moment, but understanding this connection between coordinates, transformations, and matrices is one of the most fundamental questions of linear algebra. Lessons 25, 30, and 31 are all dedicated to this important topic, and many additional lessons are simply background for these discussions.

Now that we have a good geometric description of *A*, namely, "reflect points around the $y = x$ line," we will discover its *algebraic* effect.

### Exercise 9.2

- Experiment as in Ex. 9.1 to determine the precise connection between any given point and its image under *A*. Specifically, determine the *formulas* for *u* and *v*, in terms of *x* and *y*. *Hint*: You may wish to construct a table of inputs and outputs.

## 9.3  A Menagerie of Matrices

Now that we have see how a matrix may act to transform the plane and have learned to use some tools for analyzing this type of action, we wish to examine the variety of geometric transformations that may represented as a matrix. This section will consist of a series of experiments with five different matrices. Together they will illustrate all of the "elementary" geometric transformations into which any matrix may be decomposed. We will return to these examples in Lesson 18.

In order to speed our analysis, we have provided two other commands from the "MPicts.m" package, MatrixPicture[] and MatrixPlotSquare[], which each produce a more comprehensive picture of the action of a matrix on the plane than MatrixPlot[]. In the case of the matrix *A* from Section 9.2, MatrixPicture[] produces a diagram that is similar to that of MatrixPlot[], except that it plots an entire *grid* of inputs and uses color to associate inputs and outputs:

- `MatrixPicture[A];`

In contrast, MatrixPlotSquare[] plots a labeled, colored square and its image under the given matrix transformation (i.e., the image of each of the points that comprise the original the square):

- `MatrixPlotSquare[A];`

Both pictures emphasize the fact that *A* represents the geometric reflection through the line $y = x$. We will learn that the entire transformation for a matrix is actually determined by its effect on relatively few points. That is because such a transformation is linear and will preserve straight lines. In particular, a linear transformation will always map any square into a parallelogram. Thus, MatrixPlotSquare[] actually produces the simplest, most comprehensive description of the geometric effect of a matrix.

*Warning*: Although matrices of any dimension may be viewed from this type of geometric perspective (but become difficult to visualize for dimensions greater 3), these commands work only for 2 × 2 matrices and points in the plane.

Our first example is similar to that in the previous section and is given by the matrix:

- ```
  B = {{1,0}, {0,-1}}; MatrixForm[B]
  ```

## Exercise 9.3

- Investigate the action of $B$ on points in the plane. Describe the geometric and algebraic effect of $B$, just as in Ex. 9.2.

Although it would seem natural to name our next example "$C$", the letters $C$, $D$, and $E$ have special meaning to *Mathematica* and may not be used as variable names. What do you think these letters represent to *Mathematica*? You may test your intuition with the ? command:

- ```
  ?C
  ```

- ```
  ?D
  ```

- ```
  ?E
  ```

Thus, we name our next example:

- ```
  F = {{3,0}, {0,1/2}}; MatrixForm[F]
  ```

This illustrates a rather different type of geometric effect than the first two examples.

## Exercise 9.4

- Repeat Ex. 9.2 with this matrix to give an algebraic and geometric description of $F$.

Another type of geometric transformation is illustrated by trigonometric matrices of the form:

- ```
  G = {{Cos[45 Degree], Sin[45 Degree]},
        {-Sin[45 Degree], Cos[45 Degree]}}; MatrixForm[G]
  ```

This matrix will be easier to manipulate, if we first use the N[] command to convert from symbolic to numeric entries:

- ```
  G = N[G]; MatrixForm[G]
  ```

## Exercise 9.5

- Repeat Ex. 9.2 with this matrix to give an algebraic and geometric description of $G$. *Hint*: It may be useful to first convert $\{x, y\}$ to polar coordinates $\{r, \theta\}$ and then express $\{u, v\}$ in terms of $r$ and $\theta$.

A more unusual example, often referred to as a horizontal "shear," is given by the matrix:

- ```
  H = {{1, 1}, {0, 1}}; MatrixForm[H]
  ```

## Exercise 9.6

- Repeat Ex. 9.2 with this matrix to give an algebraic and geometric description of $H$.

The term "shear" comes from physics, where one talks about stress, strain, and shearing forces.

### *Advanced Exercise 9.7*

- For each of our examples so far, infer the exact matrix that "reverses" the effect of the given matrix. You should use your observations on the geometric and algebraic nature of each matrix and a bit of creative guessing.

As a final example, we consider a matrix whose effect is qualitatively quite distinct from all of the others we have considered. All of the previous examples could be reversed in that the effect of each transformation could be undone by another transformation. We say that such transformations are "invertible." (Compare the usual definition of an invertible function). This next example illustrates an "irreversible" or "noninvertible" matrix:

- ```
  J = {{1, 0}, {0, 0}}; MatrixForm[J]
  ```

## Exercise 9.8

- Repeat Ex. 9.2 with this matrix to give an algebraic and geometric description of $J$.

*Note*: The resulting picture may look strange because some of the labels and lines have necessarily been plotted on top of one another.

## 9.4   Linear Transformations and Calculus

In Section 9.1, we mentioned the applications of various transformations. The first two examples were given by matrices, namely as a rotation (compare $G$ from Section 9.3) and a reflection (compare $B$ from Section 9.3). The final example used the following nonlinear tranformation:

- ```
  h[x_,y_] := {0.5 Sin[Pi/2(x/20 - 2)](E^y + 1/E^y),
               0.5 Cos[Pi/2(x/20 - 2)](E^y - 1/E^y)}
  ```

Although this transformation cannot be accomplished by a *single* matrix, it may be approximated by a *collection* of matrices. That is because the derivative of $h[x, y]$ provides a "locally linear" approximation to $h[x, y]$ and, as a *linear* transformation, it is given by a matrix. We are all familiar with this process for functions of a single variable. In that case, the derivative "matrix" is $1 \times 1$; that is, it is a single number, which determines the tangent line to the graph of the given function at a specific point. In this section, we show how this idea generalizes to functions of two variables.

The reader should be aware that we will be somewhat imprecise with the mathematics in this section. This material is intended only to convey the sense of how valuable matrix transformations are in practice and may be skipped without loss of continuity. If the reader is interested in pursuing this material in greater detail, she should refer to a text on vector calculus.

We begin by learning a new *Mathematica* command for plotting an arbitrary function from the $x$-$y$ plane to the $u$-$v$ plane. We then compare the behavior of nonlinear functions with the linear transformations we examined in Section 9.3. We conclude by illustrating how the derivative provides linear approximations to a nonlinear function.

Since nonlinear transformations are not as well-behaved as linear transformations, we must plot many more points to gain a good geometric understanding of the effect of any such function. Thus, we provide another command, from the "Plots.m" package, called CoordinatePlot[]. We first illustrate its use with the *linear* transformation $f[x, y]$ from Section 9.2, corresponding to the reflection matrix $A$:

- ```
  A = {{0,1},{1,0}};
  f[x_,y_] := A.{x,y};
  CoordinatePlot[f[x,y],{x,-2,2,1},{y,-2,2,1}];
  ```

CoordinatePlot[] is a cross between MatrixPicture[] and MatrixPlotSquare[]. Instead of using a single square as MatrixPlotSquare[] does, it plots the image of an entire grid of lines from the domain. It uses a coloring scheme similar to MatrixPicture[] to show the correspondence of outputs in the range to inputs. We specify the grid in the same way one specifies the domain of a Plot[] command, by giving smallest, largest, and spacing parameters for each variable.

A linear transformation such as this will preserve straight lines. In particular, the image of the rectangular grid is at worst a grid of identical parallelograms. On the other hand, a nonlinear transformation, such as $h[x, y]$ from the beginning of this section, generally distorts each rectangle in a different manner:

- ```
  CoordinatePlot[h[x,y],{x,20,60,5},{y,0,2,.4},AspectRatio->.6];
  ```

To illustrate these ideas more fully, we will use the simpler transformation:

- ```
  g[x_,y_] := {x, x^2 + y}
  ```

Geometrically, this transformation "bends" the entire $x$-$y$ plane along a parabola.

- ```
  CoordinatePlot[g[x,y],{x,-2,2,.4},{y,-2,2,1}];
  ```

Although the grid resulting from a nonlinear transformation is not as regular as that of a linear one, if we focus on one particular region in the range, the output looks *roughly* like a parallelogram. That is because even a nonlinear function may be *approximated* by a matrix. For example, if we "zoom in" on this plot near the point $\{x, y\} = \{1, 1\}$, which maps to $\{u, v\} = \{1, 2\}$, we obtain the following:

- ```
  CoordinatePlot[g[x,y],{x,1,1.1,.02},{y,1,1.1,.02},
              AspectRatio->Automatic];
  ```

Here we see that, for small changes in the inputs for values near $\{x, y\} = \{1, 1\}$, $g[x, y]$ behaves like a vertical shear (compare matrix $H$ from Section 9.3).
  ```

Because the nature of a nonlinear transformation varies from point to point, to represent this transformation by a matrix, we would need to use a matrix with *variable* entries. The matrix of partial derivatives:

$$\begin{bmatrix} \frac{\partial u}{\partial x} & \frac{\partial u}{\partial y} \\ \frac{\partial v}{\partial x} & \frac{\partial v}{\partial y} \end{bmatrix}$$

- ```
  Clear[x,y,u,v, dg]; {u,v} = g[x,y];
  dg[x_,y_] = {{D[u,x], D[u,y]},{D[v,x], D[v,y]}};
  MatrixForm[dg[x,y]]
  ```

known as the "total" derivative of $g$ and written as $D(g)$, provides the best linear approximation to $g[x, y]$ near any given point. For example, we may verify that, at the point $\{x, y\} = \{1, 1\}$, this formula produces the vertical shear we observed earlier:

- ```
  CoordinatePlot[dg[1,1].{x,y},{x,0,1,.2},{y,0,1,.2},
              AspectRatio->Automatic];
  ```

except for the fact that the origins in the domain and range have been moved.

The following exercise will review the ideas from this section by encouraging you to carry out the calculations youself.

### Exercise 9.9

Consider the transformation:

- ```
  k[x_,y_] := {(x^2 - y^2)/2, x y}
  ```

- Compute its total derivative $D(k)$ as a matrix in $x$ and $y$. Determine the specific matrix $M$ approximating $k[x, y]$ near the point $\{x, y\} = \{0, 1\}$. Enter your value for $M$ into the following cell and verify your calculation by comparing the corresponding plots.

- ```
  M = {{,},{,}};
  CoordinatePlot[M.{x,y},{x,0,.1,.02},{y,0,0.1,.02},
              AspectRatio->Automatic];
  CoordinatePlot[k[x,y],{x,0,.1,.02},{y,1,1.1,.02},
              AspectRatio->Automatic];
  ```

## 9.5   The Matrix of a Transformation

In Section 9.2, we saw how a matrix may act as a transformation through multiplication. In this section we wish to reverse the process. That is, given a transformation of the plane, we wish to discover the matrix to which it corresponds. In general, this may not be possible. Only transformations that are "linear" (i.e., that map straight lines to straight lines and map the origin to itself) may be represented as a matrix. In this section, we will see that this correspondence depends on the effect of the given transformation on a set of "standard basis points."

In the plane, this "standard basis" is given by two points:

- ```
  InPoint1 = {1,0}; InPoint2 = {0,1};
  ```
  ```

## Exercise 9.10

• Compute the image of InPoint1 under the transformation of *A* (using *Mathematica*). *Hint*: You may need to refer back to the beginning of this lesson to find the necessary *Mathematica* commands. Record your result. How does this output compare with the entries of the matrix *A* itself?

## Exercise 9.11

• Repeat Ex. 9.10 with InPoint2. Give a rule expressing *A* in terms of these two image points.

Now that you are somewhat familiar with this process, we will quickly generate the image points for all of the examples from Section 9.3 with the following sequence of commands:

• 
```
Print["A = ",MatrixForm[A]];
Print[" "];
Print["  Image of 1st point = ",A.InPoint1];
Print["  Image of 2nd point = ",A.InPoint2];
Print[" "];Print[" "];
Print["B = ",MatrixForm[B]];
Print[" "];
Print["  Image of 1st point = ",B.InPoint1];
Print["  Image of 2nd point = ",B.InPoint2];
Print[" "];Print[" "];
Print["F = ",MatrixForm[F]];
Print[" "];
Print["  Image of 1st point = ",F.InPoint1];
Print["  Image of 2nd point = ",F.InPoint2];
Print[" "];Print[" "];
Print["G = ",MatrixForm[G]];
Print[" "];
Print["  Image of 1st point = ",G.InPoint1];
Print["  Image of 2nd point = ",G.InPoint2];
Print[" "];Print[" "];
Print["H = ",MatrixForm[H]];
Print[" "];
Print["  Image of 1st point = ",H.InPoint1];
Print["  Image of 2nd point = ",H.InPoint2];
Print[" "];Print[" "];
Print["J = ",MatrixForm[J]];
Print[" "];
Print["  Image of 1st point = ",J.InPoint1];
Print["  Image of 2nd point = ",J.InPoint2];
```

### Exercise 9.12

- Give a rule relating a matrix to its two image points for InPoint1 and InPoint2 that holds for all six matrices: $A$, $B$, $F$, $G$, $H$, and $J$. *Hint*: You may need to revise your answer to Ex. 9.11 to obtain a pattern that correctly describes *every* case.

### Exercise 9.13

Using only basic geometric reasoning, determine the images of InPoint1 and InPoint2 under a counterclockwise rotation by 90°. Determine the matrix $M$ for this rotation by using the pattern you observed in Ex. 9.12. Fill in the matrix $M$ in the following cell and use MatrixPlotSquare[] to verify your answer.

- ```
M = {{,},{,}};
Print["M = ",MatrixForm[M]];
Print["  "];
Print["  Image of 1st point = ",M.InPoint1];
Print["  Image of 2nd point = ",M.InPoint2];
MatrixPlotSquare[M];
```

Conjecture how one might obtain $M$ from $G$ through multiplication alone, and then use *Mathematica* to verify your conjecture.

## 9.6  Summary

Our main goal in this lesson has been to achieve a deeper understanding of what matrices really are. The most accurate description of a matrix is as the representation of a transformation. Thus, in this lesson we have used *Mathematica* to explore various transformations of the plane to illustrate:

- how matrix multiplication produces a linear transformation,
- that such transformations include a variety of geometric operations, and
- how we may infer the matrix of a linear transformation in the plane by plotting the images of $\{1,0\}$ and $\{0,1\}$.

### Exercise 9.14

Each of our examples are commonly referred to by one of the following names: projection, reflection, rotation, shear, and rescaling of axes. Match each example to the correct name.

In the next few lessons, we will discuss how to add, subtract, multiply, and "divide" matrices. In other words, we will see how to perform arithmetic operations with matrices. This will give us a *third* interpretation of matrices — as algebraic objects that
```

generalize "ordinary" numbers. You may be more comfortable thinking of matrices notationally, as shorthand for linear systems, or algebraically, as generalized "numbers." However, the transformational point of view introduced in this lesson provides us with a powerful problem-solving tool that will allow us to expand the applicability of linear algebra into a wide variety of different settings. Ultimately, it will clarify and unite the various observations we will make throughout this text into a coherent and quite beautiful body of results.

# Matrix Arithmetic

In the last lesson, we explored the geometric significance of matrices. In this lesson, we examine their algebraic character. We will ultimately see that matrices act very much like ordinary numbers, in that they may be added, subtracted, multiplied, etc. and obey many of the usual laws of algebra, but with some important differences. In this lesson, we will only discuss *how to calculate* with matrices, that is, the "arithmetic" of matrices. We will postpone the discussion of the *patterns* of matrix arithmetic, namely, matrix "algebra," until Lesson 12.

Instead of expaining how to perform matrix arithmetic in very formal terms, we will allow you to follow your own intuition. Because you should be able to conjecture the correct rules of matrix arithmetic, we will encourage you to discover them on your own. We will then discuss how to convert your own, intuitive, verbal descriptions into the corresponding mathematical formulas.

You should keep in mind, however, that matrix arithmetic is *not* governed by "made-up" rules! Because matrices represent geometric transformations, their behavior is governed by the geometry of the physical world. Thus, the "laws" of matrix algebra are just like the "laws" of physics or chemistry or any other physical science. They do not depend on the whims of any individual, but describe *how the world works*!

## 10.1 Addition, Subtraction, "Multiplication," and Equality

### Exercise 10.1

Record what you believe the sum of the following two matrices should be:

$$\begin{bmatrix} 1 & 0 & -3 \\ -2 & 5 & 4 \\ 0 & 3 & 7 \end{bmatrix} + \begin{bmatrix} 2 & 2 & 4 \\ -3 & 7 & -2 \\ 5 & 1 & -9 \end{bmatrix} = \begin{bmatrix} & & \\ & & \\ & & \end{bmatrix}$$

Provide a clear, verbal prescription for performing matrix addition.

### Exercise 10.2

Should one be able to add the following matrices?

$$\begin{bmatrix} 1 & 0 & -3 \\ -2 & 5 & 4 \\ 0 & 3 & 7 \end{bmatrix} \text{ and } \begin{bmatrix} 2 & 2 & 4 & -1 \\ -3 & 7 & -2 & 3 \\ 5 & 1 & -9 & 2 \end{bmatrix}$$

If so, how would this be done? If not, why not?

## Exercise 10.3

Repeat Ex. 10.1 but with the following subtraction problem.

$$\begin{bmatrix} 1 & 0 & -3 \\ -2 & 5 & 4 \\ 0 & 3 & 7 \end{bmatrix} - \begin{bmatrix} 2 & 2 & 4 \\ -3 & 7 & -2 \\ 5 & 1 & -9 \end{bmatrix} = \begin{bmatrix} & & \\ & & \\ & & \end{bmatrix}$$

## Exercise 10.4

Repeat Ex. 10.1 but with the following multiplication problem.

$$3 \cdot \begin{bmatrix} 1 & 0 & -3 \\ -2 & 5 & 4 \\ 0 & 3 & 7 \end{bmatrix} = \begin{bmatrix} & & \\ & & \\ & & \end{bmatrix}$$

Because we wish to think of matrices as "numbers," we require another term to refer to ordinary real numbers. We call them "scalars" (like the "scale" on a map), and this type of multiplication is known as "scalar multiplication."

## Exercise 10.5

If we have the following equation:

$$\begin{bmatrix} x & -2 & 5 \\ 3 & 0 & 4 \end{bmatrix} = \begin{bmatrix} 1 & y & 5 \\ 3 & z & 4 \end{bmatrix}$$

what must $x$, $y$, and $z$ equal?

## Exercise 10.6

Can the following equation be true for some values of $x$ and $y$?

$$\begin{bmatrix} x & -2 & 5 \\ 3 & 0 & 4 \end{bmatrix} = \begin{bmatrix} 1 & y & 5 \\ 3 & 7 & 4 \end{bmatrix}$$

If so, what are the necessary values for $x$ and $y$? If not, why not? Explain in words how to verify a matrix equation.

Although it may seem that you have simply made-up some rules for performing matrix arithmetic, that is not the case. Because matrices represent linear transformations, they are governed by laws independent of anyone's personal whims. Once we have established the theoretical foundation of linear transformations, we could prove that you have actually *discovered* these laws. However, because these rules are so very intuitive, it does not make sense to delay their discussion for so long.

## 10.2   Translating Mathematics

So far, we have treated the subject of linear algebra informally, avoiding precise definitions and formal proofs. However, learning to read and understand mathematical definitions, theorems, and proofs is an important skill to develop and is crucial to continuing education in any scientific discipline. In particular, one must be able to translate between intuitive, operational descriptions and precise, mathematical formulas. This is not easy, but whenever one encounters a mathematical formula, definition, or theorem, one should rephrase it in English. In this way, one learns to gain an intuitive understanding of mathematical concepts. Instead of seeing a confusing jumble of symbols on a page, one may then discover the beautiful ballet of *ideas*, expressed in terms of formulas, which is the heart of mathematics.

We will practice this skill by translating each of the arithmetic operations from Section 10.1 into mathematical formulas. For example, you should have decided that two matrices are equal if and only if every pair of "corresponding entries" (i.e., entries in the same row and column) are equal. To translate this condition into a formula, we must be able to refer to any particular entry of a matrix. We do so with "subscript" notation.

**Definition 10.1**  We denote the entry of a matrix $A$ which occurs in the $i$th row and $j$th column by $(A)_{ij}$. The letters $i$ and $j$ are called "subscripts." Similarly, $(A)_{i\cdot}$ refers to the $i$th row of $A$, while $(A)_{\cdot j}$ denotes the $j$th column.

An alternate notation is to specify that $A = [a_{ij}]$ and use $a_{ij}$ in place of $(A)_{ij}$. We may also sometimes drop the parentheses and write $A_{ij}$ in place of $(A)_{ij}$. In general, it is common to use a capital letter for a matrix and the corresponding lowercase letter with subscripts to represent its entries.

Via subscript notation, our intuitive definition of matrix equality translates directly into a precise formal definition:

$$A = B \quad \Leftrightarrow \quad A_{ij} = B_{ij} \quad \text{for all } i \text{ and } j$$

Likewise, we may express matrix addition by the formula:

$$(A + B)_{ij} = A_{ij} + B_{ij}$$

A direct translation of this formula would be:

"Each entry of the sum of two matrices is obtained by adding the entries of each of the two matrices in exactly the same position."

### Exercise 10.7

Express the formal definitions for subtraction and scalar multiplication by completing the following formulas. Give a direct English translation for each formula.

(a)  $(A - B)_{ij} = $ _____ .

(b)  $(sA)_{ij} = $ _____ .

## 10.3   Matrix Multiplication

So far we have not discussed how to *multiply* two *matrices*. This is because matrix multi-plication (as opposed to scalar multiplication) is not as intuitive as the other arithmetic operations. In this section, we will give *one possible* motivation for the formulas in terms of systems of linear equations. The *best* motivation is based on the geometry of matrices and the notion of a "linear transformation," which we will not be able to explain com-pletely until Lesson 30. We will proceed in three stages, examining ever larger matrix products, to develop the three main principles governing matrix multiplication.

### Row Times Column

We begin with the simplest linear system:

$$ax = b \tag{10.1}$$

namely, one equation in one unknown $x$ and two constants $a$ and $b$. The next simplest linear system would have two variables, such as:

$$2x + 3y = 5 \tag{10.2}$$

Matrix multiplication allows us to unify these two examples by expressing Eq. 10.2 as:

$$AX = B \tag{10.3}$$

where $X$ is an unknown *matrix* and $A$ and $B$ are constant *matrices*! To write Eq. 10.2 as Eq. 10.3, there is little choice in defining $A$, $B$, and $X$. Because $B$ represents the right-hand side of our equation, it must be the matrix $\begin{bmatrix} 5 \end{bmatrix}$. Likewise, $A$ should contain the constants on the left-hand side, namely $\begin{bmatrix} 2 & 3 \end{bmatrix}$, while $X$ must be the matrix of un-knowns $X = \begin{bmatrix} x \\ y \end{bmatrix}$.

At this point, the only real choices involved were in the precise dimensions of $A$ and $X$. From the point of view of linear systems, there is no good reason to take $A$ to be $1 \times 2$ and $X$ to be a $2 \times 1$. This choice is dictated by the transformational approach to systems of equations which we will introduce in Lesson 20. Given these choices, however, there is only one possible way to define multiplication so that Eq. 10.2 and Eq. 10.3 directly correspond:

$$2x + 3y = 5 \iff [2x + 3y] = [5] \overset{???}{\iff} \begin{bmatrix} 2 & 3 \end{bmatrix} \begin{bmatrix} x \\ y \end{bmatrix} = [5]$$

$$\iff AX = B \quad \text{with} \quad A = \begin{bmatrix} 2 & 3 \end{bmatrix}, X = \begin{bmatrix} x \\ y \end{bmatrix}, \text{ and } B = [5]$$

Namely, we must take:

$$\begin{bmatrix} 2 & 3 \end{bmatrix} \begin{bmatrix} x \\ y \end{bmatrix} = \begin{bmatrix} 2x + 3y \end{bmatrix} \tag{10.4}$$

This suggests the following principle of matrix multiplication:

**Observation:**

(1) To multiply a single row with a single column, multiply corresponding entries and total the results.

## Exercise 10.8

Use this rule to multiply $\begin{bmatrix} 2 & 3 & -1 \end{bmatrix} \begin{bmatrix} 4 \\ -2 \\ 5 \end{bmatrix}$.

## Exercise 10.9

Should one be able to multiply $\begin{bmatrix} 2 & 3 & -1 \end{bmatrix} \begin{bmatrix} 4 \\ -2 \\ 5 \\ 6 \end{bmatrix}$? If so, describe precisely how this would be done. If not, explain why not.

## Rows Times Column

Now that we know how multiply a single row and column, the generalization to more rows and columns is almost immediate. We first consider the case where the first matrix consists of more than one row. As before, we consider how to rewrite the system of equations:

$$\begin{aligned} 2x + 3y - z &= 5 \\ x - 4y + 6z &= 1 \end{aligned} \qquad (10.5)$$

as $AX = B$. By analogy with the previous case, the natural choices are:

$$A = \begin{bmatrix} 2 & 3 & -1 \\ 1 & -4 & 6 \end{bmatrix}, B = \begin{bmatrix} 5 \\ 1 \end{bmatrix}, \text{ and } X = \begin{bmatrix} x \\ y \\ z \end{bmatrix}$$

As before, there is only one way to define matrix multiplication so that Eq. 10.5 and Eq. 10.3 correspond. Namely, we must have:

$$\begin{bmatrix} 2 & 3 & -1 \\ 1 & -4 & 6 \end{bmatrix} \begin{bmatrix} x \\ y \\ z \end{bmatrix} = \begin{bmatrix} 2x + 3y - z \\ x - 4y + 6z \end{bmatrix} \qquad (10.6)$$

This suggestions the following two additional rules for matrix multiplication:

**Observation:**

(2) The result of any multiplication has as many rows as the first matrix and as many columns as the second.

(3) To obtain the entry in the $i$th row, we simply multiply the $i$th row from the first matrix by the second matrix.

## Exercise 10.10

Compute $\begin{bmatrix} 2 & 3 & -1 \\ 1 & -4 & 6 \end{bmatrix} \begin{bmatrix} 1 \\ -2 \\ 3 \end{bmatrix}$.

This is the type of multiplication that *Mathematica* computed in Lesson 9 to evaluate $A$ as a linear transformation.

# Rows Times Columns

The two rules from the previous example generalize directly to the case when the second matrix has more columns, if we simply choose to multiply each column in the second matrix individually. Specifically,

**Observation:**

(2) The result of any multiplication has as many rows as the first matrix and as many columns as the second.

(3) To obtain the entry in the $i$th row and $j$th column, we simply multiply the $i$th row from the first matrix by the $j$th column in the second matrix.

This choice has no clear motivation from the point of view of systems of equations. We will see in Lesson 30 that this unusual definition precisely reflects the *composition* of linear transformations.

For example, multiplication of a matrix with two rows by a matrix with two columns yields a matrix with two rows and two columns. To obtain the entry in the first row and second column, we would multiply the first row (of the first matrix) by the second column (of the second matrix). To illustrate:

$$\begin{bmatrix} 2 & 3 & -1 \\ 1 & -4 & 6 \end{bmatrix} \begin{bmatrix} 1 & 5 \\ -2 & 0 \\ 3 & -1 \end{bmatrix} = \begin{bmatrix} & 11 \\ & \end{bmatrix} \quad \text{because} \quad \begin{bmatrix} 2 & 3 & -1 \end{bmatrix} \begin{bmatrix} 5 \\ 0 \\ -1 \end{bmatrix} = \begin{bmatrix} 11 \end{bmatrix}$$

## Exercise 10.11

Complete this multiplication by filling in the remaining three entries of the product.

Using the subscript notation from Definition 10.1, we may express our observations on matrix multiplication formally in the following definition.

**Definition 10.2** If $A = [a_{ik}]$ is an $p \times q$ matrix and $B = [b_{kj}]$ is a $q \times r$ matrix, then the product $C = AB$ is defined as the $p \times r$ matrix with entries $c_{ij}$ that are given by the formula:

$$c_{ij} = a_{i1}b_{1j} + a_{i2}b_{2j} + \cdots + a_{iq}b_{qj}$$

or:

$$(AB)_{ij} = a_{i1}b_{1j} + a_{i2}b_{2j} + \cdots + a_{iq}b_{qj}$$

or:

$$(AB)_{ij} = (A)_{i1}(B)_{1j} + (A)_{i2}(B)_{2j} + \cdots + (A)_{iq}(B)_{qj} \qquad (10.7)$$

Notice that $a_{i1}, a_{i2}, \ldots, a_{iq}$ are the entries of the $i$th row of $A$ and $b_{1j}, b_{2j}, \ldots, b_{qj}$ are the entries of the $j$th column of $B$. Thus, this formal definition expresses exactly the three rules from our previous observations, namely, take each pair of row and column, multiply corresponding entries and total the results.

Although Definition 10.2 is the usual way of phrasing matrix multiplication, an alternative approach would be to break our definition into two parts, first defining the product of a single row and column and then defining a general matrix product.

**Definition 10.3** If $R = \begin{bmatrix} r_1 & \cdots & r_q \end{bmatrix}$ is a $1 \times q$ matrix (i.e., a single row) and $C = \begin{bmatrix} c_1 \\ \vdots \\ c_q \end{bmatrix}$, a $q \times 1$ matrix (i.e., a single column), then the product is defined as the $1 \times 1$ matrix $RC$, given by the single number:

$$RC = \begin{bmatrix} r_1 & \cdots & r_q \end{bmatrix} \begin{bmatrix} c_1 \\ \vdots \\ c_q \end{bmatrix} = \begin{bmatrix} r_1c_1 + r_2c_2 + \cdots + r_qc_q \end{bmatrix}$$

More generally, if $A$ is $p \times q$ and $B$ is $q \times r$, then $AB$ is $p \times r$ and

$$(AB)_{ij} = (A)_{i\cdot}(B)_{\cdot j} \qquad (10.8)$$

Notice how $(A)_{i\cdot}$ and $(B)_{\cdot j}$ consist of a single row and column, respectively. Thus, the second part of Definition 10.3 builds on the first part. Both Definitions 10.2 and 10.3 are logically equivalent, so neither definition is "more right" than the other, but Definition 10.3 will be more convenient for proofs.

Moreover, although Definition 10.2 is more usual, because it is given by a *single* formula, one should notice that Definition 10.3 actually describes our previous observations regarding matrix multiplication more precisely. The first part of the definition corresponds to observation 1 of this section, while the second part expresses observations 2 and 3. Notice how the introduction of appropriate notation allows us to write down formulas that most closely represent our actual thought process. This is the function of notation, namely, to help us get our thoughts on paper as quickly, succinctly, and accurately as possible. In fact, the proper choice of notation will actually help us think more clearly. This then allows us to manipulate more difficult concepts and solve ever more complicated problems.

## 10.4   Transposition

There is one more simple operation that we may perform with matrices, which has no analog in "ordinary" arithmetic. As a noun, this operation is known as "transposition." In verb form, we "transpose" or "form the transpose of" a matrix, $A$, and denote the result by $A^T$. Although the value of tranposition will not become fully apparent until we study linear transformations and inner products in Lessons 27 and 28, as with most other matrix operations, computing a transpose is rather straightforward. For example, if:

$$A = \begin{bmatrix} 2 & 3 & -1 \\ 1 & -4 & 6 \end{bmatrix} \quad \text{then} \quad A^T = \begin{bmatrix} 2 & 1 \\ 3 & -4 \\ -1 & 6 \end{bmatrix}$$

In words, we simply turn all the rows into columns (maintaining the same order).

### Exercise 10.12

Consider the matrix $A = \begin{bmatrix} -1 & 2 & 5 \\ 0 & 7 & 11 \\ 3 & -9 & -6 \end{bmatrix}$.

(a) Compute $A^T$.

(b) Determine the row and column of $A^T$ where $(A)_{23} = 11$ appears.

(c) Generalize your observations to complete the formula:

$$\left(A^T\right)_{ij} = \underline{\hspace{7cm}}.$$

(d) Give a direct English translation for your formula.

We conclude this section by indicating a few of the uses of transposition. A common use of transposition is to save space. Instead of writing $\begin{bmatrix} 2 & 3 & -1 \end{bmatrix} \begin{bmatrix} 4 \\ -2 \\ 5 \end{bmatrix}$, one will often write $\begin{bmatrix} 2 & 3 & -1 \end{bmatrix} \begin{bmatrix} 4 & -2 & 5 \end{bmatrix}^T$. Although these represent the same product, the second is more compact. Alternatively, this multiplication may be written as $v^T w$, where $v$ and $w$ are the column vectors, $v = \begin{bmatrix} 2 \\ 3 \\ -1 \end{bmatrix}$ and $w = \begin{bmatrix} 4 \\ -2 \\ 5 \end{bmatrix}$. Written this way, this expresses the standard inner product formula on column vectors, which we will explore in great detail in Lessons 27 and 28. In more advanced texts, transposition is also used to define the "dual" of a given linear transformation.

## 10.5   Summary

So far we have examined the matrix operations of addition, subtraction, scalar multiplication, equality, matrix multiplication and transposition. In Lesson 11, we will discuss how to perform these operations in *Mathematica*. At this point, you may need more practice with these operations. You may even wish to verify that your conjectures about the procedures for matrix arithmetic are in fact correct. Using *Mathematica*, you will be able to practice these operations by hand and then verify your calculations on the computer. Once you become proficient with manual computation, you may then use the computer to save time, as well.

# Matrix Arithmetic with *Mathematica*

*Mathematica* Notebook

## 11.1   Introduction

In the last lesson, you discovered how to perform many basic matrix operations by hand. In this lesson, you will:

- learn how to perform these calculations using *Mathematica,* and
- gain confidence in your understanding of matrix algebra by using the computer to check your work.

We will also:

- discover a very important fact about the rank of a matrix,
- discuss some *Mathematica* pitfalls to avoid, and
- employ matrix multiplication to quickly substitute values to verify a solution to a system of linear equations.

## 11.2   Basic Matrix Arithmetic

*Mathematica* will simplify matrix arithmetic as easily as it performs ordinary arithmetic. For example, to compute:

$$\begin{bmatrix} 2 & 3 & -1 \\ 1 & -4 & 6 \end{bmatrix} \begin{bmatrix} 1 & 5 \\ -2 & 0 \\ 3 & -1 \end{bmatrix} - 3 \begin{bmatrix} 1 & 2 \\ 3 & 4 \end{bmatrix}$$

one may simply evaluate:

- 
```
A = {{2,3,-1},
     {1,-4,6}};
B = {{1,5},
     {-2,0},
     {3,-1}};
F = {{1,2},
```

```
{3,4}}; MatrixForm[A.B - 3 F]
```

Notice how we use "." to indicate matrix multiplication and "−" for subtraction. Likewise, we use a space for scalar multiplication and "+" for addition.

## Exercise 11.1

- Carry out the previous calculation by hand, showing all of your work, and compare your results with those of the computer.

In entering the previous command, we carefully formatted our input to make it more readable. This is helpful if one wishes to avoid typos and other kinds of mistakes. However, for ease in typing, one may achieve the same result with:

- ```
  Partition[{2,3,-1,1,-4,6},3].Partition[{1,5,-2,0,3,
  -1},2] - 3 Partition[{1,2,3,4},2]
  ```

Here we avoid separate definitions of each matrix, suppress all superfluous spaces, and enter each matrix with the Partition[] command. Notice how this command allows us to enter all of the entries of a matrix in a single list (only one set of brackets) and then specify how to break the input into rows by giving the number of elements in each row (i.e., the number of columns of the desired matrix).

This entry is clearly much less readable than the one in the first approach, but may be typed much more quickly. You will need to decide for yourself how to achieve a proper balance between these two objectives. A compromise between these two approaches might be:

- ```
  A = Partition[{2,3,-1,1,-4,6},3]; Print["A = ",MatrixForm[A]];
  B = Partition[{1,5,-2,0,3,-1},2]; Print["B = ",MatrixForm[B]];
  F = Partition[{1,2,3,4},2]; Print["F = ",MatrixForm[F]];
  Print["A.B - 3 F = ", MatrixForm[A.B - 3 F]];
  ```

Although this method involves more typing than both of the previous ones, the typing is easier than the first method (and may be speeded via the Copy and Paste commands) and the results are the clearest of all.

Unlike the standard arithmetic operations, transposition in *Mathematica* does not use traditional notation. For example, to compute the transpose of matrix $A$ from above, we would use the Transpose[] command:

- ```
  Transpose[A]
  ```

Of course this looks more natural in MatrixForm[]:

- ```
  MatrixForm[Transpose[A]]
  ```

## Exercise 11.2

- Compute:

$$26 \begin{bmatrix} 14 & 26 \\ 31 & 41 \\ 43 & -72 \end{bmatrix}^{T} + \begin{bmatrix} 13 & -21 & 39 \\ 5.6 & 0 & -1 \end{bmatrix} \begin{bmatrix} 2.7 & 301 & -15 \\ 0.01 & -4.8 & 63 \\ 7.5 & 0 & 37 \end{bmatrix}$$

using the computer.

## 11.3   Important Warnings

*Mathematica* will always try to compute what you say, even if it does not make mathematical sense or does not yield the computation you intend. For example, even though we may intend to multiply two matrices

- ```
  A = {{2,3}, {1,-4}}; MatrixForm[A]
  B = {{1,5}, {-2,0}}; MatrixForm[B]
  ```

if we forget to use the multiplication operator ".", then we will obtain

- ```
  MatrixForm[A B]
  ```

instead of:

- ```
  MatrixForm[A.B]
  ```

### Exercise 11.3

- From the result, infer the procedure that *Mathematica* uses to compute $AB$ and contrast that with the correct multiplication $A.B$.

This type of mistake is particularly dangerous, because it is not obvious that we have input an incorrect command, unless we actually carry out the operations by hand.

There is one more subtle type of mistake, indicated by the following command:

- ```
  A = MatrixForm[{{2,3}, {1,-4}}];
  B = MatrixForm[{{1,5}, {-2,0}}];
  Print["The sum of A = ", A," and B = ", B," is ",A + B];
  ```

Notice that *Mathematica* did not perform the indicated addition, and it does not explain why. That is because, as they have been defined in this example, *A* and *B are not matrices*! They are MatrixForm[] objects, which cannot be added. In other words, one should be careful not to use the MatrixForm[] command in assignment statements, which is designed to be used only to view results. That is, we should have entered the previous command as:

- ```
  A = {{2,3}, {1,-4}}; B = {{1,5}, {-2,0}};
  Print["The sum of A = ", MatrixForm[A]," and B = ",
      MatrixForm[B]," is ",MatrixForm[A + B]];
  ```

On the other hand, there are certain mistakes which *Mathematica* will attempt to catch. Unfortunately, *Mathematica*'s warnings are often rather cryptic. For example, if one attempts to combine matrices that do not have the proper dimensions, *Mathematica* will issue various warning messages in red.
  ```

### Exercise 11.4

• If we define $A$ and $F$ as the following:

• 
```
A = {{2,3},{1,-4}}; MatrixForm[A]
F = {{1,5},{-2,0},{3,-1}}; MatrixForm[F]
```

why should we be unable to compute $A + F$? How about $A.F$? What warnings does *Mathematica* give in each case?

### *Drill Exercise 11.5*

• Invent three more arithmetic expressions involving matrices of various dimensions and the operations of addition, subtraction, matrix and scalar multiplication, and transposition. Simplify each expression by hand and then verify your computations using the computer. You should make your expressions as varied as possible to more fully test your understanding of matrix arithmetic.

### *Drill Exercise 11.6*

Invent two more matrix expressions that are impossible to simplify according to the rules of matrix arithmetic. We then say that each expression is "undefined." Explain why each of your expressions is undefined, according to the rules discussed in the last lesson.

## 11.4   A Fact About Rank

We have described the rank of a matrix $A$ as the number of leading 1's in any row echelon form of the matrix (i.e., after Gaussian elimination), and we have seen that linear systems may be characterized qualitatively based on certain rank conditions. In future lessons, we will discuss the significance of rank in greater detail. For now, there is one important fact that we will not be able to *prove* for quite a while, but which is very useful, easy to remember, and easy to see through experimentation. Specifically, there is a direct connection between the rank of a matrix $A$ and the rank of its transpose $A^T$.

### Exercise 11.7

• Compare the ranks of the following matrices with the ranks of their transposes:

(a) $\begin{bmatrix} 1 & 1 & -8 & -14 \\ 3 & -1 & -3 & 0 \\ 2 & -1 & -7 & -10 \end{bmatrix}$
(b) $\begin{bmatrix} 1 & -2 & 1 & 4 \\ 2 & -3 & -1 & 2 \end{bmatrix}$
(c) $\begin{bmatrix} 1 & 2 & 3 & 4 & 7 \\ 0 & 5 & 7 & 6 & 8 \\ 0 & 5 & 2 & 4 & 7 \\ 0 & 0 & 2 & 1 & 0 \end{bmatrix}$
(d) $\begin{bmatrix} 4 & -1 & 2 & 6 \\ -1 & 5 & -1 & -3 \\ 3 & 4 & 1 & 3 \end{bmatrix}$

Hypothesize a general relationship between rank($A$) and rank(Transpose[$A$]).

On the surface, it is quite surprising that there should be any connection at all, since the specific steps involved in Gaussian elimination on $A$ and Transpose[$A$] will, in general, be quite different.

## 11.5   A Short Cut Using Matrices

We already know how to verify a proposed solution to a system of equations: Substitute each value for its corresponding variable, simplify each equation, and verify that each resulting equality is valid. In this section, we observe that there is another technique which exploits matrix multiplication. This new approach is quite convenient, in that it may be accomplished quickly and easily with *Mathematica*.

We introduced matrix multiplication by highlighting its close connection with systems of equations. Namely, a system such as

$$2x + 3y - z = 5$$
$$x - 4y + 6z = 1$$

may be rewritten as:

$$\begin{bmatrix} 2 & 3 & -1 \\ 1 & -4 & 6 \end{bmatrix} \begin{bmatrix} x \\ y \\ z \end{bmatrix} = \begin{bmatrix} 5 \\ 1 \end{bmatrix}$$

Substitution in the original matrix leads to a number of separate multiplications and additions, while substitution into the matrix form of the system only requires a single matrix multiplication.

### Exercise 11.8

Determine if $x = -7$, $y = 9$, and $z = 7$ is a solution by:

(a)  plugging into the original equations by hand.

(b)  using the computer to multiply

$$\begin{bmatrix} 2 & 3 & -1 \\ 1 & -4 & 6 \end{bmatrix} \begin{bmatrix} -7 \\ 9 \\ 7 \end{bmatrix}$$

and comparing the entries of the product with the right-hand sides of the original equations. Which method is easier?

### Exercise 11.9

Determine if $x = -106$, $y = 101$, and $z = 85$ is a solution by either method. Explain why you used the method you chose.

## 11.6   Summary

In this lesson, we have seen that the computer is very helpful for performing matrix arithmetic. However, one must be careful to:

   – check that all matrices are the right "shape" so that the necessary operations are well-defined, and

– always use the "." for matrix multiplication.

By this point, you should be confident in your ability to perform basic matrix arithmetic by hand. If not, you should continue to create and solve drill questions, using *Mathematica* (or a friend) to verify your answers, until you become proficient.

You should also have:

– learned that the rank of a matrix and its transpose are always equal, and

– seen how to use matrix multiplication to verify a potential solution to a system of linear equations.

One might argue that these two results are rather theoretical. As we will see over and over in this text, theory often has practical uses. In general, a little bit of thought is usually worth a lot of computation. Although we will not see the practical use of the first result until much later, we have already seen how the second result is very helpful in dealing with solutions to large systems or solutions with unusual values. In the next two lessons, we will investigate some more "theoretical" facts concerning matrix algebra that will be of great practical help throughout the remainder of the text.

# Properties of Matrix Algebra

When a scientist observes regular patterns in nature, he will try to describe those patterns in very precise language (often involving mathematical formulas) and refer to these patterns as "natural laws." When a mathematician notices patterns, she will do the same thing, except she will usually refer to these patterns as "theorems." When a scientist hears about the results of a colleague regarding some "natural law," if it is not too difficult, he will often try to reproduce the results for himself by performing careful experiments. A mathematician will usually do this as well; however, she will also attempt to *prove* that these patterns hold quite generally.

In Lesson 10, we saw how to perform arithmetic operations with matrices. In this lesson, we will observe that matrices obey many of the ordinary laws of algebra. We will perform mathematical "experiments" to discover and verify for ourselves these "natural laws" of matrix algebra. Because matrices satisfy many of the familiar laws of algebra, such as the associative, commutative, and distributive laws, we may consider matrices as numbers and carry out algebraic computations with them, such as solving equations, essentially as usual.

Although matrix algebra is quite similar to ordinary algebra, there are a number of important differences. Thus, in the first two sections, we will contrast all of the properties of matrix algebra that are similar to ordinary algebra with those ways in which it is different. Along the way, we will learn to act like mathematicians and provide simple proofs to generalize our observations.

"Transposition" of matrices has no analog in ordinary algebra. We will dedicate the final two sections to discussing the patterns relating transposition to the other arithmetic operations. We will also formally introduce the notion of a "partitioning" and discover a simple theorem between a partitioning and transposition. This will serve as precursor to the important Principle of Partitioned Matrices to which we will dedicate Lesson 13.

## 12.1  Similar Properties

In this section, we will investigate those properties of ordinary algebra that hold for matrix algebra as well, by examining some particular examples. We also discuss how one may verify that these properties hold in general.

# Associativity

## Exercise 12.1

Compute:

$$\begin{bmatrix} 1 & 0 & -3 \\ -2 & 5 & 4 \\ 0 & 3 & 7 \end{bmatrix} + \left( \begin{bmatrix} 2 & 2 & 4 \\ -3 & 7 & -2 \\ 5 & 1 & -9 \end{bmatrix} + \begin{bmatrix} 0 & 2 & 6 \\ 4 & -2 & 1 \\ -1 & 0 & 3 \end{bmatrix} \right)$$

by first adding inside the parentheses. Now perform the addition in a different order:

$$\left( \begin{bmatrix} 1 & 0 & -3 \\ -2 & 5 & 4 \\ 0 & 3 & 7 \end{bmatrix} + \begin{bmatrix} 2 & 2 & 4 \\ -3 & 7 & -2 \\ 5 & 1 & -9 \end{bmatrix} \right) + \begin{bmatrix} 0 & 2 & 6 \\ 4 & -2 & 1 \\ -1 & 0 & 3 \end{bmatrix}$$

In each case, record the result of the addition in parentheses and then the total result. Do your results tend to confirm or deny the associative law of matrix addition, as expressed formally below?

$$A + (B + C) = (A + B) + C$$

# Constructing a Proof

When a scientist states a "natural law," he makes no guarantee that it will *always* hold true. He can only say that it has never been observed to fail. After conducting experiments, a mathematician is held to a higher standard. She should justify her observations with a logical explanation of why her theorem will *always* hold. It is this higher standard of accountability that makes mathematics one of the most challenging, useful, and rewarding of academic disciplines, the so-called Queen of the Sciences.

In fact, *every* statement that a mathematician utters must withstand the eternal question: Why is this true? To provide such justifications, mathematicians are trained to construct careful, logical "proofs." Although such proofs may sometimes involve lots of notation and equations, fundamentally a proof should simply be a clearly worded explanation that convinces its audience. The most important requirement for a "proof" (which distinguishes it from simply an "experiment") is that a "proof" must not contain any "loopholes," in that it must address every possible contingency.

Mathematicians do so by utilizing very general arguments that appeal to commonly accepted definitions and general properties. Likewise, they will avoid reference to the specific properties of any particular example. For example, we could "prove" the associative law for matrix addition with the following argument:

We know (by definition) that matrix addition is done by adding corresponding entries. We know that the associative law holds for ordinary numbers (known property) and therefore in each entry of the matrices $A$, $B$, and $C$. Because any matrix equality may be verified entry-by-entry, the associative law holds for matrices themselves.

Notice that we did *not* refer to any *specific* numbers which would limit the generality of our observations. To convert this argument into a "formal" proof, we simply introduce appropriate notation and more clearly state the relevant definitions and theorems that we use. As it stands, this reasoning is what is known as an "informal" argument. This is the type of argument that mathematicians most frequently use "between friends." Hopefully, we are on a friendly basis, and so we will rely on informal arguments much of the time. However, it is important to be able to understand and construct formal arguments, as well.

To illustrate this process, we will convert the previous argument into a formal proof.

**Proof:**

To prove that $A+(B+C) = (A+B)+C$, we must show that both sides have precisely the same entries. Using subscript notation, we must show that $(A + (B + C))_{ij} = ((A + B) + C))_{ij}$ for every row $i$ and column $j$. In Section 10.2, we saw that matrix addition is defined by the formula $(X + Y)_{ij} = X_{ij} + Y_{ij}$. Thus, we may make the following computation:

$$(A + (B + C))_{ij} = A_{ij} + (B + C)_{ij} = A_{ij} + (B_{ij} + C_{ij}) \qquad (12.1)$$

Here we have used the addition formula of Section 10.2 twice. The first time, we let $X = A$ and $Y = B + C$, while the second time $X = B$ and $Y = C$.

Notice that the $+$ signs on the left of Eq. 12.1 represent *matrix* addition, while those on the right represent *ordinary* addition. Thus, at this point, we may apply the associative law of *ordinary* arithmetic to give:

$$A_{ij} + (B_{ij} + C_{ij}) = (A_{ij} + B_{ij}) + C_{ij}$$

We may again use the definition of matrix addition, but in reverse, to obtain:

$$(A_{ij} + B_{ij}) + C_{ij} = (A + B)_{ij} + C_{ij} = ((A + B) + C)_{ij}$$

Putting these three equations together gives our desired result. Q.E.D.

*Note*: "Q.E.D." is an abbreviation for the Latin phrase *quod erat demonstrandum*, which means "Thus, it is proved." It is traditional to put this at the end of a proof to indicate its completion. Compare this proof with the informal proof given earlier and you should observe that they involve essentially the same ideas. In this case, however, the argument was restructured into a single, extended calculation.

From this example, one may observe the structure of a formal proof. As in most formal proofs, we begin by clarifying what it is we must prove and introducing appropriate notation. In this case, we specified a particular row and column. However, we did not limit our argument by giving precise values for $i$ or $j$. Because this theorem was given by a single equation, the proof consisted of a computation, namely a sequence of equalities. Each equality follows by substitution, using an equation from a known theorem or definition.

As is often the case, the first and last few equalities in our computation consisted of a direct appeal to a definition. The key idea of the proof occurs in the central equality, where we apply the known theorem of ordinary arithmetic, called the associative law of addition. Ninety percent of the construction of a simple proof, such as this, consists of obtaining a clear understanding of what one is trying to prove, establishing clear notation and applying known definitions. Once one has done this, the heart of the proof becomes visible and the proof may be completed by direct algebraic manipulation, substitution, and application of known results.

## Commutativity of Addition

### Exercise 12.2

Compute:

$$\begin{bmatrix} 1 & 0 & -3 \\ -2 & 5 & 4 \\ 0 & 3 & 7 \end{bmatrix} + \begin{bmatrix} 2 & 2 & 4 \\ -3 & 7 & -2 \\ 5 & 1 & -9 \end{bmatrix} \quad \text{and} \quad \begin{bmatrix} 2 & 2 & 4 \\ -3 & 7 & -2 \\ 5 & 1 & -9 \end{bmatrix} + \begin{bmatrix} 1 & 0 & -3 \\ -2 & 5 & 4 \\ 0 & 3 & 7 \end{bmatrix}$$

Do your results tend to confirm or deny the commutative law of matrix addition, as expressed formally below?

$$A + B = B + A$$

### Exercise 12.3

Provide an informal argument to support your observations of Ex. 12.2, then translate your argument into a formal proof, as we did for associativity.

## Scalar Multiplication Distributes over Addition

### Exercise 12.4

Compute:

$$3 \cdot \left( \begin{bmatrix} 2 & 2 & 4 \\ -3 & 7 & -2 \\ 5 & 1 & -9 \end{bmatrix} + \begin{bmatrix} 1 & 0 & -3 \\ -2 & 5 & 4 \\ 0 & 3 & 7 \end{bmatrix} \right)$$

by first performing the addition. Now compute:

$$3 \cdot \begin{bmatrix} 2 & 2 & 4 \\ -3 & 7 & -2 \\ 5 & 1 & -9 \end{bmatrix} + 3 \cdot \begin{bmatrix} 1 & 0 & -3 \\ -2 & 5 & 4 \\ 0 & 3 & 7 \end{bmatrix}$$

by first multiplying, recording all intermediate steps. Do your results tend to confirm or deny the distributive law of scalar multiplication, as expressed formally below?

$$x(A + B) = xA + xB$$

### Exercise 12.5

Provide an informal argument to support your observations of Ex. 12.4, then translate your argument into a formal proof.

## Matrix Multiplication Distributes over Addition

### Exercise 12.6

Compute:

$$\begin{bmatrix} 2 & 3 & -1 \\ 1 & -4 & 6 \end{bmatrix} \left( \begin{bmatrix} 1 & 5 \\ -2 & 0 \\ 3 & -1 \end{bmatrix} + \begin{bmatrix} 0 & -6 \\ 1 & 7 \\ -2 & 3 \end{bmatrix} \right)$$

by first performing the addition. Now compute:

$$\begin{bmatrix} 2 & 3 & -1 \\ 1 & -4 & 6 \end{bmatrix} \begin{bmatrix} 1 & 5 \\ -2 & 0 \\ 3 & -1 \end{bmatrix} + \begin{bmatrix} 2 & 3 & -1 \\ 1 & -4 & 6 \end{bmatrix} \begin{bmatrix} 0 & -6 \\ 1 & 7 \\ -2 & 3 \end{bmatrix}$$

by first multiplying, recording all intermediate steps. Do your results tend to confirm or deny the distributive law of matrix multiplication, as expressed formally below?

$$C(A+B) = CA + CB$$

Because matrix multiplication is not "entry-wise," we cannot give the same kind of informal justification for this property as before. This is a case where the *formal* proof is much clearer than a verbal explanation. That is because we have established clear and concise notation for expressing matrix multiplication in Definition 10.3.

**Proof:**

As usual, we must show that the entries of both sides are equal, by appealing to Definition 10.3, the definition of matrix addition, and the laws of ordinary arithmetic. Because Definition 10.3 is stated in two parts, it makes sense to approach the proof in two stages.

*Case I*

First, we will assume that $C$ consists of a single row and $A$ and $B$ each consist of a single column. That is, $C = (C)_{1\cdot}$ is a $1 \times q$ matrix, while $A = (A)_{\cdot 1}$ and $B = (B)_{\cdot 1}$ are $q \times 1$ matrices. This means that $C(A+B)$ is a $1 \times 1$ matrix, given by the single number:

$$\begin{aligned} (C(A+B))_{11} &= (C)_{11}(A+B)_{11} + \cdots + (C)_{1q}(A+B)_{q1} \\ &= (C)_{11}\{(A)_{11} + (B)_{11}\} + \cdots + (C)_{1q}\left\{(A)_{q1} + (B)_{q1}\right\} \end{aligned}$$

As usual, we begin by applying the definitions of matrix multiplication and addition. Notice that we are using Eq. 10.7 from Definition 10.2, because it is more convenient. Now that all terms in this equation are scalars, we may apply the distributive law of *ordinary* multiplication to obtain:

$$\begin{aligned} &(C)_{11}\{(A)_{11} + (B)_{11}\} + \cdots + (C)_{1q}\left\{(A)_{q1} + (B)_{q1}\right\} \\ =\ &(C)_{11}(A)_{11} + (C)_{11}(B)_{11} + \cdots + (C)_{1q}(A)_{q1} + (C)_{1q}(B)_{q1} \\ =\ &\left\{(C)_{11}(A)_{11} + \cdots + (C)_{1q}(A)_{q1}\right\} + \left\{(C)_{11}(B)_{11} + \cdots + (C)_{1q}(B)_{q1}\right\} \end{aligned}$$

Notice how we needed to use the commutative and associative laws, as well, to regroup terms.

As usual, it remains to apply the definitions of matrix multiplication again, in reverse:

$$\left\{ (C)_{11}\,(A)_{11} + \cdots + (C)_{1q}\,(A)_{q1} \right\} + \left\{ (C)_{11}\,(B)_{11} + \cdots + (C)_{1q}\,(B)_{q1} \right\}$$
$$= \ (CA)_{11} + (CB)_{11} = (CA + CB))_{11}$$

Since we have shown that the one and only entry of the matrices $C(A + B)$ and $CA + CB$ are equal, we are done with this case.

*Case II*

Now we assume that $A$, $B$, and $C$ are *arbitrary* compatible matrices. The proof in this general case follows almost immediately from the previous case and Definition 10.3. As usual, we examine an arbitrary entry from the matrix $C(A + B)$ and equate this to the corresponding entry of $CA + CB$. From the definitions, we have:

$$(C(A + B))_{ij} = (C)_{i \cdot}\,(A + B)_{\cdot j} = (C)_{i \cdot} \left\{ (A)_{\cdot j} + (B)_{\cdot j} \right\}$$

Because $(C)_{i \cdot}$ is a single row while $(A)_{\cdot j}$ and $(B)_{\cdot j}$ are single columns, we may apply the distributive law, which we have proven in Case I. Using this and the definitions of multiplication and addition, we obtain:

$$(C)_{i \cdot} \left\{ (A)_{\cdot j} + (B)_{\cdot j} \right\} \ = \ (C)_{i \cdot}\,(A)_{\cdot j} + (C)_{i \cdot}\,(B)_{\cdot j}$$
$$= \ (CA)_{ij} + (CB)_{ij} = (CA + CB)_{ij}$$

Putting these equations together gives $(C(A + B))_{ij} = (CA + CB)_{ij}$. Since $i$ and $j$ were arbitrary, we have shown that every entry of $C(A + B)$ equals the corresponding entry of $CA + CB$, so that $C(A + B) = CA + CB$. Since we did not put any special restrictions on $A$, $B$, or $C$, this equality holds generally for any three matrices for which these operations are defined. Q.E.D.

Notice how one case in this proof built on the other. It often happens that one may begin by proving a specialized version of a result, in this example by restricting the dimensions of the matrices involved, since special cases are often easier to analyze. If one is very fortunate, as we were here, the general case follows with only a little more work. Besides being a good proof technique, examining special cases often helps one gain a better understanding of any given theorem or definition. In general, one should illustrate for oneself any theorem, definition, or proof by exploring a particular example.

## 12.2  Unusual Properties

Now we examine some of the properties of matrix arithmetic that distinguish matrix algebra from ordinary algebra. One of the most important differences has to do with its unusual multiplication.

## Matrix Multiplication Is Not Commutative

In Ex. 12.2, we saw that addition is commutative. Now we examine the corresponding property for multiplication.

### Exercise 12.7

If $A = \begin{bmatrix} 2 & 3 & -1 \end{bmatrix}$ and $B = \begin{bmatrix} 4 \\ -2 \\ 5 \end{bmatrix}$, determine whether the commutativity equation $AB = BA$ holds. Explain your conclusions.

This is *not* a particularly unusual example.

### Exercise 12.8

Repeat Ex. 12.7 with $A = \begin{bmatrix} 1 & 2 \\ 3 & 4 \end{bmatrix}$ and $B = \begin{bmatrix} 5 & 6 \\ 7 & 8 \end{bmatrix}$.

In fact, it is unusual if the multiplication *is* commutative for a given pair of matrices.

## The "Zero" Matrix

Just as with numbers, there is a matrix which performs like 0, in that:
$$A + 0 = A \tag{12.2}$$
for any matrix $A$, with a slight difference. Because matrices must have the same dimensions for the addition to be defined, there must be a "zero" matrix *of every possible shape*. It may seem confusing to use the same symbol 0 to indicate *all* of the possible zero matrices, as well at the scalar 0. However, the exact meaning of the symbol "0" is usually quite easy to determine from its context.

### Exercise 12.9

Determine the $3 \times 3$ zero matrix. Verify that your candidate for 0 satisfies Eq. 12.2 with $A = \begin{bmatrix} 2 & 2 & 4 \\ -3 & 7 & -2 \\ 5 & 1 & -9 \end{bmatrix}$. *Hint*: We should have the equation $A - A = 0$.

## The "One" Matrix

Likewise, there exist matrices that perform as 1, in that:
$$1A = A \tag{12.3}$$
for any matrix $A$. Unlike zero matrices, we must be more careful with our notation. It would be difficult to determine whether "1" stands for a matrix or a scalar from its context alone. Thus, we use the letter $I$ to indicate a "one" matrix, better known as an "identity matrix." As with zero matrices, there must be more than one possible identity

matrix. Notice that Eq. 12.3 requires an identity matrix to be square. A typical identity matrix is of the form:

$$I = \begin{bmatrix} 1 & 0 \\ 0 & 1 \end{bmatrix}$$

## Exercise 12.10

Determine the $3 \times 3$ identity matrix. Verify that your candidate for $I$ satisfies Eq. 12.3 with $A = \begin{bmatrix} 2 & 2 & 4 \\ -3 & 7 & -2 \\ 5 & 1 & -9 \end{bmatrix}$. Verify that $AI = A$, as well.

Stated in formal terms:

**Definition 12.1** We say that $I$ is both a left- and a right-sided *identity* matrix iff

$$IA = A \qquad \text{and} \qquad AI = A \tag{12.4}$$

In this definition, the term "iff" is short-hand for "if and only if." We use this term in definitions, since the defined terms are logically equivalent to the defining conditions.

## Matrix "Division"

Division is rather more difficult, even in ordinary arithmetic. While "long division"is a difficult algorithm, matrix "division" is even more involved. For example, in ordinary division, there is only one number that is off-limits, in that you may not divide by $0$. In contrast, *most* matrices are off-limits in this way. We say that a matrix with which you cannot divide is "singular" or "non-invertible." Moreover, it is rather difficult to determine whether or not a matrix is non-singular. Thus, we will postpone further discussion of invertibility until Lessons 14 – 16.

## 12.3   Transposition and Matrix Arithmetic

Transposition is an operation that is unique to matrix arithmetic. However, the transposition operator interacts well with the other more usual arithmetic operations. We may describe what we mean by this either intuitively in words or formally in symbols.

**Theorem 12.1** *For any matrices of the appropriate dimensions, A and B, the following equalities hold:*

(a) $\left(A^T\right)^T = A$

(b) $(A + B)^T = A^T + B^T$

(c) $(AB)^T = B^T A^T$

While these equations are clear and easy to remember, it is also helpful to rephrase each part of this theorem in words:

a. The transpose operator "undoes" itself. We say that transposition is "self-invertible".

b. The transpose operator "commutes" with addition. That is, the order of operations does not matter — add, then transpose or vice versa. This type of property is common in linear algebra. We this behavior by saying that the transposition operator "preserves" addition.

c. The transpose *almost* preserves multiplication, except we must reverse the order of the terms in the product.

Notice that parts 12.1b and 12.1c each resemble a kind of distributive law, which suggests that it may be used in a manner similar to the ordinary distributive law. Verbal descriptions, such as this, often provide insight into the meaning and usefulness of a theorem.

We will supply a formal proof of Theorem 12.1c, which is the most difficult part, leaving the remaining parts of the theorem for you to explore in subsequent exercises. As with the distributive law for matrix multiplication, this result may be easily proven using the approach of a "proof by cases."

**Proof:**

*Case I*

The simplest case is when $A$ is a single row and $B$ is a single column. Since the result is a $1 \times 1$ matrix, the transposition on the left side has no effect, i.e., $(AB)^T = AB$. By the definition of matrix multiplication:

$$AB = (AB)_{11} = (A)_{1\cdot}(B)_{\cdot 1} = (A)_{11}(B)_{11} + \cdots + (A)_{1n}(B)_{n1}$$

and

$$\begin{aligned} B^T A^T &= (B^T A^T)_{11} = (B^T)_{1\cdot}(A^T)_{\cdot 1} = (B^T)_{11}(A^T)_{11} + \cdots + (B^T)_{1n}(A^T)_{n1} \\ &= (B)_{11}(A)_{11} + \cdots + (B)_{n1}(A)_{1n} \end{aligned}$$

These two expressions are equal by commutativity of *ordinary* multiplication.

*Case II*

Now we prove the general case, where $A$ is $n \times q$ and $B$ is $q \times m$. As usual, we begin on the left-hand side of the equation, examine a general entry and try to rewrite it until we obtain the corresponding entry from the right-hand side of the original equation. By the definitions of transposition and matrix multiplication, we have:

$$\left((AB)^T\right)_{ij} = (AB)_{ji} = (A)_{j\cdot}(B)_{\cdot i}$$

Now transpose both sides of this equation and we obtain:

$$\left((AB)^T)_{ij}\right)^T = \left((A)_{j\cdot}(B)_{\cdot i}\right)^T$$

Because the left-hand side is a single number, it will remain unchanged and we have:

$$((AB)^T)_{ij} = ((A)_{j\cdot}(B)_{\cdot i})^T$$

The right-hand side, however, is an example of a row times a column, which we have already examined. Thus, using Case I and the definition of multiplication, we have:

$$((AB)^T)_{ij} = ((B)_{\cdot i})^T ((A)_{j\cdot})^T = (B^T)_{i\cdot}(A^T)_{\cdot j} = (B^T A^T)_{ij}$$

Here we have used the fact that $((B)_{\cdot i})^T = (B^T)_{i\cdot}$, or in other words, the transpose of the $i$th column of $B$ is the $i$th row of the transpose of $B$.

In every case, we have shown that each entry of $(AB)^T$ equals the corresponding entry of $B^T A^T$. Thus, these two matrices must be equal. Q.E.D.

In the next few exercises, we verify Theorem 12.1 by applying it to specific matrices.

## Exercise 12.11

Taking $A = \begin{bmatrix} 2 & 3 & -1 \\ 1 & -4 & 6 \end{bmatrix}$ and $B = \begin{bmatrix} 7 & -2 & 5 \\ 3 & 0 & 4 \end{bmatrix}$, compute $B^T_\cdot A^T + B^T_\cdot A + B$, and $(A + B)^T_\cdot$ and compare the results.

## Exercise 12.12

Using $A$ from Ex. 12.11, compute $A^T$ and then $(A^T)^T$ and compare the results.

## Exercise 12.13

Take $A = \begin{bmatrix} 2 & 3 & -1 \\ 1 & -4 & 6 \end{bmatrix}$ and $B = \begin{bmatrix} 7 & 3 \\ -2 & 0 \\ 5 & 4 \end{bmatrix}$ and compute $AB$, $(AB)^T_\cdot A^T_\cdot B^T_\cdot$ and $B^T A^T_\cdot$ Compare your results with Theorem 12.1c.

In the next exercise, we will discover our own theorem.

## Exercise 12.14

Use $A$ from Ex. 12.11 to compute $2A$ and $(2A)^T$. Compare your results with $2(A^T)$. Generalize this example to give a new theorem. State your theorem both formally in symbols and informally in English.

## Exercise 12.15

Explain in words why parts 12.1a and 12.1b of Theorem 12.1, as well as your theorem from Ex. 12.14, should be true. Convert each of your explanations into formal proofs by writing down the appropriate calculations.

## 12.4  Partitioning and Transposition

In the next lesson we will discuss an important technique known as the Principle of Partitioned Matrices (compare Theorem 13.1). It describes how the operation of matrix multiplication interacts with partitioning. In this section, we discover a similar principle for the transposition operator, namely, that transposition "preserves" partitioning in a nice way. This section serves to provide a gradual introduction to such partitioning theorems.

Formally, we "partition" a matrix by simply drawing vertical and horizontal lines through the matrix. We then consider the original matrix as a *matrix of smaller matrices*. This may be accomplished in many different ways for the same matrix. For example, the matrix $M = \begin{bmatrix} 1 & 2 & 3 & 4 \\ 5 & 6 & 7 & 8 \end{bmatrix}$ may be partitioned as $\begin{bmatrix} 1 & 2 & 3 & 4 \\ 5 & 6 & 7 & 8 \end{bmatrix}$. In this way, we consider $M$ to be the $1 \times 2$ matrix $\begin{bmatrix} A & B \end{bmatrix}$, where $A = \begin{bmatrix} 1 & 2 \\ 5 & 6 \end{bmatrix}$ and $B = \begin{bmatrix} 3 & 4 \\ 7 & 8 \end{bmatrix}$.
We refer to $A$ and $B$ as "submatrices" of $M$.

In order to examine the Partitioning Principle for Transposition, we will partition $M$ as $M = \begin{bmatrix} 1 & 2 & 3 & 4 \\ 5 & 6 & 7 & 8 \end{bmatrix}$. In this way, we may consider $M = \begin{bmatrix} A & B \\ C & D \end{bmatrix}$ as a matrix of matrices, with $A = \begin{bmatrix} 1 & 2 \end{bmatrix}$, $B = \begin{bmatrix} 3 & 4 \end{bmatrix}$, $C = \begin{bmatrix} 5 & 6 \end{bmatrix}$, and $D = \begin{bmatrix} 7 & 8 \end{bmatrix}$. If we transpose $M$ *with* this partitioning, we obtain $M^T = \begin{bmatrix} 1 & 5 \\ 2 & 6 \\ 3 & 7 \\ 4 & 8 \end{bmatrix}$. If we introduce the variables for each submatrix, we begin to observe a pattern:

$$\begin{bmatrix} A & B \\ C & D \end{bmatrix}^T = M^T = \begin{bmatrix} A^T & \phantom{x} \\ & \phantom{x} \end{bmatrix} \tag{12.5}$$

### Exercise 12.16

Fill in the blanks in the last matrix of Eq. 12.5 to complete the pattern. Based on this example, generalize to give a description of a Partitioning Principle for Transposition:

"To compute the transpose of a partitioned matrix, such as $\begin{bmatrix} A & B \\ C & D \end{bmatrix}$, one should . . ."

## 12.5  Summary

At this point, we should make a number of observations. Besides offering a notational convenience or representing geometric transformations, matrices are also algebraic objects that we may treat as a new kind of number. Because they satisfy most of the familiar rules of algebra, we may solve equations as usual, such as solving $2X + 3A = 5B$ for $A$, to obtain $A = (5/3)B - (2/3)X$, even though $X$, $A$, and $B$ may be matrices.

For the most part, the procedures and practices we have learned for ordinary algebra hold true for matrix algebra, as well. However, we must be a bit careful when manipulating matrices, because matrices differ from ordinary numbers in some very important ways:

- matrix multiplication is not commutative (i.e., order matters),

- not every two matrices can be added, subtracted, or multiplied,

- there is more than one type of multiplication for matrices,

- there is more than one 0 matrix,

- there is more that one 1 matrix,

- more than just the 0 matrices are off-limits for division, and

- we may transpose matrices.

From a historical point of view, matrices formed one of the first systems of numbers beyond ordinary, real numbers. For many years mathematicians used the term "real" to refer to the so-called real number system, because they believed this system to have a unique position corresponding to physical reality. The discovery of matrices forced mathematicians to re-examine their own biased and narrow conception of a number, mathematics in general, and the very foundations of mathematics.

At this point, you may wonder what matrices *really* are. Are they notation, geometric transformations, or numbers? They are all of these and more! This ability to see the same thing from many points of view is crucial to mathematics, in general, and linear algebra, in particular. Changing our perspective is often the crucial step in solving any particular problem. We have already seen this type of thinking, for example, when solving linear systems. The problems in Lesson 3 were originally phrased in words, in the context of a specific discipline. We first translated each problem into a mathematical setting, as equations. To employ Gaussian elimination, we would then convert these equations into matrices. To use *Mathematica*, we would then translate each matrix into *Mathematica* notation and use the computer to perform Gaussian elimination and back-addition. We would ultimately translate the results from the computer back into the context of the original problem.

Throughout the remainder of the text, we will discover even *more* ways of viewing matrices. All of them will prove useful at one time or another. In general, you will need to become very flexible in your thinking. You will learn to rephrase and reformulate problems in a variety of different ways, each of which will lead to different insights and problem-solving strategies.

The technique of partitioning provides one such method for examining a matrix from multiple points of view. In the next lesson, we will investigate how matrix multiplication relates to partitioning. This will provide a powerful technique for analysis, known as the Principle of Partitioned Matrices, which we will use often in the remainder of the text. We will immediately observe some of its uses for analyzing systems of equation.

# Matrix Algebra and Block Matrices

In Lessons 9 – 12, we discussed matrix multiplication and some of its important properties. In this lesson, we will discuss an important property of matrix multiplication that will allow us to perform the same matrix multiplication in many different ways, via partitioning. We will use it repeatedly throughout the text to prove most properties of matrices, especially those involving multiplication.

Partitioning also allows us to *view* matrix multiplication in many different ways. One of those alternative viewpoints concerns the idea of a "linear combination" of "column vectors," which is foundational for the entire theoretical development of linear algebra. Computationally, partitioning allows us to divide a large matrix product into many smaller pieces. We may then compute each piece "in parallel." This technique is used, for example, on modern supercomputers to greatly speed calculations.

As you progress to become an independent learner, you will need to be able to read and write formal mathematics, because this is how mathematical concepts are communicated in college and beyond. In Lesson 12, we introduced subscript notation that allowed us to provide formal proofs to buttress our intuition. You were asked to mimic the proofs from the text to provide simple proofs of your own. We will continue this process by giving a formal proof of the Principle of Partitioned Matrices, where you will need to fill in part of the proof for yourself.

Remember that a mathematical proof is simply a way of trying to convince a skeptical audience of the truth of certain observations. A formal proof may not always be the best way to do this. For example, you may notice that we will not state the Principle of Partitioned Matrices in very formal terms. That is because this would involve a lot of extra notation that could ultimately obscure the actual meaning of the theorem. Throughout this text, you will never need to take anything "on faith" or "because I say so." The justfication may be in the form of a formal proof. Occasionally, although a formal proof could be given, you will be asked to settle for some well-chosen, convincing examples. This is because sometimes a careful proof may be actually *less* convincing. Most often, we will give an informal argument embodied in an extended "conversation" that we will have as you work through examples and respond to the text. Although you may need to wait a while for some explanations, we will eventually give some sort of justification for every definition and theorem in this book. This is what makes mathematics such an exciting and satisfying discipline: There are always reasons for everything, and given enough time and effort, you will always be rewarded for your patient inquiry.

## 13.1    Introduction to the Principle of Partitioned Matrices

Partitioning provides a powerful tool for analyzing matrix multiplication, based on the Principle of Partitioned Matrices, expressed in the following theorem.

**Theorem 13.1 (The Principle of Paritioned Matrices)** *If two matrices $X$ and $Y$ are partitioned as matrices with matrix entries, in such a way that all necessary products are defined, the product $XY$ may be computed by the same formulas as if the submatrices were numbers.*

We will see in the following exercise that, because there is only the most minimal restriction on the manner of partitioning, this Principle implies that the same product may be computed through a variety of different intermediate calculations.

For example, take $X = \begin{bmatrix} 1 & 2 \\ 5 & 6 \end{bmatrix}$ and $Y = \begin{bmatrix} 3 & 4 \\ 7 & 8 \end{bmatrix}$, and partition the product $XY$ as $\begin{bmatrix} A & B \end{bmatrix} \begin{bmatrix} C \\ D \end{bmatrix}$. This means that we must take $A = \begin{bmatrix} 1 \\ 5 \end{bmatrix}$, $B = \begin{bmatrix} 2 \\ 6 \end{bmatrix}$, $C = \begin{bmatrix} 3 & 4 \end{bmatrix}$, and $D = \begin{bmatrix} 7 & 8 \end{bmatrix}$. The Principle of Partitioned Matrices says that we may compute $XY$ by the formula $\begin{bmatrix} A & B \end{bmatrix} \begin{bmatrix} C \\ D \end{bmatrix} = \begin{bmatrix} AC + BD \end{bmatrix}$. At first, this may not seem surprising, since this formula simply expressed the rule for multiplying a row and column given in Eq. 10.4. However, that rule was originally stated assuming that the entries were *scalars* not *matrices*!

### Exercise 13.1

Verify the formula $\begin{bmatrix} A & B \end{bmatrix} \begin{bmatrix} C \\ D \end{bmatrix} = \begin{bmatrix} AC + BD \end{bmatrix}$ in the previous example by computing $AC$, $BD$, and $AC + BD$ and comparing this with $XY$.

Because there are three other possible partitionings for $XY$, the Principle of Partitioned Matrices implies that there are three alternative formulas for this product.

### Exercise 13.2

Let $X$ and $Y$ be as in Ex. 13.1. For each part, you should:

- identify how the variables *must* be defined so that the product on the left represents a partitioning of the product $XY$,
- compute the corresponding products on the right-hand side, and
- compare the result of simplifying the right-hand side as written, with the value of $XY$.

You should be aware that the definitions of the variables $A$, $B$, $C$, and $D$, are *not* the same as in Ex. 13.1 and are different for all three parts.

(a) $\begin{bmatrix} A \end{bmatrix} \begin{bmatrix} B & C \end{bmatrix} = \begin{bmatrix} AB & AC \end{bmatrix}$

(b) $\begin{bmatrix} A \\ B \end{bmatrix} \begin{bmatrix} C \end{bmatrix} = \begin{bmatrix} AC \\ BC \end{bmatrix}$

$$\text{(c)} \quad \begin{bmatrix} A \\ B \end{bmatrix} \begin{bmatrix} C & D \end{bmatrix} = \begin{bmatrix} AC & AD \\ BC & BD \end{bmatrix}$$

*Note*: Since matrix multiplication is not commutative, it is important to be careful about the order of all multiplications.

Notice how the Principle of Partitioned Matrices provides us with a number of ways of viewing the same multiplication. For example, while Ex. 13.2c is essentially a restatement of the second part of Definition 10.3, Ex. 13.2a says that we may perform a multiplication by "distributing" the multiplication over the *columns* of the *second* matrix. Similarly, Ex. 13.2b suggests that we may "distribute" the multiplication over the *rows* of the first matrix. Although the formula from Ex. 13.1 looks rather innocuous, it is probably the most subtle of the four formulas we have examined. It is known as the "outer product" and is used to speed matrix calculations on a supercomputer, as mentioned earlier.

### Drill Exercise 13.3

Construct your own example and verify Theorem 13.1, as in Ex. 13.1. Specifically, you will need to:

- choose two matrices $X$ and $Y$, which may be multiplied,
- partition both matrices, making sure that the position of the row partitions in the first is the same as that of the column partitions in the second,
- replace each of the submatrices created by this partitioning with variables to obtain two matrices with variable entries,
- multiply out these matrices symbolically, and
- compare the result of simplifying that result as written with the product $XY$.

This deceptively simple principle will show up again and again during our study of linear algebra, because it usually makes computations and theorems so much shorter and easier to follow. Because we will rely on it so heavily, you should make sure that you are comfortable with the principle before going on, and be careful to watch for its use.

## 13.2 Linear Combinations and Column Vectors

In this section, we will discover a new algebraic and geometric perspective on systems of equations by using "column vectors." In Lesson 10, we defined matrix multiplication to allow us to write a system of equations, such as:

$$\begin{aligned} 2x - y - z &= 1 \\ -x + 2y + z &= 2 \end{aligned} \qquad (13.1)$$

as the matrix equation:

$$\begin{bmatrix} 2 & -1 & -1 \\ -1 & 2 & 1 \end{bmatrix} \begin{bmatrix} x \\ y \\ z \end{bmatrix} = \begin{bmatrix} 1 \\ 2 \end{bmatrix}$$

Using the Principle of Matrix Multiplication, we may reformulate this equation. If we partition $A$ into columns and $X$ into rows, we have:

$$AX = \begin{bmatrix} a_1 & \cdots & a_n \end{bmatrix} \begin{bmatrix} x_1 \\ \vdots \\ x_n \end{bmatrix} = a_1 x_1 + \cdots + a_n x_n$$

Since $X$ was originally a single column, the $x_i$'s are $1 \times 1$ matrices, the matrix product $a_i x_i$ is the same as the scalar product $x_i a_i$, and we have:

$$AX = x_1 a_1 + \cdots + x_n a_n \tag{13.2}$$

This means that we may view Eq. 13.1 as:

$$x \begin{bmatrix} 2 \\ -1 \end{bmatrix} + y \begin{bmatrix} -1 \\ 2 \end{bmatrix} + z \begin{bmatrix} -1 \\ 1 \end{bmatrix} = \begin{bmatrix} 1 \\ 2 \end{bmatrix} \tag{13.3}$$

This simple change is one of the most important techniques in linear algebra. It is so important that we will develop an entire language for describing equations such as Eq. 13.3. For instance, we usually refer to a matrix with only one column as a "column vector," and we describe Eq. 13.3 as expressing the column vector $\begin{bmatrix} 1 \\ 2 \end{bmatrix}$ as a "linear combination" of the column vectors $\begin{bmatrix} 2 \\ -1 \end{bmatrix}$, $\begin{bmatrix} -1 \\ 2 \end{bmatrix}$, and $\begin{bmatrix} -1 \\ 1 \end{bmatrix}$. We use the term "linear combination," because we have combined the given column vectors using only the linear operations of addition and scalar multiplication (compare Lesson 2). We refer to the numbers $x$, $y$, and $z$ in Eq. 13.3 as the "coefficients" of this linear combination.

Solving Eq. 13.1, we discover that $x = 2$, $y = 1$, $z = 2$ is a solution. At this point we may now view this single solution in three distinct ways.

1. Lesson 2 emphasized how substitution of the values $x = 2$, $y = 1$, $z = 2$ into Eq. 13.1, would produce two valid equalities:

$$\begin{aligned} 2(2) - (1) - (2) &= 1 \\ -(2) + 2(1) + (2) &= 2 \end{aligned} \tag{13.4}$$

2. Lesson 11 stressed that multiplication of $\begin{bmatrix} 2 \\ 1 \\ 2 \end{bmatrix}$ and $\begin{bmatrix} 2 & -1 & -1 \\ -1 & 2 & 1 \end{bmatrix}$ will give $\begin{bmatrix} 1 \\ 2 \end{bmatrix}$, that is:

$$\begin{bmatrix} 2 & -1 & -1 \\ -1 & 2 & 1 \end{bmatrix} \begin{bmatrix} 2 \\ 1 \\ 2 \end{bmatrix} = \begin{bmatrix} 1 \\ 2 \end{bmatrix} \tag{13.5}$$

3. Now we may also interpret this solution as saying that the column vectors $\begin{bmatrix} 2 \\ -1 \end{bmatrix}$, $\begin{bmatrix} -1 \\ 2 \end{bmatrix}$, and $\begin{bmatrix} -1 \\ 1 \end{bmatrix}$ may be combined in a linear fashion, using the coefficients 2, 1, and 2, respectively, to achieve the column vector $\begin{bmatrix} 1 \\ 2 \end{bmatrix}$, or as a vector equation:

$$2 \begin{bmatrix} 2 \\ -1 \end{bmatrix} + 1 \begin{bmatrix} -1 \\ 2 \end{bmatrix} + 2 \begin{bmatrix} -1 \\ 1 \end{bmatrix} = \begin{bmatrix} 1 \\ 2 \end{bmatrix} \tag{13.6}$$

## Exercise 13.4

Show that $x = 2$, $y = 1$, $z = 2$ is in fact a solution of Eq. 13.1 by verifying Eq. 13.4–13.6. Show all intermediate calculations.

Now that we know how to solve systems of equations, we wish to understand the solution process more completely. In particular, we want to learn why some systems of equations have more than one solution, while others have no solutions. Although each point of view has its uses, the notion of "linear combination" will provide the clearest answer to this question.

The concept of a linear combination also allows us to view the process of solving systems of equations from a *geometric* viewpoint. For example, we may describe Eq. 13.6 by the following picture.

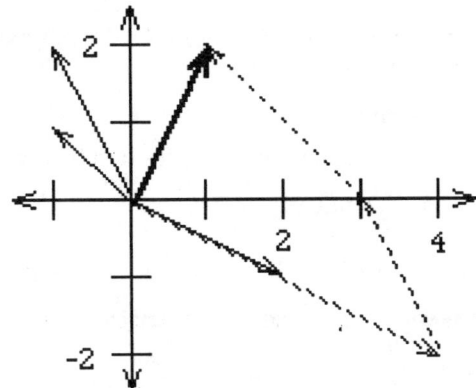

In this picture, the bold arrow corresponds to the right-hand side of Eq. 13.6. Likewise, the three solid arrows represent the column vectors on the left-hand side, while the three dotted arrows represent the process of taking a linear combination of those vectors.

## Exercise 13.5

Infer how the arrows correspond to the "column vectors" in Eq. 13.6. Deduce how the dotted arrows correspond to the scalar multiplication and addition on the left-hand side of Eq. 13.6.

## Exercise 13.6

Determine another solution to Eq. 13.1 and verify that your solution works by each of the three methods that we have discussed. Draw a picture to represent the corresponding linear combination.

Throughout this text, we will explore the *geometric* perspective of each concept. It is often said that "a picture is worth a thousand words." That is because our brains are "hardwired" to maneuver in two and three dimensions. By grounding linear algebra concepts on our geometric intuition, we gain an automatic head start on the subject. The theorems of linear algebra so closely reflect our innate sense of geometric reality, most of them will make intuitive sense, once you learn to view them geometrically.

## 13.3   The Proof of the Principle of Partitioned Matrices

In this section, we give a formal proof of Theorem 13.1. Although a proof must be written in general terms, when reading a proof like this, it sometimes helps to pick specific examples for all of the variables mentioned and follow the steps of the proof with numbers in place of the symbols. That was the intent of Ex. 13.1, to help us understand the statement of Theorem 13.1. This process of substituting specific examples can help to bring an abstract proof down-to-earth.

**Proof:**

Assume that $A$ and $B$ are $p \times q$ and $q \times r$, respectively. The general case will follow from three special cases.

*Case I: Horizontal–None*

Assume that $A$ has been partitioned horizontally into two parts and $B$ was left untouched. Specifically, let $A = \begin{bmatrix} C \\ D \end{bmatrix}$ where $C$ is $n \times q$ and $D$ is $(p-n) \times q$. The Principle of Partitioned Matrices says that $AB = \begin{bmatrix} C \\ D \end{bmatrix} B = \begin{bmatrix} CB \\ DB \end{bmatrix}$. To compare entries of these two matrices, there are two natural cases to consider.

*Case 1:*

When $i \leq n$, $(AB)_{ij} = (A)_i.(B)_{.j} = (C)_i.(B)_{.j} = (CB)_{ij}$, which is the $(i,j)$th entry of $\begin{bmatrix} CB \\ DB \end{bmatrix}$.

*Case 2:*

Similarly, if $i > n$, $(AB)_{ij} = (A)_i.(B)_{.j} = (D)_{(n-i).}(B)_{.j} = (DB)_{(n-i)j}$, which is the $(i,j)$th entry of $\begin{bmatrix} CB \\ DB \end{bmatrix}$.

*Case II: None–Vertical*

This case is very similar to the previous one. We leave this part for Ex. 13.7.

*Case III: Vertical–Horizontal*

If we partition $A = \begin{bmatrix} C & D \end{bmatrix}$, where $C$ is $p \times n$ matrix and $D$ is $p \times (q-n)$, then we must partition $B$ horizontally into $\begin{bmatrix} E \\ F \end{bmatrix}$, where $E$ is $n \times r$ and $F$ is $(q-n) \times r$.

The Principle of Partitioned Matrices says that $AB = \begin{bmatrix} C D \end{bmatrix} \begin{bmatrix} E \\ F \end{bmatrix} = CE + DF$.

In this case:

$$
\begin{aligned}
(AB)_{ij} &= (A)_{i1}(B)_{1j} + \cdots + (A)_{in}(B)_{nj} \\
&\quad + (A)_{i(n+1)}(B)_{(n+1)j} + \cdots + (A)_{iq}(B)_{qj} \\
&= (C)_{i1}(E)_{1j} + \cdots + (C)_{in}(E)_{nj} \\
&\quad + (D)_{i1}(F)_{1j} + \cdots + (C)_{i(q-n)}(F)_{(q-n)j} \\
&= (CE)_{ij} + (DF)_{ij} = (CE + DF)_{ij}
\end{aligned}
$$

*General Case*

A general partitioning of $A$ and $B$ may be achieved by a sequence of partitions of each of the three previous types. Since we can prove that the Principle of Partitioned Matrices holds for each step, it must hold in general. Q.E.D.

### *Advanced Exercise 13.7*

Repeat the proof of Case I of Theorem 13.1, making the appropriate changes, to give a formal proof of Case II.

## 13.4  Summary

The ability to change our perspective is the key to solving problems in linear algebra, and in mathematics in general. Much of linear algebra is simply a matter of rephrasing a problem, until one achieves a perspective from which the solution becomes clear. In this lesson, we have seen how the Principle of Partitioned Matrices allows one to view the same matrix multiplication in a variety of ways. In particular, we have seen how multiplication of a matrix and a column vector, and hence solutions to linear systems, may be interpreted in terms of a linear combination of column vectors. We have also visualized this process geometrically. In Lessons 14 – 16, we will use this principle to investigate matrix "division."

# Matrix Inverses: Definitions and Basic Properties

## Accompanied by Notebook

In Lesson 12, we originally claimed that we would discuss "matrix division" at this point. It would be somewhat more accurate to say that we will consider matrix "reciprocals." From ordinary algebra, we know how division is intimately related to both fractions and reciprocals, as indicated, for example, by the equation:

$$2 \div 3 = \frac{2}{3} = 2 \cdot \frac{1}{3} = 2 \cdot 3^{-1}$$

In this way, division by 3 is the same as multiplication by the reciprocal of 3. In the same way, for a (nonsingular) matrix $A$, we will define the "reciprocal" of $A$ and denote it by $A^{-1}$, and we will refer to this as the "inverse" of $A$.

Unlike ordinary numbers, since matrix multiplication is noncommutative, there is no way to clearly define division for matrices. Specifically, there is no reasonable way to choose whether $A \div B$ should be defined as $A \cdot B^{-1}$ or $B^{-1} \cdot A$. Even the defining property of a reciprocal

$$3 \cdot 3^{-1} = 1$$

must be slightly modified. That is, we must logically distinguish between three different types of inverses. We will see these three types of inverses correspond directly to the three different types of solutions to linear systems. Thus, a careful exploration of inverses will yield deep insights into our original problem of solving linear equations.

Unlike study in many other disciplines, mathematics allows one to explain *why* one's specific observations are true in general. Thus, we will:

- discuss the precise definition for each type of inverses,

- present a precise, formal proof for most statements, which carefully explain the general outline of the proof, as well as each specific step, and

- encourage and guide the reader to construct proofs that have been omitted or merely outlined.

*Important*: Do not spend a *lot* of time struggling with any proof in this lesson. Consider each proof until you understand *what* must be shown, and then devote a solid 10 minutes of thought into reading or composing a proof. If you do not understand the proof in that amount of time, go on to something else. Return to the proof at a later time and you may discover that the proof comes more easily.

## 14.1 Matrix Inverses: The Formal Definition

In elementary school, you learned to compute reciprocals, such as $\frac{1}{3} = .33\overline{3}$, using a rather complicated algorithm, commonly known as "long division." This is typically the first really difficult arithmetic operation one learns, and is usually offered with little explanation of its effectiveness, because any such explanation is usually beyond the comprehension of the audience. Likewise, computing matrix inverses is a rather difficult operation and the proof of the correctness of the algorithm is rather involved. However, with the computer this algorithm becomes much easier, and we will eventually be able to discover why our algorithm will always yield a correct result.

The fundamental property of an inverse is its "cancelation" property. For ordinary numbers, such as 3, this property is expressed by the equation:

$$\frac{1}{3} \cdot 3 = 1$$

It is this property which allows us to solve linear equations, such as $3x = 2$. One simply multiplies both sides by the inverse of 3:

$$3x = 2 \Rightarrow \frac{1}{3} \cdot 3x = \frac{1}{3} \cdot 2 \Rightarrow 1 \cdot x = .33\overline{3} \cdot 2 \Rightarrow x = .66\overline{6}$$

This suggests that if we define the inverse of $A$ in a similar way, we will be able to solve equations of the form $AX = B$ in *exactly* the same way:

$$AX = B \Rightarrow A^{-1} \cdot AX = A^{-1} \cdot B \Rightarrow I \cdot X = A^{-1} \cdot B \Rightarrow X = A^{-1} \cdot B \qquad (14.1)$$

Specifically, this implies that the equation

$$A^{-1} \cdot A = I$$

should be true for the inverse. In other words, multiplication by an inverse (on the left) cancels out multiplication by the original matrix. Because matrix multiplication is non-commutative, we must specify cancelation on the right separately. In this way, we have just *discovered* the formal definition of the inverse of a matrix:

**Definition 14.1** A matrix $B$ is called the (two-sided) *inverse* of $A$ and denoted by $A^{-1}$ iff $B$ satisfies the two equations:

$$B \cdot A = I \qquad \text{and} \qquad A \cdot B = I$$

### Exercise 14.1

Show that $\begin{bmatrix} 1 & -1.25 \\ -1 & 1.5 \end{bmatrix}$ is the inverse of $A = \begin{bmatrix} 6 & 5 \\ 4 & 4 \end{bmatrix}$. Make sure to explain your reasoning by appealing to Definition 14.1.

## Exercise 14.2

• Use the results of Ex. 14.1 to solve the system of equations:

$$6x + 5y = 3$$
$$4x + 4y = 8$$

*Hint*: Write this as a matrix equation and imitate the calculation in Eq. 14.1.

## Exercise 14.3

• Determine if $A = \begin{bmatrix} 6 & 5 & -1 \\ 4 & 4 & 2 \end{bmatrix}$ and $B = \begin{bmatrix} 8 & .5 \\ -9 & -.5 \\ 2 & .5 \end{bmatrix}$ are inverses. Make sure to explain your reasoning by appealing to Definition 14.1.

This definition provokes a few observations. First, notice that we will always have the cancelation equations:

$$A^{-1} \cdot A = I \qquad \text{and} \qquad A \cdot A^{-1} = I$$

Because we have cancelation on the left and on the right, we refer to such an inverse as "two-sided." Also, we should say that $B$ is *an* inverse, instead of *the* inverse. That is because, at this point, we do not know if there may be *more than one* matrix $B$ that satisfies these equations.

This leads us to consider the following theorem.

**Theorem 14.1** *A matrix $A$ has at most one two-sided inverse.*

### Proof:

This proof will differ from those which we have given so far, in that we will be able to reason by matrix algebra alone. That is, we will not need to examine particular entries of any matrix. Thus, this type of proof is even easier.

Assume that we have two *potentially* different inverses of $A$, call them $B$ and $C$. By Definition 14.1, this means that we have the equations

$$B \cdot A = I \qquad \text{and} \qquad A \cdot B = I$$

as well as:

$$C \cdot A = I \qquad \text{and} \qquad A \cdot C = I$$

We will show that $B = C$, so that there is only one actual inverse.

The argument may be summarized by the following calculation:

$$B = B \cdot I = B \cdot (A \cdot C) = (B \cdot A) \cdot C = I \cdot C = C$$

The first equality follows from the definition of an identity matrix, the second by substitution (using one of our given equations), and the next by the Associative Law of Matrix Multiplication. The last two steps are similar to first two but in reverse order. Q.E.D.

Ultimately, for any theorem "the proof is in the pudding." That is, we will see the truth of this, and every other theorem we will examine, as we look at specific examples and see them *work*! For example, in the next lesson we will discuss how to calculate the inverse $A^{-1}$ of a given matrix $A$ by solving a certain system of linear equations. We will see that such systems are never underdetermined. In particular, there will be at most one possible inverse.

Although we can now be sure that there is at most one inverse $B$ for any specific matrix $A$, we do not know if there is *any* matrix $B$ satisfying the conditions of Definition 14.1! In other words, although the process of finding an inverse is not underdetermined, it may be *overdetermined*. Just as the number 0 has no reciprocal (because 0 times any number is always 0 and can never be 1), there are many matrices that do not have a two-sided inverse. If the matrix $A$ does have an inverse, we say that $A$ is "invertible" (or "nonsingular"). If the matrix $A$ does not have an inverse, we say that $A$ is "noninvertible" or "singular."

In Lessons 16 and 23, we will learn a number of different conditions on a matrix $A$ that will guarantee that $A$ is invertible. For now, we will demonstrate how the invertibility of certain matrices will guarantee the invertibility of others.

**Theorem 14.2** *Assuming that the formulas on the right-hand side of each equation are well-defined and that the given inverses exist, each inverse on the left-hand side exists and can be computed by the given formula:*

(a) $(AB)^{-1} = B^{-1}A^{-1}$

(b) $(A^T)^{-1} = (A^{-1})^T$

(c) $(A^{-1})^{-1} = A$

- *Note*: One may wish to explore the truth of this theorem by experimenting in the accompanying *Mathematica* Notebook, before reading on to the formal proofs of these results. We will prove the first two parts of this theorem, leaving the third for practice.

**Proof:**

*Proof of Theorem 14.2a*

First, we should clarify what we must demonstrate. The first equation says that if $C = AB$, then $C^{-1} = B^{-1}A^{-1}$. In other words, if we let $D = B^{-1}A^{-1}$, we must show that $D$ satisfies the equations for an inverse for $C$, namely:

$$D \cdot C = I \qquad \text{and} \qquad C \cdot D = I$$

Plugging in our temporary definitions for $C$ and $D$, this means that we must verify the following two equations:

$$(B^{-1}A^{-1}) \cdot (AB) = I \qquad \text{and} \qquad (AB) \cdot (B^{-1}A^{-1}) = I$$

Using the basic rules of matrix algebra, this is very easy. For example,

$$(B^{-1}A^{-1}) \cdot (AB) = (B^{-1}(A^{-1} \cdot A)) B = (B^{-1} \cdot I) \cdot B = B^{-1} \cdot B = I$$

Here we use the associative law of matrix multiplication, the definition of an inverse (twice), and the defining property of the identity matrix $I$. We leave the proof of the second equation as an exercise.

*Proof of Theorem 14.2b*

To prove the second part of the theorem, we will take a slightly different approach. This equation says that if $B = A^T$, then $B^{-1} = (A^{-1})^T$. In other words, if we let $C = (A^{-1})^T$, we must show that $C$ satisfies the equations for an inverse for $B$:

$$C \cdot B = I \quad \text{and} \quad B \cdot C = I$$

Plugging in our temporary definitions for $B$ and $C$, this means that we must prove that:

$$(A^{-1})^T \cdot A^T = I \quad \text{and} \quad A^T \cdot (A^{-1})^T = I \tag{14.2}$$

Instead of proving these equations one at a time, we produce them simultaneously from other known equations.

From the defining equations for $A^{-1}$, we have:

$$A^{-1} \cdot A = I \quad \text{and} \quad A \cdot A^{-1} = I$$

Applying the transposition operator to both sides of each equation preserves both equalities and produces the equations:

$$(A^{-1} \cdot A)^T = I^T \quad \text{and} \quad (A \cdot A^{-1})^T = I^T \tag{14.3}$$

We have seen in Theorem 12.1 that transposition "commutes with" multiplication, as long as we reverse the order of multiplication. This means that:

$$(A^{-1} \cdot A)^T = A^T \cdot (A^{-1})^T \quad \text{and} \quad (A \cdot A^{-1})^T = (A^{-1})^T \cdot A^T \tag{14.4}$$

Also, if one writes out a specific example, it becomes clear that $I^T = I$. Substituting this and Eq. 14.4 into Eq. 14.3 yields:

$$A^T \cdot (A^{-1})^T = I \quad \text{and} \quad (A^{-1})^T \cdot A^T = I$$

which are just the desired equations from Eq. 14.2 (in reverse order).

The proof of Theorem 14.2c is the easiest of all three, so it is left as an exercise. In fact, you may be surprised by how little actual work is required, once you clearly state what you need to show. Q.E.D.

To obtain a more intuitive understanding of this theorem, it is helpful to translate each part of the theorem into English. For example, the first part of the theorem is very reminicent of the corresponding result for transposes. It says that taking inverses "preserves" matrix multiplication, as long as we reverse the order of multiplication. Phrased in this way, you may also find this result easier to remember.

## Exercise 14.4

Express the remaining parts of Theorem 14.2 in English. *Hint*: You may wish to re-examine the comments following Theorem 12.1.

### Exercise 14.5

Complete the proof of Theorem 14.2a. That is, simplify the left-hand side of

$$(AB) \cdot (B^{-1}A^{-1}) = I$$

until it is clear that equality holds. Be sure to explain *why* each of your steps is true.

### Exercise 14.6

Verify the third part of Theorem 14.2. In other words, assume that $B = A^{-1}$ and show that $B^{-1} = A$. *Hint*: Imitate the proof of either Theorem 14.2a or 14.2b. Do not be concerned if your proof seems very short, because it should be.

## 14.2   One-sided Inverses

Looking back at Eq. 14.1, we notice that *both* cancelation properties of the inverse are unnecessary, but rather we only require cancelation on the *left*. This suggests two new definitions:

**Definition 14.2** A matrix $B$, is called:

-   a *left-inverse* of $A$ iff $B$ satisfies the equation $B \cdot A = I$,
-   a *right-inverse* of $A$ iff $B$ satisfies the equation $A \cdot B = I$.

### Exercise 14.7

Determine if $B = \begin{bmatrix} -2 & 1 \\ 1 & \frac{1}{2} \\ 0 & -\frac{1}{2} \end{bmatrix}$ is a right-inverse for $A = \begin{bmatrix} 1 & 3 & 5 \\ 2 & 4 & 6 \end{bmatrix}$. Is it a (two-sided) inverse? Explain your answers. As usual, you may use *Mathematica* to perform any necessary calculations, unless specifically directed otherwise, as long as you describe the commands that you use, the resulting computer output, and how you *interpret* those results.

### Exercise 14.8

Show that $C = \begin{bmatrix} -1 & \frac{5}{2} \\ -1 & -\frac{5}{2} \\ 1 & 1 \end{bmatrix}$ is also a right-inverse to the matrix $A$ from Ex. 14.7.

We pause to make two rather subtle observations. We have seen that if a (two-sided) inverse exists, then there is only one possible inverse. Thus, there is no ambiguity in our notation of $A^{-1}$ for this (unique) inverse. On the other hand, the last two exercises show that there may be *more* than one possible right-inverse. Likewise, there can be more than one possible left-inverse. In particular, this means that we cannot invent a clear notation for left- or right-inverses. Likewise, we must refer to a one-sided inverse as a left- or right-inverse, as opposed to the (two-sided) inverse.

We have already observed how left-inverses are valuable for solving systems of equations. We conclude this lesson by suggesting how a right-inverse may help us solve a linear system, as well.

### Exercise 14.9

•   Using $A$ and $B$ from Ex. 14.7, show that $X = B \begin{bmatrix} 1 \\ 2 \end{bmatrix}$ is a solution to:

$$AX = \begin{bmatrix} 1 \\ 2 \end{bmatrix} \tag{14.5}$$

in two ways:

(a) by multiplication, to compute $X$, and substitution, and

(b) by substitution (*without* multiplication) and the definition of a right-inverse.

### Exercise 14.10

•   Determine *another* solution to Eq. 14.5. *Hint*: $C$ is also a right-inverse for $A$.

## 14.3   Summary

In this lesson, one should have learned:

- the definitions of left-, right-, and two-sided inverses,

- how inverses give rise to solutions to systems of equations,

- the definition of a singular *vs.* a nonsingular matrix, and

- that left- and right-inverses are not unique, in general, while two-sided inverses always are.

Notice that we have yet to discuss how to actually compute inverses. In the next lesson, we will explore that question and devise two different ways of computing left- or right-inverses. A complete discussion of two-sided inverses, however, will be postponed until Lesson 16.

# Computing Matrix Inverses

## Accompanied by Notebook

In Lesson 14, we learned the definitions of left-, right-, and two-sided inverses. We saw that not all matrices have inverses and others have more than one. However, we did see that if a matrix $A$ has a two-sided inverse, then it has only one, which we denote by $A^{-1}$.

In this lesson, we address the question of actually computing left- and right-inverses. We will supply two different methods for computing inverses: one that follows directly from the definitions and a faster one that uses the Principle of Partitioned Matrices. Along the way, we will learn how to solve "parallel" systems of equations. We will also expand on the connection between inverses and solving systems of equations, as well as informally observe some basic facts about inverses.

## 15.1  Computing Matrix Inverses from the Definitions

It is very easy to compute inverses, using the computer, once we write out the definition. For example, take the matrix $A = \begin{bmatrix} 1 & 3 & 5 \\ 2 & 4 & 6 \end{bmatrix}$ from Ex. 14.7. If we think carefully about the definition, we can come up with necessary restrictions on any possible right-inverse $B$. For example, we know that we must have $AB = I$. In particular, the multiplication must be possible, which implies that $B$ must have three rows. Since $I$ is a square matrix and we know that $AB$ has the same number of rows as $A$, $I$ must be $2 \times 2$. Similarly, since $AB$ has the same number of columns as $B$, $B$ must have two columns.

Thus, we know that a "generic" right-inverse must be of the form $B = \begin{bmatrix} a & b \\ c & d \\ e & f \end{bmatrix}$ and satisfy the equation $AB = I$.

Multiplying out gives the matrix equation:

$$AB = \begin{bmatrix} a + 3c + 5e & b + 3d + 5f \\ 2a + 4c + 6e & 2b + 4d + 6f \end{bmatrix} = I = \begin{bmatrix} 1 & 0 \\ 0 & 1 \end{bmatrix}$$

Equating the entries leads to a system of linear equations in $a$ through $f$:

$$
\begin{array}{rcrcrcl}
a & + & 3c & + & 5e & = & 1 \\
b & + & 3d & + & 5f & = & 0 \\
2a & + & 4c & + & 6e & = & 0 \\
2b & + & 4d & + & 6f & = & 1
\end{array}
$$

Now that we know Gaussian elimination, we can easily solve this system of four equations in six unknowns. We would not expect to get a unique solution and, as we saw in Ex. 14.7 –14.8, there will be more than one right-inverse for $A$. Using *Mathematica*'s RowReduce[] command, the only difficulty is in writing down the correct matrix:

$$
\begin{bmatrix}
1 & 0 & 3 & 0 & 5 & 0 & 1 \\
0 & 1 & 0 & 3 & 0 & 5 & 0 \\
2 & 0 & 4 & 0 & 6 & 0 & 0 \\
0 & 2 & 0 & 4 & 0 & 6 & 1
\end{bmatrix}
$$

• It is already typed into the accompanying Notebook inside a RowReduce[] command; evaluate that command now.

From the resulting reduced row echelon form we see that $a$, $b$, $c$, and $d$ are basic variables, while $e$ and $f$ are free. This is what we would expect with only four equations.

### Exercise 15.1

Record a specific solution to this system of equations, and the corresponding right-inverse. Be sure that your solution is different from either of the two right-inverses that have already been given in Ex. 14.7–14.8. Verify that your right-inverse does in fact work.

Now we will look for a left-inverse of $A$. This would be a matrix $B$ such that $BA = I$.

### Exercise 15.2

Explain why a left-inverse $B$ must be of the same form as a right-inverse, namely
$$
B = \begin{bmatrix} a & b \\ c & d \\ e & f \end{bmatrix}.
$$

### Exercise 15.3

• Write out the system of equations corresponding to the condition $BA = I$ for the same matrix $A$, as before. How many equations do you get? What do you expect the solution set to look like? Use RowReduce[] to check your hypothesis and give a left-inverse (if possible).

### Exercise 15.4

Can $A$ have a two-sided inverse? Explain your answer. *Hint*: A two-sided inverse must also be a left-inverse.

### Exercise 15.5

Explain why $A = \begin{bmatrix} 1 & -1 & 1 \\ -3 & 3 & -3 \end{bmatrix}$ has neither a left- nor a right-inverse.

### Exercise 15.6

Explain why $A = \begin{bmatrix} 1 & -1 \\ 1 & 1 \\ 2 & 3 \end{bmatrix}$ has no right-inverse, and every possible left-inverse

of $A$ is given by the formula $B = \begin{bmatrix} \frac{1+c}{2} & \frac{1-5c}{2} & c \\ \frac{-1+f}{2} & \frac{1-5f}{2} & f \end{bmatrix}$.

### Exercise 15.7

If $A$ is an $n \times m$ matrix, explain why we should *not* expect to find a left-inverse for $A$, if $n < m$. Likewise, explain why we should not expect to find a right-inverse for $A$, if $n > m$. *Hint*: In each case, compare the number of variables one would need to solve for with the number of equations that one would have.

*Aside*: In Lesson 16, we will be able to show that the intuitively plausible statements from Ex. 15.7 are in fact true. Commonly, mathematics and mathematical proofs help us to:

– confirm our intuition when it is correct, *and*

– correct our intuition when it leads us astray.

In general, education both confirms and corrects our intuition about the nature of "reality" and enables us to function independently in the world.

### Exercise 15.8

Assuming that the statements from Ex. 15.7 are true, explain why only "square" matrices (i.e., same number of rows and columns) can have (two-sided) inverses.

## 15.2   A Faster Way: The Method

The systems that we must solve, to compute a left- or right-inverse for a given matrix $A$, have very striking patterns. With a bit of practice, one may learn to determine an appropriate matrix to RowReduce[] without writing out the equations explicitly, as we have done so far. However, this process soon becomes unwieldy, as these matrices tend to grow very large as $A$ grows larger. In this section, we will see how to exploit these patterns to come up with a more compact way of computing left- and right-inverses. We will use a general technique that works for solving any set of "parallel" systems of equations.

Whenever we have a number of similar systems to solve like $AX = C$ and $AY = D$, we could perform Gaussian elimination on the partitioned matrix $\begin{bmatrix} A & C \end{bmatrix}$ to find $X$, and then on $\begin{bmatrix} A & D \end{bmatrix}$ to find $Y$. You should notice, however, that we would be doing *exactly* the same row operations in both cases, since the choice of row operations is determined by the left-most entries, which will be the same for the two systems. Thus, this would be a rather inefficient way to proceed. Another approach would be to perform Gaussian elimination *once* on $\begin{bmatrix} A & C & D \end{bmatrix}$, and then read off the solutions $X$ and $Y$, *as if* we had done each problem separately. In this way, we solve both systems at the same time, in "parallel." This technique is useful when computing matrix inverses. For example, look at the equations in the last section. They may be grouped together into two similar systems:

$$
\begin{array}{rcrcrcl}
a & + & 3c & + & 5e & = & 1 \\
2a & + & 4c & + & 6e & = & 0
\end{array}
\quad \text{and} \quad
\begin{array}{rcrcrcl}
b & + & 3d & + & 5f & = & 0 \\
2b & + & 4d & + & 6f & = & 1
\end{array}
$$

Notice that the left-hand side of both systems is given by the matrix $A = \begin{bmatrix} 1 & 3 & 5 \\ 2 & 4 & 6 \end{bmatrix}$ that we have been studying. This means that we have two "parallel" systems with $X = \begin{bmatrix} a \\ c \\ e \end{bmatrix}, C = \begin{bmatrix} 1 \\ 0 \end{bmatrix}, Y = \begin{bmatrix} b \\ d \\ f \end{bmatrix}$, and $D = \begin{bmatrix} 0 \\ 1 \end{bmatrix}$. Using the trick mentioned previously, we may solve both systems at the same time by performing Gaussian elimination on:

$$
\begin{bmatrix} A & C & D \end{bmatrix} = \begin{bmatrix} 1 & 3 & 5 & 1 & 0 \\ 2 & 4 & 6 & 0 & 1 \end{bmatrix}
$$

Notice the special form of this matrix — it looks like $\begin{bmatrix} A & I \end{bmatrix}$. Using *Mathematica*, we may RowReduce[] this to solve for the two columns of the right inverse, given by $X$ and $Y$, respectively. Determining the correct matrix to RowReduce[] using this technique is much easier than that of Section 15.1. The tradeoff is that our answer is slightly more difficult to interpret. With a little practice, you will find that it is not very difficult, and *it is an important skill to develop for future work.*

- You should perform the corresponding RowReduce[] command now (again, it is pretyped in the accompanying Notebook). You should obtain:

$$
\begin{bmatrix} 1 & 0 & -1 & -2 & \frac{3}{2} \\ 0 & 1 & 2 & 1 & -\frac{1}{2} \end{bmatrix}
$$

Covering up the last column and focusing on the first augmented column, we can see that the result of doing Gaussian elimination on $\begin{bmatrix} A & C \end{bmatrix}$ would have been:

$$
\begin{bmatrix} 1 & 0 & -1 & -2 \\ 0 & 1 & 2 & 1 \end{bmatrix}
$$

We can read off the solution for $X = \begin{bmatrix} a \\ c \\ e \end{bmatrix}$, namely:

$$
\begin{array}{rcl}
a & = & -2 + e \\
c & = & 1 - 2e \\
e & = & \text{anything}
\end{array}
\quad \text{or} \quad
X = \begin{bmatrix} -2 + e \\ 1 - 2e \\ e \end{bmatrix}
$$

Similarly, by covering the next-to-last column and focusing on the second augmented column, we see that the result of doing Gaussian elimination on $\begin{bmatrix} A & D \end{bmatrix}$ would have been:

$$\begin{bmatrix} 1 & 0 & -1 & \frac{3}{2} \\ 0 & 1 & 2 & -\frac{1}{2} \end{bmatrix}$$

This means that $Y = \begin{bmatrix} b \\ d \\ f \end{bmatrix}$ is given by:

$$\begin{array}{rcl} b & = & \frac{3}{2} + f \\ d & = & -\frac{1}{2} - 2f \\ f & = & \text{anything} \end{array} \qquad \text{or} \qquad Y = \begin{bmatrix} \frac{3}{2} + f \\ -\frac{1}{2} - 2f \\ f \end{bmatrix}$$

We may obtain particular values for $a$ through $f$, and thus a particular right-inverse, by choosing values for the free variables $e$ and $f$. If we take $e = 0$ and $f = -\frac{1}{2}$, we obtain the right-inverse $B$ of Ex. 14.7. Likewise, $e = 1 = f$ yields matrix $C$ of Ex. 14.8. Using the general solution of each system, we obtain a general formula for all possible right-inverses of $A$, namely:

$$B = \begin{bmatrix} -2 + e & \frac{3}{2} + f \\ 1 - 2e & -\frac{1}{2} - 2f \\ e & f \end{bmatrix} = \begin{bmatrix} -2 & \frac{3}{2} \\ 1 & -\frac{1}{2} \\ 0 & 0 \end{bmatrix} + e \begin{bmatrix} 1 & 0 \\ -2 & 0 \\ 1 & 0 \end{bmatrix} + f \begin{bmatrix} 0 & 1 \\ 0 & -2 \\ 0 & 1 \end{bmatrix}$$

## 15.3   A Faster Way: The Proof

We will now verify that the method of Section 15.2 will always work via the Principle of Partitioned Matrices. Specifically, assume that $A$ is an $n \times m$ matrix. Then by the same argument as in Section 15.1, any right-inverse must be $m \times n$. Take a potential right-inverse, call it $R$, and the identity matrix $I$ and partition them into columns, so that $R = \begin{bmatrix} r_1 & \cdots & r_n \end{bmatrix}$ and $I = \begin{bmatrix} e_1 & \cdots & e_n \end{bmatrix}$.

*Aside*: Since we know what $I$ looks like, we may say that $e_i$ is a column of 0's with a single 1 in the $i$th row; this is a *very important* set of column vectors, which we have already exploited in Lesson 9 and which will arise repeatedly in the future (called the "standard basis" for Euclidean space — more on this later).

Using this notation, we may conclude that $R$ is a right-inverse for $A$ if and only if:

$$\begin{bmatrix} Ar_1 & \cdots & Ar_n \end{bmatrix} = A\begin{bmatrix} r_1 & \cdots & r_n \end{bmatrix} = AR = I = \begin{bmatrix} e_1 & \cdots & e_n \end{bmatrix}$$

The first equality holds by the Principle of Partitioned Matrices, the second by the definition of $r_i$, the third by the definition of a right-inverse, and the last by the definition of $e_i$. Setting the columns equal gives the parallel set of systems:

$$Ar_i = e_i \qquad \text{for} \qquad i = 1, \ldots, n$$

To solve this collection of systems, the method of Section 15.2 would have us perform Gaussian elimination on $\begin{bmatrix} A & e_1 & \cdots & e_n \end{bmatrix} = \begin{bmatrix} A & I \end{bmatrix}$. This confirms our experience with the calculation in Section 15.2. As in that example, the solution corresponding to each of the augmented columns $e_i$ gives the corresponding column of $R$, $r_i$.

### Exercise 15.9

• Use this method to find a right-inverse (if possible) for each of the following matrices. Check your answers by multiplication.

(a) $\begin{bmatrix} 2 & 4 & 2 \\ 4 & 2 & -14 \\ 2 & 6 & 11 \end{bmatrix}$
(b) $\begin{bmatrix} 3 & 6 & -3 \\ 2 & 5 & 1 \end{bmatrix}$
(c) $\begin{bmatrix} 2 & -3 \\ -8 & 12 \end{bmatrix}$

Do any of these matrices have a two-sided inverse? Explain.

## 15.4   The Connection Between Right- and Left-Inverses

Although the corresponding argument does not work for left-inverses, a simple observation will allow us to reduce the problem of computing a left-inverse to computing a right-inverse, so that we may again exploit the algorithm of Section 15.2. Consider the equations for a left-inverse $B = \begin{bmatrix} a & b \\ c & d \\ e & f \end{bmatrix}$ from Ex. 15.3. Notice that they group into three parallel systems:

$$
\begin{array}{rcl}
a + 2b &=& 1 \\
3a + 4b &=& 0 \\
5a + 6b &=& 0
\end{array}
\quad
\begin{array}{rcl}
c + 2d &=& 0 \\
3c + 4d &=& 1 \\
5c + 6d &=& 0
\end{array}
\quad \text{and} \quad
\begin{array}{rcl}
e + 2f &=& 0 \\
3e + 4f &=& 0 \\
5e + 6f &=& 1
\end{array}
$$

which we may solve by performing Gaussian elimination on:

$$
\begin{bmatrix}
1 & 2 & 1 & 0 & 0 \\
3 & 4 & 0 & 1 & 0 \\
5 & 6 & 0 & 0 & 1
\end{bmatrix}
$$

Notice that this looks like we are computing a *right*-inverse for $\begin{bmatrix} 1 & 2 \\ 3 & 4 \\ 5 & 6 \end{bmatrix} = A^T$. Moreover, the three columns of that right-inverse form the *rows* of our desired left-inverse. This reflects the following theorem:

**Theorem 15.1** *A matrix $L$ is a left-inverse of $A$ if and only if $L^T$ is a right-inverse for $A^T$.*

### Proof:

To prove an "if and only if" statement, we must show that if we assume the first statement the second necessarily follows, and vice versa. We start with the "forward" implication and treat the "reverse" implication separately. *Note*: This proof is very similar to that of Theorem 14.2b. If matrix $L$ is a left-inverse of $A$, then we know that $LA = I$. Taking transposes of both sides gives $(LA)^T = I^T$. We know that $I^T = I$

and $(LA)^T = A^T L^T$. Thus, we have the equation $A^T L^T = I$. Since this is the defining equation of a right-inverse, we have that $L^T$ is a right-inverse for $A^T$.

Conversely, starting with the fact that $L^T$ is a right-inverse for $A^T$, so that $A^T L^T = I$, we may transpose both sides again to obtain $I = \left(A^T L^T\right)^T = \left(L^T\right)^T \left(A^T\right)^T = LA$. Here we have used Theorem 12.1a, which says two transposes "cancel out." Thus, we have $I = LA$, or $L$ is a left-inverse of $A$. Q.E.D.

In other words, letting $B = L^T$, if we compute a right-inverse $B$ for $A^T$, then $L = \left(L^T\right)^T = B^T$ a left-inverse for $A$.

## Exercise 15.10

Prove the following:

**Theorem 15.2** *A matrix $R$ is a right-inverse of $A$ if and only if $R^T$ is a left-inverse for $A^T$.*

*Hint*: You can use the preceding proof as a model, but try to put it in your own words.

These last two theorems describe a very close connection between left- and right-inverses. Since we have a fast algorithm for computing right-inverses, we may easily compute left-inverses as well. To summarize the results of this section, we have seen that:

- we may compute the columns of the general right-inverse of $A$ by performing Gaussian elimination on $\begin{bmatrix} A & I \end{bmatrix}$ and interpreting the general solution corresponding to each of the augmented columns as the corresponding column of the right-inverse;

- we may compute the general left-inverse of $A$ by computing the *right*-inverse for $A^T$ and transposing the result.

## Exercise 15.11

Use this method to find a left-inverse (if possible) for each of the following matrices. Verify your answers by multiplication.

$$\text{(a)} \begin{bmatrix} 2 & -4 & -6 \\ 4 & -6 & -16 \\ -6 & 8 & 29 \end{bmatrix} \qquad \text{(b)} \begin{bmatrix} 2 & 4 \\ -3 & -4 \\ 4 & 5 \end{bmatrix} \qquad \text{(c)} \begin{bmatrix} 2 & -6 \\ -1 & 3 \\ 3 & -9 \end{bmatrix}$$

Do any of these matrices have a two-sided inverse? Explain.

## Exercise 15.12

Do the results of Ex. 15.9 and Ex. 15.11 tend to confirm or deny the remarks in Ex. 15.7 and Ex. 15.8? Explain.

## Exercise 15.13

Do the results of Ex. 15.9 and Ex. 15.11 suggest any other facts about matrix inverses? *Hint*: When will a one-sided inverse also be a two-sided inverse?

## 15.5   Special Matrices

There are collections of matrices that are singled out for special consideration because of their particular form.

**Definition 15.1** A matrix $A$ is said to be *symmetric* if and only if $A^T = A$. In particular, a symmetric matrix must be square. An $n \times n$ matrix $B$ is *lower-triangular* if and only if $(B)_{ij} = 0$ for all $1 \leq i \leq j \leq n$, while $C$ is *upper-triangular* iff $C^T$ is lower-triangular.

### Exercise 15.14

Write down a $3 \times 3$ example of each of these three types of matrices. Verify that the corresponding definition holds for each of your examples.

### Exercise 15.15

Compute a right-inverse for $A = \begin{bmatrix} 1 & 2 & 3 \\ 2 & 4 & 5 \\ 3 & 5 & 6 \end{bmatrix}$. Verify that you obtain the unique two-sided inverse $A^{-1}$. What do you observe about $A^{-1}$? Generalize your observation and state it as a theorem about symmetric matrices.

### *Advanced Exercise 15.16*

Give a formal proof of your theorem from Ex. 15.15. *Hint*: Use Theorem 14.2.

### Exercise 15.17

Repeat Ex. 15.15 with $B = \begin{bmatrix} 1 & 0 & 0 \\ 2 & 3 & 0 \\ 4 & 5 & 6 \end{bmatrix}$ to obtain a theorem about lower-triangular matrices.

*Note*: A formal proof of this is difficult without the explicit, general formula for $B^{-1}$, which we derive in Lesson 19.

## 15.6   Summary

In this lesson, one should have learned:

- that any inverse of a $p \times q$ matrix must be a $q \times p$ matrix,

- how to compute a left-, right-, or two-sided inverse by translating the definition directly into a system of equations,

- how to solve a set of systems of equations with the same left-hand sides and different right-hand sides in parallel,

- how to set up and solve for left- or right-inverses, using this technique of parallel systems, and

- the definitions of symmetric and triangular matrices and how these matrices behave with respect to inversion.

Notice that we have *not* yet given an efficient way to compute two-sided inverses. In Lesson 16 we will discuss how the three types of inverses are intimately connected with the three types of systems of Definition 8.1. As a result, we will learn how the algorithm of Section 15.2 will always suffice to produce two-sided inverses.

# Matrix Inverses and Systems of Equations

## Accompanied by Notebook

In Lessons 14 – 15, we defined three different types of inverses and showed how to use Gaussian elimination to quickly calculate left- and right-inverses. In this lesson, we will see:

- how these three types of inverses of a matrix $A$ correspond directly to the three different types of solutions to a system of linear equations,

- how we may distinguish these three cases using the rank of $A$,

- how to compute two-sided inverses, and

- a new function, called the "determinant," which will detect invertibility.

Along the way, we will gain more practice in reading and doing proofs.

## 16.1 Inverses and Systems of Equations

In this section, we discuss the detailed connection between inverses of $A$ and the nature of solutions to systems of equations involving $A$. In fact, these three cases correspond to an entire sequence of ideas that weave through all of linear algebra, related to the concepts of "existence" and "uniqueness." As we emphasized in the preface To the Student, the study of linear algebra is not simply about solving particular problems but also seeing the *connections* between the various concepts.

We begin by reviewing the results of Lesson 8 in the following theorem:

**Theorem 16.1** *We may characterize a system of $n$ linear equations in $m$ variables, written in matrix form as $AX = B$, in terms of the rank of $A$ and the partitioned matrix $A' = \begin{bmatrix} A & B \end{bmatrix}$. Specifically, the system is:*

(a) *overdetermined if and only if $rank\,(A') > rank(A)$,*

(b) *underdetermined if and only if $rank\,(A') = rank(A) < m$, or*

(c) *uniquely determined if and only if $rank\,(A') = rank(A) = m$.*

## Exercise 16.1

Under the assumptions of Theorem 16.1, explain why we cannot have $rank\,(A') < rank(A)$ or $rank(A) > m$, so that every system of linear equations falls into one of these three categories.

Theorem 16.1 refers to a matrix $A$ and a particular right-hand side $B$. We may connect these three cases to invertibility by considering *every* possible system with a given $A$ on the left-hand side.

This suggests that we establish the following definitions:

**Definition 16.1** A matrix $A$ is said to be:

(a) "onto" or *surjective* iff $AX = B$ has at *least* one solution for every possible $B$;

(b) "one-to-one" or *injective* iff $AX = B$ has at *most* one solution for every possible $B$; or

(c) a "one-to-one correspondence" or *bijective* iff $AX = B$ has *exactly* one solution for every possible $B$.

In these terms, we have the following conditions for invertibility:

**Theorem 16.2** *A matrix A has:*

(a) *a right-inverse iff A is surjective.*

(b) *a left-inverse iff the equation A is injective.*

(c) *a (two-sided) inverse iff the equation A is bijective.*

**Proof:**

*Proof of 16.2a*

In Ex. 14.9, we observed that if $A$ has a right-inverse, $R$, then $RB$ is a solution to $AX = B$. Check: If $X = RB$, then $AX = ARB = IB = B$. That is, a right-invertible matrix is surjective. Conversely, we saw in Section 15.3 that if we can solve this equation for the case where $B$ is any of the columns of the identity matrix $e_i$, then we may compute a right-inverse.

*Proof of 16.2b*

We begin with the forward direction. Assume that $A$ has a left-inverse $L$. For a given $B$, assume also that we have two *potentially* different solutions $X$ and $Y$ (compare the proof of Theorem 14.1), so that $AX = B = AY$. We may then multiply both sides by $L$ to obtain $LAX = LAY$, or $X = Y$, since $LA = I$. This means that we actually only had (at most) one solution. We leave the proof of the reverse direction to Section 16.2.

*Note*: Although it may be unsettling to leave this proof incomplete, this is very much how working mathematicians operate. They will often take a result that seems plausible and *assume* that it is true, until they can supply a formal proof for it (which is independent of any results proven from it).

*Proof of 16.2c*

The forward direction of this part is easy. If $A$ has an inverse, it is unique (by Theorem 14.1), and we may solve explicitly for $X$ by multiplying on the left of both sides of the equation to obtain $X = A^{-1}B$. Conversely, if $AX = B$ has exactly one solution for every $B$, we may use Theorem 16.2a to produce a right-inverse, $R$. It remains to show that $R$ is in fact a left-inverse as well. The Principle of Partitioned matrices implies that:

$$\begin{bmatrix} Ae_1 & \cdots & Ae_n \end{bmatrix} = A \begin{bmatrix} e_1 & \cdots & e_n \end{bmatrix} = AI = A = \begin{bmatrix} a_1 & \cdots & a_n \end{bmatrix}$$

so that:

$$Ae_i = a_i \quad \text{for} \quad i = 1, \ldots, n$$

or, in other words, $e_i$ is a solution to $Ae_i = a_i$ for $i = 1, \ldots, n$. Notice that this calculation is similar to that in Section 15.3, except that this time we partition the matrix $A$ into columns. The central equality follows, since $I$ is the multiplicative identity for matrix multiplication.

We may verify that $Ra_i$ is *also* a solution to $AX = a_i$ for $i = 1, \ldots, n$ (compare the proof of Theorem 16.2a). Since $A$ has a left-inverse, solutions to such an equation are unique (by the forward part of Theorem 16.2b), which we *have* proven). This means that $Ra_i = e_i$ for $i = 1, \ldots, n$. Using the Principle of Partitioned Matrices again, this implies that $RA = I$. Specifically:

$$RA = R \begin{bmatrix} a_1 & \cdots & a_n \end{bmatrix} = \begin{bmatrix} Ra_1 & \cdots & Ra_n \end{bmatrix} = \begin{bmatrix} e_1 & \cdots & e_n \end{bmatrix} = I$$

so that $RA = I$ and $R$ is a left-inverse as well. Q.E.D.

Using these results we may now prove the statements in Ex. 15.7–15.8, thus rigorously confirming our intuition.

**Corollary 16.3** *Given an $n \times m$ matrix $A$:*

(a) *If $A$ has more columns than rows (i.e., $n < m$), it cannot have a left-inverse.*

(b) *If $A$ has more rows than columns (i.e., $n > m$), it cannot have a right-inverse.*

(c) *If $A$ is not square (i.e., $n \neq m$), it cannot have a two-sided inverse.*

**Proof:**

*Proof of 16.3a*

Assume that $A$ has more rows than columns. For any column matrix $B$, there will be free variables after we perform Gaussian elimination to solve $AX = B$. Therefore, $A$ will *not* have unique solutions to equations, and hence cannot have a left-inverse, by the forward implication in Theorem 16.2b (which we have proven).

*Proof of 16.3b*

If $A$ has more rows than columns, we will obtain at least one zero row during Gaussian elimination on $A$. This implies that we may produce a column vector $B$ such that $AX = B$ will have no solution. Specifically, compute the reduced row echelon form of $A$, remembering the precise sequence of row operations used, augment the resulting matrix with a column vector with a 1 in one of the zero rows, then apply the row operations again in *reverse*. This will produce an augmented matrix $[A\,|B]$ which, by construction, corresponds to an overdetermined system $AX = B$. Theorem 16.2a then implies $A$ cannot have a right-inverse.

*Proof of 16.3c*

This part follows immediately from either Theorem 16.3a or 16.3b, since a two-sided inverse is, in particular, a left- and right-inverse. Q.E.D.

*Note*: We use the term "corollary" to refer to a theorem when we want to indicate that it follows from another well-known theorem.

You may find these results a little hard to remember at first. That is one reason to prove more theorems — to try and discover more intuitive and easy-to-remember characterizations of invertibility. We will see in Lesson 20 that the results of Theorem 16.2 are actually very general results about functions and solving equations that apply to *any* set of equations, not simply linear ones. For the time being, we will examine the connection between rank and invertibility. Once we learn to view rank and the general action of a matrix geometrically, these facts will become more intuitive, based on simple geometric reasoning.

## 16.2   Inverses and Rank

In Theorem 16.1, we have given the connection between rank and the nature of solutions to a system of equations. In this section, we extend those observations and relate them to the three types of inverses. Corollary 16.3 is actually a special case of a more general set of results concerning the rank of a matrix.

**Theorem 16.4** *Given an $n \times m$ matrix $A$:*

*(a) A matrix has a left-inverse iff its rank equals its number of columns.*

*(b) A matrix has a right-inverse iff its rank equals its number of rows.*

*(c) A matrix has an inverse iff it is square and its rank equals its number of columns (rows).*

**Proof:**

*Proof of 16.4a*

It is clear that a matrix does not yield multiple solutions iff its rank equals its number of columns. That is because, it would have a leading 1 in every column in row echelon form and no free variables. The result would then follow from Theorem 16.2b. Since the proof of the theorem is still incomplete, this part will not be complete until we finish the proof of Theorem 16.2b.

### Proof of 16.4b

By the same argument as in Corollary 16.3b, it is clear that a matrix always yields solutions iff its rank equals its number of rows. This means that it will never have any zero rows in its row echelon form that could lead to an over-determined situation.

### Proof of 16.4c

This is a direct consequence of Theorems 16.4a and 16.4b taken together.

Q.E.D.

In Lesson 15, while computing left- and right-inverses, one should have noticed a surprising pattern. Re-examining Ex. 15.9 and Ex. 15.11, one observes that each right- or left-inverse of a square matrix also happens to be a two-sided inverse. This phenomena, which we describe precisely in the following corollary, also known as the "Fredholm Alternative," is completely general and may be proven as a simple consequence of the previous theorem:

**Corollary 16.5 (The Fredholm Alternative)** *If A is any square matrix A, the following conditions are equivalent:*

(a) *A is injective/has a left-inverse,*

(b) *A is surjective/has a right-inverse, and*

(c) *A is bijective/has a two-sided inverse.*

**Proof:**

Since the number of rows equals the number of columns, we see that these three possibilities all occur at precisely the same time, when we consider the rank of $A$ and apply Theorem 16.4. Q.E.D.

To put this theorem into perspective, so far, we have seen three different threads weaving through linear algebra. We have seen that systems of equations can have no solution, infinitely many solutions, or exactly one solution. We have also seen that a matrix can have a right-inverse, a left-inverse, or a two-sided inverse. Each case can be described in terms of the rank of matrix either equaling the number rows or columns or both. The Fredholm Alternative says that all three strands come together in the case of square matrices.

We stop to consider just how surprising and amazing this result is. To compute a right-inverse for an $n \times n$ matrix from the definition, there are $n^2$ equations in $n^2$ unknowns. To compute a left-inverse, there is a *completely different* set of $n^2$ equations in $n^2$ unknowns. The corollary says if the first system of equations has a solution, the second must as well (and vice versa). Moreover, to compute a two-sided inverse, there are $2n^2$ equations in $n^2$ unknowns ($n^2$ equations to guarantee left-cancelation and another $n^2$ to guarantee right-cancelation) — twice as many equations as unknowns. In general, you would not expect such a situation to have a solution, but Corollary 16.5 hints that the left-cancelation equations and the right-cancelation equations are *equivalent*, even though they look very different on the surface.

## Exercise 16.2

Consider the matrix $A = \begin{bmatrix} 1 & 2 \\ 3 & 4 \end{bmatrix}$. Record the equations for a two-sided inverse of $A$ (*Hint*: These are just the equations of a left-inverse and right-inverse together, giving eight equations in four unknowns) as an augmented matrix. This will yield a $(2n^2) \times (n^2 + 1) = 8 \times 5$ matrix. After Gaussian elimination, how many nonzero equations remain? Does this tend to confirm or contradict the previous paragraph? Explain.

## Exercise 16.3

Construct an example of a $3 \times 3$, noninvertible matrix. *Hint*: Use Corollary 16.5.

With the results of this section, we may now give a complete algorithm for computing a two-sided inverse for any given matrix $A$:

1. If $A$ is not square, Corollary 16.3c implies that $A^{-1}$ does not exist.

2. If $A$ is square, we may use the algorithm of Section 15.2 to compute a right-inverse $R$. When we perform Gaussian elimination on $M \equiv \begin{bmatrix} A & I \end{bmatrix}$ either:

   (i) $A$ does not have full rank (i.e., its rank is less than its dimension), we will obtain zero rows on the left of $M$, and we may conclude, that $A^{-1}$ does not exist, by Theorem 16.4c.

   (ii) $A$ has full rank, in which case, Theorems 16.4b and 16.4c imply that $A$ has both a right-inverse $R$ (which we may construct from the reduced row echelon form of $M$) and a (two-sided) inverse $B$. A simple calculation shows that $B$ and $R$ must be equal (*Check*: $B = BI = B(AR) = (BA)R = IR = R$). In other words, the matrix $R$ that we originally thought was only a right-inverse is, in fact, the unique, two-sided inverse of $A$.

It only remains to prove the reverse direction of Theorem 16.2b. We conclude this section by indicating how the proof will eventually be constructed. We will require the following fact:

$$rank(A) = rank(A^T) \tag{16.1}$$

We verified this fact *experimentally* in Ex. 11.7, but an iron-clad proof must wait until we can give a more geometric description of rank. We will be able to prove this fact directly, as Theorem 25.7b, without appealing to any of the previous theorems or corollaries (thus, avoiding circular reasoning). We will even supply a *second*, more geometric proof of this important fact, in Lesson 28.

**Proof:** ( of the remainder of Theorem 16.2b)

Remember what remains to be shown. Assuming that $AX = B$ has at most one solution for every possible value of $B$, we must show that $A$ has a left-inverse. If $AX = B$ has at most one solution for every $B$, then $\begin{bmatrix} A & B \end{bmatrix}$ cannot have any free variables after Gaussian elimination. In particular, $rank(A)$ will equal the number of columns of $A$. Trusting that we may supply an independent proof of Theorem 25.7b, we may conclude that $rank(A^T)$ equals the number of columns of $A$, which is the number of *rows* of $A^T$. By Theorem 16.4b, we may conclude that $A^T$ has a right-inverse. Theorem 15.1 then implies that $A$ has a left-inverse. Q.E.D.

We should still put the word "Proof" in quotes, because this will not be a *complete* proof until we can show that the rank of a matrix (its "row rank") and the rank of its transpose (its "column rank") are equal. Until then, we can reap the practical benefits of the theorems of this section, in the full confidence that an indisputably rigorous proof is coming.

## 16.3 Measuring Invertibility

We now have a complete test for invertibility: a matrix is invertible iff it is square and has maximal rank. This relatively straightforward test *still* requires Gaussian elimination, which is rather time-consuming to perform by hand. It would be convenient if there were a function defined on square matrices which would *tell* us whether or not a given matrix is invertible without a great deal of computation. Theoretically, there is such a function, called the "determinant" function. It takes in a square matrix $A$ and gives out a real number that, in some sense, measures how invertible a matrix is (as well as supplying certain geometric information). We usually denote the determinant of $A$ by $|A|$ (just like the absolute value function on real numbers), although *Mathematica* uses Det[$A$]. We will indicate in Lesson 19 how it was discovered, how it is defined, and how to calculate it. For now, we will simply allow *Mathematica* to compute it for us.

### Exercise 16.4

- If $A = \begin{bmatrix} 1 & 2 \\ 3 & 4 \end{bmatrix}$, use *Mathematica* to calculate Det[$A$]. *Note:* The necessary commands are pretyped in the accompanying Notebook.

### Exercise 16.5

- For the following matrices:
    - determine whether or not they are invertible (*Hint:* Use Theorem 16.4 and RowReduce[]),
    - compute their determinant, and
    - formulate a conjecture relating invertibility of a matrix with the value of its determinant.

$$\text{(a) } A = \begin{bmatrix} 1 & 2 \\ 3 & 4 \end{bmatrix} \quad \text{(b) } B = \begin{bmatrix} 1 & 2 \\ 2 & 4 \end{bmatrix} \quad \text{(c) } F = \begin{bmatrix} 1 & 2 & 3 \\ 4 & 5 & 6 \\ 7 & 8 & 9 \end{bmatrix} \quad \text{(d) } G = \begin{bmatrix} 1 & 2 & 3 \\ 4 & 5 & 6 \\ -1 & 8 & 9 \end{bmatrix}$$

Construct more examples, if necessary, until you discover the connection.

One of the most important properties of the determinant function is that it "preserves" products, that is:

$$Det[AB] = Det[A]Det[B], \tag{16.2}$$

for every choice of $A$, and $B$. *Note*: With our other notation, this would read $|AB| = |A||B|$, just like ordinary absolute values for real numbers. This property is fundamental to both theoretical and practical calculations. Although the formal definition of the determinant formula will be a bit messy, in Lesson 19 this property will allow us to compute the determinant function quickly and easily (by Gaussian elimination).

### Exercise 16.6

- Construct two square $2 \times 2$ matrices $A$ and $B$ and verify Eq. 16.2. Record $A$, $B$, $AB$, and the value of all determinants. Repeat this for a pair of $3 \times 3$ matrices.

## 16.4   Summary

In this lesson, one should have learned:

- how to compute two-sided inverses;

- how right-inverses relate to the existence of solutions to systems of equations, while left-inverses are connected with uniqueness;

- the connection between rank and inverses; and

- the connection between determinants and inverses.

Using the concept of inverses, we may re-examine Gaussian elimination more carefully in terms of matrix multiplication. This will allow us to show that the reduced row echelon form a matrix is unique. In particular, this means that our definition of the rank as its number of leading 1's is a well-defined number (i.e., it does not depend on the particular steps we took to put the matrix in row echelon form). We will *not*, however, be able to prove our key remaining fact about ranks, namely, Eq. 16.1. This must wait until we discuss the notions of "vector space" and "linear transformation." We will then supply a clear geometric characterization of rank, from which Eq. 16.1 will follow relatively easily.

# Systems from Two Advanced Viewpoints

# Elementary Matrices and the $P^T LU$ Decomposition

*Mathematica* Notebook

## 17.1 Introduction

So far we have seen that matrices may be used to represent systems of equations, they may act as generalized "numbers," or they may represent geometric transformations. In this lesson, we will see that matrices may also be used to perform algebraic transformations, namely, the elementary row operations of Gaussian elimination. Once we discover the connection between row operations and matrix multiplication, we will be able to view Gaussian elimination in an entirely new way, as a sequence of matrix multiplications. In the process, we will discover that all of Gaussian elimination may be summarized as a *single* matrix equation:

$$A = P^T LU$$

This simple little formula has a number of important consequences. We will use it in Lesson 18 to solve systems of equations (using *only* backsubstitution), as well as to furnish a simpler proof of the correctness of our algorithm on p. 136 for computing two-sided inverses. In Lesson 19, we will also learn to compute the determinant of $A$ quickly by exploiting this crucial formula. All of this will further emphasize the value of being able to change our point of view.

## 17.2 Row Operations as Matrices

We first illustrate the connection between elementary row operations and matrix multiplication by performing Gaussian elimination on:

$$\begin{array}{rcl} -2x + y + z & = & -2 \\ x - 2y + z & = & 1 \\ x + y - 2z & = & 1 \end{array}$$

As usual, we first express this as a partitioned matrix:

•    `A = {{-2, 1, 1,-2},{ 1,-2, 1, 1},{ 1, 1,-2, 1}}; MatrixForm[A]`

where we must remember the partitioning ourselves. The first row operation of Gaussian elimination would be to multiply the first row by $-\frac{1}{2}$ and leave the remaining rows unchanged. We contend that this may be accomplished by multiplication by a certain matrix $M$. Using the Principle of Partitioned Matrices, we may infer the correct matrix $M$ so that the product $MA$ yields the desired result. Although this requires more effort than simply performing Gaussian elimination, it will lead to a deeper understanding of Gaussian elimination, as well as the decomposition result mentioned in the Introduction.

We begin by observing that, if we partition $M$ into rows, $MA$ may be viewed as a collection of row products:

$$M A = \begin{bmatrix} m_1 \\ \vdots \\ m_n \end{bmatrix} A = \begin{bmatrix} m_1 A \\ \vdots \\ m_n A \end{bmatrix}$$

This means that by considering the desired rows of $MA$, in turn, we may determine each row of $M$ *individually*.

For example, because we wish to multiply the first row of $A$ by $-\frac{1}{2}$, we take the first row of $M$ to be:

•    `M1 = {-1/2, 0, 0}`

We may verify that this does produce the desired first row:

•    `M1.A`

Notice that we have yet to assign a new value to $A$. We must still determine the remaining rows of $M$. In fact, unlike our computations in Lesson 7, we will never actually change $A$ itself. Instead we will save each intermediate matrix in a separate variable.

To determine the second row of $M$, we observe that the second row of $MA$ should be identical to the second row of $A$. This suggests that we take:

•    `M2 = {0, 1, 0}`

Again, we should verify that this performs as advertised:

•    `M2.A`

Notice how this simply picks out the second row of $A$.

## Exercise 17.1

•    Determine, by analogy, the correct third row of $M$. Enter your conjecture into the following cell and verify that it is correct. Notice that, because we do not actually change $A$, you may experiment freely, without restoring the original value of $A$.

•    `M3 = { , , }; M = {M1, M2, M3};`
       `MatrixForm[M.A]`

Now that we have correctly determined $M$, we actually perform this first step of Gaussian elimination and save the resulting matrix as $A_1$:

• `A1 = M.A; MatrixForm[A1]`

This matrix $M$ is called an "elementary matrix," because it corresponds to an elementary row operation. In Lesson 18, we will see that, just as all molecules are composed of elementary particles, all matrices are composed of elementary matrices. We will save $M$, as well, for use in Section 17.4:

• `ElementaryMatrix[1] = M; MatrixForm[M]`

*Note*: In case it is not possible to complete this lesson in a single *Mathematica* session, it would be wise to save all the ElementaryMatrix[$i$]'s that one determines in this section in a separate Notebook.

The next operation requires us add $-1$ times the first row of $A1$ to the second row of $A1$ and place the result in the second row. We will proceed as before, by experimentally determining the rows of our next ElementaryMatrix[]. Again using the Principle of Partitioned Matrices, we notice that multiplying a row $m$ with a matrix $A$ is the same as forming a linear combination of the *rows* of $A$:

$$
mA = \begin{bmatrix} m_1 & \cdots & m_n \end{bmatrix} \begin{bmatrix} A_1 \\ \vdots \\ A_n \end{bmatrix} = m_1 A_1 + \ldots + m_n A_n
$$

This suggests that we use the row:

• `{-1, 1, 0}.A1`

as part of our next elementary matrix. Since the sum should be placed in the second row of the result, this should form the second row of ElementaryMatrix[2].

## Exercise 17.2

• Determine, by analogy, the remaining rows of ElementaryMatrix[2]. *Hint*: The other rows should remain *unchanged* (compare Ex. 17.1). Enter your conjecture into the following cell and verify that it is correct. *Note*: The result should be the same as the result of Gaussian elimination after the second step.

• ```
ElementaryMatrix[2] = {{  ,  ,  },
                       {-1, 1, 0},
                       {  ,  ,  }};
A2 = ElementaryMatrix[2].A1;MatrixForm[A2]
```

Gaussian elimination may now be completed without row interchanges. Thus, the remaining sequence of elementary matrices are similar to the two we have just seen.

## Exercise 17.3

• Determine the corresponding elementary matrices for the remaining three row operations. Enter your conjectures into each of the following cells and verify that they are correct:
```

- ```
  ElementaryMatrix[3] =        ;
  A3 = ElementaryMatrix[3].A2; MatrixForm[A3]
  ```

- ```
  ElementaryMatrix[4] =        ;
  A4 = ElementaryMatrix[4].A3; MatrixForm[A4]
  ```

- ```
  ElementaryMatrix[5] =        ;
  A5 = ElementaryMatrix[5].A4; MatrixForm[A5]
  ```

You should obtain the same result as Gaussian elimination:

- ```
  MatrixForm[GaussEliminate[A]]
  ```

## 17.3    Elementary Matrices

In the previous section, we saw how to perform two of the three possible types of two operations via matrix multiplication. In this section, we discuss elementary matrices in detail. In particular, we will discuss those elementary matrices that perform row interchanges, known as transpositions, and discover that every elementary matrix possesses an easily determined inverse. Throughout this section, one may find it helpful to assume that we are referring to some system of three equations in three unknowns. Once one has examined such a system in detail, one should be able to generalize our observations to an arbitrary system of equations.

First, we observe that the inverse of an elementary matrix may be inferred by considering the corresponding row operations. For example, the operation:

(a)  multiply the first row by 2

may be undone by the operation:

(b)  multiply the first row by $\frac{1}{2}$.

Thus, one would expect that the matrix of this operation would be invertible and its inverse would be the matrix corresponding to the second operation.

### Exercise 17.4

- Verify this assertion by determining the matrices $X$ and $Y$ for operations a and b and showing that $X$ and $Y$ are inverses of one another.

  Similarly, the operation:

  (c)  add 2 times the first row to the third

may be reversed by the operation:

(d)  add $-2$ times the first row to the third.

### Exercise 17.5

Record a system of three equations in three unknowns (i.e., actual equations with variables, not simply the matrix representation of the system). Perform operation c and record the result. Verify that operation d does recover your original system.

## Exercise 17.6

• Determine the matrices $X$ and $Y$ for operations c and d and verify that $X$ and $Y$ are inverses.

## Exercise 17.7

• Determine the inverses of each of the ElementaryMatrix[$i$]'s that we obtained in Section 17.2.

Now it remains to determine the elementary matrices corresponding to row interchanges. For example, consider the matrix:

• `X ={{1,2,3},{4,5,6},{7,8,9}}; MatrixForm[X]`

## Exercise 17.8

• Experiment with the following cell to determine the matrix $M$ which will interchange the first and third rows of $X$.

• `M = { , , }; X1 = M.X; MatrixForm[X1]`

*Hint*: The second row should remained unchanged. Also, the first row comes from picking out the third row.

## Exercise 17.9

• What operation will "undo" this? What is the corresponding elementary matrix? Verify your answers, showing all work.

Such an elementary matrix for a row interchange will always have a very special form and is called a "transposition." Geometrically, a transposition $T$ corresponds to a reflection. Algebraically, it satisfies the equation:

$$T = T^{-1} = T^T$$

In other words, a transpose is symmetric (equals its transpose) and equals its own inverse. If we form a product of transpositions, the result will no longer be symmetric. This lack of symmetry is easily verified by the following computation:

$$(T_1 T_2 T_3)^T = T_3^T T_2^T T_1^T = T_3 T_2 T_1$$

since transposition reverses products.

However, because inverses *also* reverse products, we still have the equality of the transpose and the inverse:

$$(T_1 T_2 T_3)^T = T_3^T T_2^T T_1^T = T_3^{-1} T_2^{-1} T_1^{-1} = (T_1 T_2 T_3)^{-1}$$

Such a product of transpositions is called a "permutation matrix," because in mathematical parlance a permutation is any transformation that changes the order of things — in this case, the rows of a matrix. In general, we make the following definitions:

**Definition 17.1** A square matrix $T$ is a *transposition* iff it is the result of a single row interchange on the identity matrix. If $P$ is a product of transpositions, we call it a *permutation*.

Our previous calculations show that any transposition is symmetric and self-invertible (i.e., it is its own inverse), while the transpose of permutation equals its inverse. As equations, these conditions are

$$T^T = T = T^{-1} \qquad \text{and} \qquad P^T = P^{-1}$$

respectively.

### Exercise 17.10

Determine the permutation matrix $P$ so that $Y = PX$, where:

- ```
X = {{1,2,3},{4,5,6},{7,8,9}};Y = {{4,5,6},{7,8,9},{1,2,3}};
Print["X = ", MatrixForm[X], "   and   Y = ", MatrixForm[Y]];
```

May $P$ be taken to be a transposition? Explain.

We conclude this section with a final observation. One may quickly determine the matrix of any elementary row operation.

#### Observation:

To determine the elementary matrix of any row operation, simply apply that operation to the appropriately sized identity matrix.

One may easily verify that all of the elementary matrices we have used in this lesson conform to this principle.

## 17.4  The $LU$ Decomposition

Now that we understand the connection between each *step* of Gaussian elimination and matrix multiplication, we move on to discuss how the *entire process* may be summarized as a single multiplication, which will lead to our important decomposition formula. Up to this point, we have characterized Gaussian elimination algebraically as a sequence of multiplications by elementary matrices:

$$A_1 = E_1 A \implies A_2 = E_2 A_1 = E_2 E_1 A \implies \dots \implies A_5 = E_5 E_4 E_3 E_2 E_1 A$$

It is a simple matter to write this as a single multiplication, by defining:

- ```
M = ElementaryMatrix[5].ElementaryMatrix[4].
    ElementaryMatrix[3].ElementaryMatrix[2].
    ElementaryMatrix[1]; MatrixForm[M]
```

We may easily verify that this matrix performs as advertised:

- ```
  A5 == M.A
  ```

  This matrix is not particularly helpful, because it has no special properties, other than being lower-triangular:

- ```
  MatrixForm[M]
  ```

  However, we have seen that each of these elementary matrices is invertible. This implies that their product $M$ is also invertible. If we look at the inverse of $M$, call it $L$, we notice that it possesses a rather striking property:

- ```
  L = Inverse[M]; MatrixForm[L]
  ```

Specifically, all of the multipliers that we employed during Gaussian elimination appear in $L$. Moreover, their positions precisely indicate the corresponding row operations. In particular, once we examine this pattern in greater detail, we will be able to determine $L$ *without ever computing any of the corresponding elementary matrices!* This result partially explains our concern over the precise order of operations in Gaussian elimination, because this pattern only holds if we perform all row operations in the correct order.

## Exercise 17.11

The entries on the diagonal of this matrix come from (the inverses of) those elementary matrices that divide a row by a constant. Indicate precisely which operation of Gaussian elimination corresponds to each diagonal entry. Indicate how you would determine the diagonal entries of $L$ while performing Gaussian elimination.

*Hint*: It may be helpful to make a chart of all of the row operations, as we did in Ex. 4.1.

## Exercise 17.12

The off-diagonal entries of $L$ come from (the inverses of) the second type of elementary matrix which subtracts a multiple of one row from another. Indicate precisely which operation of Gaussian elimination corresponds to each off-diagonal entry. Indicate how you would determine these entries of $L$ while performing Gaussian elimination.

Although $L$ does have a nice form, we must determine what it represents mathematically. We do this by introducing a bit of notation. Because the row echelon form for square matrices is upper-triangular, it is common to rename the resulting row echelon matrix $A_5$ as $U$. *Note*: This is why the command GaussianElimination[] used the letter $U$. Likewise, because the inverse of $M$ is lower-triangular, we have called it $L$. Now if we perform the simple calculation:

$$A_5 = MA \Rightarrow U = MA \Rightarrow M^{-1}U = M^{-1}MA \Rightarrow LU = A$$
  ```

we see that $A$ may be written as the product of $L$ and its row echelon form. This type of equation is referred to as a *decomposition* of $A$, because we have broken $A$ up into simpler pieces, namely a lower- and upper-triangular matrix.

Notice that because this decomposition is completely described by Gaussian elimination, and there was no freedom in choosing the steps of elimination, there is only one possible value for $L$ and $U$. In other words,

**Observation:**

When there are no row interchanges, the $LU$ decomposition for $A$ is unique.

This fact allows us to easily create practice problems with "built-in" solutions and a guarantee that the calculations will only involve integer values. We indicate how to do so in the following exercise.

### Exercise 17.13

Choose a $3 \times 3$ lower-triangular matrix $X$, with integer entries and nonzero entries on the diagonal. Similarly, create a $3 \times 5$ row echelon matrix $Y$ with integer entries. Form the product $A = XY$ and then perform Gaussian elimination to convert $A$ to its row echelon form $U$. Use the patterns that you observed in the previous two exercises to record the matrix $L$, which corresponds to the entire set of row operations. How do $L$ and $U$ compare with the original matrices $X$ and $Y$?

## 17.5   Using Row Interchanges

The previous example was somewhat unusual, in that it did not require any row interchanges. Computing the correct decomposition becomes slightly more difficult when there are row interchanges. For example, consider the system:

$$\begin{aligned} 2x + 4y - 2z &= 6 \\ 3x + 6y + 2z &= 4 \\ -x + y + 4z &= 9 \end{aligned}$$

After the first three row operations, we are left with:

$$\begin{aligned} x + 2y - z &= 3 \\ 5z &= -5 \\ 3y + 3z &= 12 \end{aligned}$$

At this point, we must interchange the second and third equations. We will repeat the analysis of Section 17.4 with this example to determine the precise effect of this interchange on the decomposition that we have already observed.

To this point, the elementary matrices that we have used are:

```
ElementaryMatrix[1] = {{1/2,0,0},{0,1,0},{0,0,1}};
ElementaryMatrix[2] = {{1,0,0},{-3,1,0},{0,0,1}};
ElementaryMatrix[3] = {{1,0,0},{0,1,0},{1,0,1}};
```

and the corresponding inverse:

- ```
  M = ElementaryMatrix[3].ElementaryMatrix[2].
  ElementaryMatrix[1]; MatrixForm[Inverse[M]]
  ```

conforms to the pattern we observed earlier. Namely, the divisor of the first row appears in the first diagonal entry, while the multipliers for each of the remaining operations appear in the corresponding rows.

The remaining three operations would then be

- ```
  ElementaryMatrix[4] = {{1,0,0},{0,0,1},{0,1,0}};
  ElementaryMatrix[5] = {{1,0,0},{0,1/3,0},{0,0,1}};
  ElementaryMatrix[6] = {{1,0,0},{0,1,0},{0,0,1/5}};
  ```

to interchange the last two rows and divide the diagonal entries to obtain leading 1's. Now we may observe the effect of the row interchange. If we calculate $L$ as before, we would obtain the matrix:

- ```
  M = ElementaryMatrix[6].ElementaryMatrix[5].
      ElementaryMatrix[4].ElementaryMatrix[3].
      ElementaryMatrix[2].ElementaryMatrix[1]; MatrixForm[Inverse[M]]
  ```

Notice that this is no longer lower-triangular. Naively, it seems that the rows are somewhat "mixed up." If we multiply *again* by the row interchange matrix, the rows will "straighten out" to yield a lower-triangular matrix $L$:

- ```
  L = ElementaryMatrix[4].Inverse[M]; MatrixForm[L]
  ```

Although we have recovered a lower-triangular matrix, we should stop to analyze what this implies about our original decomposition. As before, we have defined $M$ to be the matrix that converts $A$ to its row-echelon form $U$, so that we have the equation $MA = U$. A bit of algebra gives:

$$MA = U \;\Rightarrow\; A = M^{-1}U \;\Rightarrow\; E_4A = E_4M^{-1}U = LU$$

In order to have a "decomposition," we must solve this final equation for $A$. By convention, we relabel ElementaryMatrix[4] as $P$ (for "permutation"). Earlier we observed such a permutation matrix is invertible and its inverse equals its transpose. Thus, we we may solve for $A$, as follows:

$$PA = E_4A = LU \;\Rightarrow\; A = P^{-1}LU = P^{T}LU$$

We have finally produced the decomposition that we promised in the Introduction. Although this example only dealt with a single interchange, this analysis generalizes directly to the case where there are more row interchanges. In general, $P$ will be the product of all interchanges, in their original order. In the next section, we will work through a more general example, focusing more on the practical details of how to carry out the computation by hand.

Because this may seem to be a long lesson, we pause to remember the benefits of our efforts. First, because of the patterns we have observed relating the matrices $L$ and $P$ to the operations of Gaussian elimination, it is possible to determine the decomposition of $A$, *while* performing elimination, with little additional effort. We discuss this in detail in the next section. We have seen how this decomposition allows one to create convenient exercises to practice elimination. We will see how this decomposition provides the most effective techniques for computing determinants and for repeatedly solving $AX = B$ for various right-hand sides $B$. Theoretically, this decomposition formula will allow us to see clearly why the qualitative nature of a system of equations is determined by its rank, and to view any matrix as a sequence of simple geometric transformations. Thus, all of our effort in this lesson will be repaid with interest.

## 17.6   An Example of a $P^T LU$ Decomposition

In this section, we work slowly through an example and describe in detail how to compute a permutation matrix $P$, a lower-triangular matrix $L$, and a row echelon matrix $U$ for a given matrix $A$ so that

$$A = P^T LU$$

For this example, we will take:

$$A = \begin{bmatrix} 3 & -12 & 9 & 9 & -9 \\ -2 & 8 & -6 & -5 & 7 \\ -3 & 10 & -7 & -3 & 19 \end{bmatrix}$$

Since we will obtain $U$ as the row echelon form of $A$ after Gaussian elimination, it remains to focus on how to construct $P$ and $L$. Both matrices will be square with the same number of rows as $A$. As suggested by the examples in the previous two sections, $L$ will contain all of the multipliers used during elimination and $P$ will be the product of all elemetary interchange matrices that arise during Gaussian elimination. Instead of computing $P$ explicitly by multiplication, we may obtain the same result by simply performing the sequence of row interchanges on the identity matrix.

To begin, write out $A$, a $3 \times 3$ identity matrix and an empty $3 \times 3$ matrix. These last two matrices will *eventually* contain $P$ and $L$. To save writing, we may omit the 0 entries of $P$. Thus, we would have:

$$A = \begin{bmatrix} 3 & -12 & 9 & 9 & -9 \\ -2 & 8 & -6 & -5 & 7 \\ -3 & 10 & -7 & -3 & 19 \end{bmatrix} \quad P = \begin{bmatrix} 1 & & \\ & 1 & \\ & & 1 \end{bmatrix} \quad L = \begin{bmatrix} & & \\ & & \\ & & \end{bmatrix}$$

Now we carry out Gaussian elimination and record each of the steps appropriately in $P$ and $L$.

In this example, the first three steps of Gaussian elimination are:

$$\xrightarrow{R_1 \div 3} \begin{bmatrix} 1 & -4 & 3 & 3 & -3 \\ -2 & 8 & -6 & -5 & 7 \\ -3 & 10 & -7 & -3 & 19 \end{bmatrix} \quad P = \begin{bmatrix} 1 & & \\ & 1 & \\ & & 1 \end{bmatrix} \quad L = \begin{bmatrix} 3 & & \\ & & \\ & & \end{bmatrix}$$

$$\xrightarrow{R_2 - (-2) \times R_1} \begin{bmatrix} 1 & -4 & 3 & 3 & -3 \\ 0 & 0 & 0 & 1 & 1 \\ -3 & 10 & -7 & -3 & 19 \end{bmatrix} \quad P = \begin{bmatrix} 1 & & \\ & 1 & \\ & & 1 \end{bmatrix} \quad L = \begin{bmatrix} 3 & & \\ -2 & & \\ & & \end{bmatrix}$$

$$\xrightarrow{R_3 - (-3) \times R_1} \begin{bmatrix} 1 & -4 & 3 & 3 & -3 \\ 0 & 0 & 0 & 1 & 1 \\ 0 & -2 & 2 & 6 & 10 \end{bmatrix} \quad P = \begin{bmatrix} 1 & & \\ & 1 & \\ & & 1 \end{bmatrix} \quad L = \begin{bmatrix} 3 & & \\ -2 & & \\ -3 & & \end{bmatrix}$$

Here we should observe that whenever we use operation 4.1a from Definition 4.1, we record the divisor in the diagonal entry of $L$ on the same row. Now it becomes clear why we were so careful to phrase this operation in terms of division (as opposed to multiplication by $\frac{1}{3}$).

Similarly, when we use operation 4.1c, we record the multiplier in the same row of $L$ as the row from which we are subtracting. At first, it may seem that we record the multiplier in the same column as that which we are eliminating, sometimes called the "pivot column." However, with experience, one will see that the column in which we record the multiplier is the same as the *row* in which the current leading 1 resides. Notice also that although one may be tempted to think of these row operations as "add a multiple" (and may in fact compute the results that way), in order to get the correct sign on the corresponding entry of $L$, one must phrase each such operation in terms of *subtraction*. Finally, observe that if one records the entries correctly, each column of $L$, from the diagonal entry on down, looks identical to the corresponding pivot column *before* elimination in that column. This pattern provides an easy check on one's computation.

Notice that, after most steps, we simply fill in the correct entry of $L$ as we go, without entirely rewriting $P$ and $L$. However, in the next step we must use operation 4.1b, which requires more significant modifications. In this case, we simply perform the *same* switch in *all three matrices*.

$$\xrightarrow{R_2 \leftrightarrow R_3} \begin{bmatrix} 1 & -4 & 3 & 3 & -3 \\ 0 & -2 & 2 & 6 & 10 \\ 0 & 0 & 0 & 1 & 1 \end{bmatrix} \quad P = \begin{bmatrix} 1 & & \\ & & 1 \\ & 1 & \end{bmatrix} \quad L = \begin{bmatrix} 3 & & \\ -3 & & \\ -2 & & \end{bmatrix}$$

The next step is similar to the first, except that it occurs in the second row.

$$\xrightarrow{R_2 \div -2} \begin{bmatrix} 1 & -4 & 3 & 3 & -3 \\ 0 & 1 & -1 & -3 & -5 \\ 0 & 0 & 0 & 1 & 1 \end{bmatrix} \quad P = \begin{bmatrix} 1 & & \\ & & 1 \\ & 1 & \end{bmatrix} \quad L = \begin{bmatrix} 3 & & \\ -3 & -2 & \\ -2 & & \end{bmatrix}$$

Although finished with elimination, we must record something in the $(3, 2)$ and $(3, 3)$ entries of $L$. We may act as if we performed the following two operations:

$$\xrightarrow{R_3 - (0) \times R_2} \begin{bmatrix} 1 & -4 & 3 & 3 & -3 \\ 0 & 1 & -1 & -3 & -5 \\ 0 & 0 & 0 & 1 & 1 \end{bmatrix} \quad P = \begin{bmatrix} 1 & & \\ & & 1 \\ & 1 & \end{bmatrix} \quad L = \begin{bmatrix} 3 & & \\ -3 & -2 & \\ -2 & 0 & \end{bmatrix}$$

$$\xrightarrow{R_3 \div 1} \begin{bmatrix} 1 & -4 & 3 & 3 & -3 \\ 0 & 1 & -1 & -3 & -5 \\ 0 & 0 & 0 & 1 & 1 \end{bmatrix} \quad P = \begin{bmatrix} 1 & & \\ & & 1 \\ & 1 & \end{bmatrix} \quad L = \begin{bmatrix} 3 & & \\ -3 & -2 & \\ -2 & 0 & 1 \end{bmatrix}$$

since subtraction by 0 and division by 1 have no effect.

At this point, the matrix on the left is the desired matrix $U$, and we have obtained the actual values of $P$ and $L$ for our $P^T LU$ decomposition. This implies that we have the decomposition formula:

$$A = P^T L U = \begin{bmatrix} 1 & 0 & 0 \\ 0 & 0 & 1 \\ 0 & 1 & 0 \end{bmatrix} \begin{bmatrix} 3 & 0 & 0 \\ -3 & -2 & 0 \\ -2 & 0 & 1 \end{bmatrix} \begin{bmatrix} 1 & -4 & 3 & 3 & -3 \\ 0 & 1 & -1 & -3 & -5 \\ 0 & 0 & 0 & 1 & 1 \end{bmatrix}.$$

The reader should simplify the right-hand side of this equation to verify that this multiplication will recover the original matrix $A$. Notice how we filled all of the "missing" entries of $L$ and $P$ with 0's. Notice also how the decomposition formula requires us to form the *transpose* of $P$ before multiplying. In this case, because $P$ happened to be symmetric, one could neglect the transpose without consequence. In general, $P$ will *not* be symmetric and one should be careful to transpose $P$ to obtain a valid decomposition.

To summarize, to compute $P$, $L$, and $U$ for a given matrix $A$, one starts with an identity matrix for $P$ and an empty square matrix for $L$, both with the same number of rows as $A$. One then performs Gaussian elimination, recording each operation in $P$ and $L$, as indicated by the following table.

| Operation | Record |
|---|---|
| Divide row $i$ by a nonzero number. | Put the divisor in entry $(i, i)$ of $L$. |
| Switch two rows. | Switch the same two rows of $L$ and $P$. |
| Subtract a multiple of row $j$ from row $i$. | Put the multiplier in entry $(i, j)$ of $L$. |

After completing Gaussian elimination, one then fills in the remaining entries of $L$ and $P$ with 0's. The row echelon form of $A$ gives $U$, yielding the decomposition formula:

$$A = P^T LU$$

where $P$ has been transposed.

## 17.7   Constructing and Solving Your Own Examples

As with an $LU$ decomposition, even if we have row interchanges, we obtain a decomposition of $A$, which means that we may verify our answer by multiplication. For instance, the example which we computed in Section 17.6 yielded:

- 
```
A = {{ 2, 4,-2, 6},{ 3, 6, 2, 4},{-1, 1, 4, 9}};
P = {{ 1, 0, 0},{ 0, 0, 1},{ 0, 1, 0}};
L = {{ 2, 0, 0},{-1, 3, 0},{ 3, 0, 5}};
U = {{ 1, 2,-1, 3},{ 0, 1, 1, 4},{ 0, 0, 1, -1}};
Print["P  =  ",MatrixForm[P],"    L  =  ",MatrixForm[L]];
Print["U  =  ",MatrixForm[U],"    A  =  ",MatrixForm[A]];
```

To verify this, we would compute:

- 
```
MatrixForm[Transpose[P].L.U]
```

and compare the result with $A$. Notice how we transposed $P$ before multiplying.

Unlike an $LU$ decomposition (i.e., without row interchanges), the choice of row interchanges prevents this decomposition from being unique. That is, two people may obtain different decompositions that are *both* correct. That is why we provide the PtLU[] command, defined in the "Decomps.m" package, to compute $P$, $L$, and $U$ automatically. As with the GaussEliminate[] command, it will always choose the *first possible* *row*, when doing interchanges, and will switch rows only when absolutely necessary. In particular, if one chooses to switch rows when it is not absolutely necessary (i.e., when there is not a 0 in the place of a leading 1), or does not switch with the first possible row, one will obtain very different answers than the computer.

If we have remembered to open "Start.ma", we may reproduce the results from Section 17.6 with the following commands:

- 
```
{P,L,U} = PtLU[A,Verbose->True];
Print["P  =  ",MatrixForm[P],"    L  =  ",MatrixForm[L]];
Print["U  =  ",MatrixForm[U],"    A  =  ",MatrixForm[A]];
```

This defines $P$, $L$, and $U$ to be those matrices in a decomposition of $A$ and then prints them out in a readable format. The Verbose->True option instructs *Mathematica* to print out some of the intermediate steps, just like GaussEliminate[]. In fact, GaussEliminate[] is defined via the PtLU[] command. You may leave out the Verbose option (or set it to False), if you only desire the final results.

Although one may create an exercise by hand (as we did in Ex. 17.13), one may not recover the original matrices $P$, $L$, and $U$. Thus, we have also provided a command, in the "Matrices.m" package, called "MakeNiceMatrix[n,m]," that will create an $n \times m$ matrix with an integer-valued $P^T LU$ decomposition.

This command has a number of options that allow one to specify the rank of $A$, as well as control the size and type of the entries of $A$. For example:

- `A = MakeNiceMatrix[5,6, Rank->4]; MatrixForm[A]`

will create a matrix with rank 4, 5 rows, and 6 columns.

### Exercise 17.14

- Use the previous MakeNiceMatrix[] command to generate a matrix $A$. Compute its decomposition by hand, recording each row operation used and the resulting decomposition. You should only use PtLU[] to check your work.

### *Drill Exercise 17.15*

- Generate three more matrices as in Ex. 17.14 and compute their decompositions. Your matrices should be at least $3 \times 3$ of various sizes and of various ranks.

## 17.8   Summary

At this point, one should know how each step of Gaussian elimination corresponds to multiplication by an elementary matrix, and how the entire process corresponds to a decomposition of the form:

$$A = P^T LU$$

where $P$ is a permutation matrix, $L$ is lower-triangular, and $U$ is row echelon. One should also be able to quickly compute $P$, $L$, and $U$ while performing Gaussian elimination, how to create additional practice problems, and how to verify one's answers. In the next lesson, we will discuss some of applications of this important formula.

# Applications of the $P^T LU$ Decomposition

## Accompanied by Notebook

In Lesson 17, by viewing Gaussian elimination in terms of matrix multiplication, we saw how any matrix $A$ may be decomposed into $A = P^T LU$. In this lesson, we will continue to see applications of this point of view and this decomposition formula. In particular, we will:

- prove that the solution set does not change as a result of Gaussian elimination,

- discover a quick way of solving $AX = B$ for different $B$'s, after performing Gaussian elimination *only once*,

- prove that any square, triangular matrix with nonzero elements down the diagonal is invertible, and

- learn how any matrix is composed of elementary matrices.

From these results one should begin to recognize the value of this important matrix decomposition. Throughout this text, we will see that many important results may be phrased in terms of a decomposition into specialized types of matrices.

## 18.1 Solutions and Gaussian Elimination

We have been using Gaussian elimination since Lesson 4 for solving systems of equations. Although our experiments in Section 2.5 indicated that the elementary operations of Definition 4.1 will not change the solution set, we have yet to provide a rigorous proof of this fact. In this section, we will use matrix multiplication to provide a simple proof. To solve a system of equations such as our example from Lesson 4:

$$
\begin{array}{rrrrrrrrr}
2x & - & 4y & - & 2z & + & 6w & = & 8 \\
x & - & 2y & & & + & 7w & = & 4 \\
-3x & + & 6y & + & 3z & - & 4w & = & -7 \\
4x & - & 5y & + & 5z & + & 6w & = & 10
\end{array}
\tag{18.1}
$$

we represent this as $AX = B$, where:

$$A = \begin{bmatrix} 2 & -4 & -2 & 6 \\ 1 & -2 & 0 & 7 \\ -3 & 6 & 3 & -4 \\ 4 & -5 & 5 & 6 \end{bmatrix} \quad \text{and} \quad B = \begin{bmatrix} 8 \\ 4 \\ -7 \\ 10 \end{bmatrix}$$

and perform Gaussian elimination on the partitioned matrix $A' = \begin{bmatrix} A & B \end{bmatrix}$. If we can show that the solutions do not change at each *step* of Gaussian elimination, then elimination as a whole cannot change the solution set of the system.

In Lesson 17, we observed that each step of Gaussian elimination may be achieved by multiplication of $A'$ by an elementary row operation $E$. By the Principle of Partitioned Matrices, we have:

$$EA' = E \begin{bmatrix} A & B \end{bmatrix} = \begin{bmatrix} EA & EB \end{bmatrix}$$

which corresponds to the system of equations $EAX = EB$. So we must show that the solutions $X$ to the original system $AX = B$ are the same as those of the transformed system $EAX = EB$.

If $X$ is a solution to the original system $AX = B$, then we may multiply both sides of this equation by $E$ to get $EAX = EB$, showing that $X$ is also a solution of the transformed system. Conversely, $X$ is a solution to the transformed system, so that $EAX = EB$, we may multiply both sides by the inverse of $E$ to obtain:

$$E^{-1}EAX = E^{-1}EB \quad \Rightarrow \quad AX = B$$

Thus, both systems have exactly the same solutions. *Note*: It is the *reversibility* (i.e., invertibility) of the row operations that makes this proof work.

## 18.2  Solving Multiple Systems

We have already given a technique for solving a set of parallel systems (compare Section 15.2), such as:

$$AX = B_1, \qquad AX = B_2, \quad \dots \quad AX = B_n$$

by performing Gaussian elimination on $\begin{bmatrix} A & B_1 & B_2 & \cdots & B_n \end{bmatrix}$. But it could be that we are not given all of the $B$'s *at the same time*. For example, suppose we must solve the recursive collection of equations $AB_{n+1} = B_n$ for n = 1, 2, .... Since we will not know $B_2$ until we have already performed Gaussian elimination, we cannot solve for $B_3$ simultaneously. Although this problem may seem far-fetched, it arises in numerous practical engineering problems, for example, when designing a car's suspension system. We might describe this type of problem as solving a set of systems "in series." Although we cannot use our "parallel" algorithm, we may *still* solve this (infinite!) collection of problems by performing Gaussian elimination *only once*. Instead of employing partitioned matrices, we exploit the $P^T LU$ decomposition.

Given a $P^T LU$ decomposition for $A$, we may quickly solve $AX = B$ by repeated backsubstituion. Specifically, to solve the system $B = AX = P^T LUX$, we may:

1. compute $PB \equiv B'$,

2. solve $Lu = B'$ *by backsubstitution*, for some $u$,

3. solve $UX = u$ for $X$, using the $u$ from step 2, again by backsubstitution.

*Note*: The symbol "$\equiv$" is often used when an equation is used to *define* one of the variables involved. Thus, our first equation is meant to define $B'$ as the product $PB$, giving this product a name for future reference. To verify that the $X$ resulting from this procedure is actually a solution to $AX = B$, we simply work backwards:

$$UX = u \quad \Rightarrow \quad LUX = Lu = B'$$
$$\Rightarrow \quad AX = P^T LUX = P^T B' = P^{-1} B' = B.$$

Here we use the fact that $P$ is a permutation matrix so that $P^T = P^{-1}$.

For example, if we wish to solve Eq. 18.1 and we know that the left-hand side of this system decomposes as:

$$A = \begin{bmatrix} 2 & -4 & -2 & 6 \\ 1 & -2 & 0 & 7 \\ -3 & 6 & 3 & -4 \\ 4 & -5 & 5 & 6 \end{bmatrix}$$

$$= P^T LU = \begin{bmatrix} 1 & 0 & 0 & 0 \\ 0 & 0 & 1 & 0 \\ 0 & 0 & 0 & 1 \\ 0 & 1 & 0 & 0 \end{bmatrix} \begin{bmatrix} 2 & 0 & 0 & 0 \\ 4 & 3 & 0 & 0 \\ 1 & 0 & 1 & 0 \\ -3 & 0 & 0 & 5 \end{bmatrix} \begin{bmatrix} 1 & -2 & -1 & 3 \\ 0 & 1 & 3 & -2 \\ 0 & 0 & 1 & 4 \\ 0 & 0 & 0 & 1 \end{bmatrix}$$

we would first compute $B' = PB = \begin{bmatrix} 1 & 0 & 0 & 0 \\ 0 & 0 & 0 & 1 \\ 0 & 1 & 0 & 0 \\ 0 & 0 & 1 & 0 \end{bmatrix} \begin{bmatrix} 8 \\ 4 \\ -7 \\ 10 \end{bmatrix} = \begin{bmatrix} 8 \\ 10 \\ 4 \\ -7 \end{bmatrix}$. We would

then solve $Lu = B'$ by backsubstitution.

## Exercise 18.1

Use backsubstitution, working from the top down, to show that $u = \begin{bmatrix} 4 \\ -2 \\ 0 \\ 1 \end{bmatrix}$ is the

unique solution to $Lu = B'$. *Hint*: Plugging in the definitions for $L$ and $B'$ yields the equation:

$$\begin{bmatrix} 2 & 0 & 0 & 0 \\ 4 & 3 & 0 & 0 \\ 1 & 0 & 1 & 0 \\ -3 & 0 & 0 & 5 \end{bmatrix} u = \begin{bmatrix} 8 \\ 10 \\ 4 \\ -7 \end{bmatrix}$$

Using $u$ from Ex. 18.1, we may then solve for $X$ by applying backsubstitution to the equation $UX = u$.

## Exercise 18.2

Plug in the values for $U$ and $u$ into the equation $UX = u$ and solve the resulting matrix equation by backsubstitution. Conclude that the unique solution to Eq. 18.1

is $X = \begin{bmatrix} 21 \\ 12 \\ -4 \\ 1 \end{bmatrix}$.

## Exercise 18.3

Now solve

$$\begin{array}{rrrrrrrrrr} 2x & - & 4y & - & 2z & + & 6w & = & 34 \\ x & - & 2y & & & + & 7w & = & 33 \\ -3x & + & 6y & + & 3z & - & 4w & = & -31 \\ 4x & - & 5y & + & 5z & + & 6w & = & 38 \end{array}$$

using the technique given above. Make sure to check your work. *Hint*: Because this system has the same left-hand side as Eq. 18.1, you will not need to perform Gaussian elimination.

Now we can explain how to solve a recursive set of equations: $AB_{n+1} = B_n$ for $n = 1, 2, \ldots$, by using the algorithm on p. 157. This may be considered as a set of equations "in series," in that the "output" (i.e., solution) of one system is the "input" (i.e., right-hand side) of the next. For example, take the matrix $A = \begin{bmatrix} -10 & 220 \\ -10 & 120 \end{bmatrix}$ and $B_1 = \begin{bmatrix} 100 \\ 0 \end{bmatrix}$. We will proceed to solve for $B_n$ for $n \geq 2$ inductively in Ex. 18.4–18.6.

## Exercise 18.4

Verify that $A = \begin{bmatrix} -10 & 0 \\ -10 & -100 \end{bmatrix}\begin{bmatrix} 1 & -22 \\ 0 & 1 \end{bmatrix}$ is a valid *LU* decomposition.

Now we may easily solve $AB_2 = B_1$ for $B_2$.

## Exercise 18.5

Solve $AB_2 = B_1$ by repeated backsubstitution, showing all work. You should obtain $B_2 = \begin{bmatrix} 12 \\ 1 \end{bmatrix}$ as a final solution.

Now we simply repeat this process to solve for $AB_3 = B_2$, etc.

### Exercise 18.6

Solve $AB_3 = \begin{bmatrix} 12 \\ 1 \end{bmatrix}$ for $B_3$. Repeat the process to solve $AB_4 = B_3$ for $B_4$ and then $AB_5 = B_4$ for $B_5$. To save time, you may use the commands in the accompanying *Mathematica* Notebook.

### Exercise 18.7

Do you notice any pattern to the solutions $B_2, \ldots, B_5$? *Hint*: Examine your solutions in decimal form. Use your pattern to estimate $B_6$. Verify your conjecture by showing that your estimate of $B_6$ is actually the solution to $AB_6 = B_5$.

*Note*: Although some aspects of this pattern are limited to this particular example, there are two features of the pattern that are very general. These features of the solutions are determined by an "eigenvalue" for $A$ and its associated "eigenvector." We will discuss eigenvalues and eigenvectors more fully in Lessons 32–33.

## 18.3   Some Criteria for Invertibility

So far it has been difficult to determine whether or not a matrix is invertible. In this section we provide two simple tests that work in *some* cases. The proofs are easy, given our understanding of Gaussian elimination (without interchanges) in terms of an $LU$ decomposition.

First observe that since the $L$ from an $LU$ decomposition is the product of invertible, elementary matrices, it is invertible (compare Theorem 14.2a). Also notice that such an $L$ has nonzero elements down its diagonal (look at any of the examples from Lesson 17). This suggestions the following theorem.

**Theorem 18.1** *Any square, lower-triangular matrix is invertible iff all of its diagonal elements are nonzero.*

### Proof:

We first prove the foward direction, that is, we will will consider a square, invertible, lower-triangular matrix $A$ and we will show that it cannot have a zero element on its diagonal. We will use a new proof technique, known as proof by contradiction. Specifically, we will assume that $A$ *does* have a zero element along its diagonal. We will then reason until we reach a logical contradiction. Assuming that we have not made a mistake in our reasoning, we must conclude that one of our original assumptions, namely that there was a zero element on the diagonal of $A$, must be incorrect.

Assume that a square, invertible, lower-triangular matrix $A$ has a 0 in the $(i, i)$ entry. By the time we reach the $i$th columns in Gaussian elimination, the $i$th row will consist entirely of 0's. This implies that we will be left with 0 rows at the conclusion of Gaussian elimination. In particular, $A$ will not have full rank and, by Theorem 16.4c, $A$ cannot be invertible. However, $A$ *is* invertible, so we have reached a logical contradiction. Thus, $A$ must have all nonzero elements on its diagonal.

Conversely, given any square lower-triangular matrix $A$ with nonzero elements down its diagonal, we may perform Gaussian elimination without row interchanges to obtain $A = LU$. Moreover, since $A$ is square, we will obtain $U = I$ (the identity matrix), so that $A = LI = L$, which we have already observed is invertible. Q.E.D.

**Corollary 18.2** *Any square, upper-triangular matrix with nonzero elements down the diagonal is invertible.*

### Proof:

If $A$ is upper-triangular, then its transpose $A^T$ is lower-triangular. Applying Theorem 18.1 to $A^T$, we have that $A^T$ is invertible. By Theorem 14.2b, the transpose of this $\left(A^T\right)^T$ is invertible. But $\left(A^T\right)^T = A$. Q.E.D.

Of course, this does not mean that these are the *only* invertible matrices. We have seen a number of examples of invertible matrices already that are not of this form. However, these theorems provide a simple test for invertibility.

### Observation:

*If* a matrix is triangular, it is invertible precisely when all the diagonal elements are nonzero.

When it applies, this observation saves us a great deal of time over the methods we have learned to this point.

## 18.4  The Structure of Invertible Matrices

In this section, we re-examine our algorithm for computing the inverse of a matrix $A$ (compare p. 136). That algorithm said to form $\begin{bmatrix} A & I \end{bmatrix}$, perform Gaussian elimination, and (if the matrix is invertible) one obtains $\begin{bmatrix} I & R \end{bmatrix}$, where $R$ is a right-inverse of $A$. We reasoned that this was also a left-inverse, by appealing to certain theorems and a simple computation. Now we can give a straightforward proof of this fact, which will have another interesting consequence.

To this point, we have viewed Gaussian elimination as multiplication by a *sequence* of elementary matrices, $E_k, \ldots, E_1$. If we define $E \equiv E_k \cdots E_1$, we can instead consider elimination as multiplication by a *single* invertible matrix $E$. In particular, we may view the computation of $R$ as the multiplication:

$$E \begin{bmatrix} A & I \end{bmatrix} = \begin{bmatrix} EA & EI \end{bmatrix} = \begin{bmatrix} EA & E \end{bmatrix} = \begin{bmatrix} I & R \end{bmatrix}$$

In particular, we see that $R = E$ and $EA = I$, so that $RA = I$ and $R$ is also a left-inverse and $R = A^{-1}$. Moreover, $A^{-1} = R = E$ is a product of elementary matrices. In fact, since $A^{-1}$ is also invertible, this argument may be repeated to show that:

### Observation:

Any invertible matrix $A$ is a product of elementary matrices.

This observation provides a very powerful description of invertible matrices. Loosely speaking, this suggests that if one understands elementary matrices, then one has a fairly complete understanding of invertible matrices.

## 18.5   The Rank Decomposition

Just as one may decompose an *invertible* matrix into a sequence of elementary matrices, *any* matrix may be decomposed into two invertible matrices and one other very special type of matrix. Specifically:

**Theorem 18.3** *Given an arbitrary matrix $A$, we may compute a decomposition of the form $A = M_1 J M_2$, known as the rank decomposition of $A$, where $J$ is referred to as the rank matrix of $A$ and the matrices $M_1$ and $M_2$ are both invertible.*

We illustrate how this decomposition may be computed with a specific matrix in the accompanying Notebook. We may also provide a formal proof.

   **Proof:**

   The decomposition is determined by the equations:

$$M_1 = P_1^T L_1, \quad M_2 = L_2^T P_2 \quad \text{and} \quad J = U_2^T, \quad \text{where}$$
$$A = P_1^T L_1 U_1 \quad \text{and} \quad U_1^T = P_2^T L_2 U_2$$

   That is, we first compute a $P^T LU$ decomposition to derive $P_1$, $L_1$, and $U_1$. We then compute a $P^T LU$ decomposition of $U_1^T$ to determine $P_2$, $L_2$, and $U_2$. Finally, we define $M_1$, $M_2$, and $J$, by the given equations. Since $M_1$ and $M_2$ are the product of invertible matrices, they are invertible. The given decomposition formula may be verified by simple algebra:

$$M_1 J M_2 = M_1 U_2^T L_2^T P_2 = M_1 \left(P_2^T L_2 U_2\right)^T = M_1 \left(U_1^T\right)^T = M_1 U_1 = P_1^T L_1 U_1 = A.$$

   Q.E.D.

One should note that although the $M_1$ and $M_2$ of a rank decomposition are not generally unique, the associated rank matrix $J$ *is* completely determined by the dimensions of $A$ and its rank.

### Exercise 18.8

   • Experiment in the accompanying Notebook until you can form a conjecture describing how the rank matrix $J$ of an arbitrary matrix $A$ compares with $A$ itself.

   From a practical standpoint, this decomposition shows why the rank of a matrix determines its behavior to such a large degree. The reason is that $J$ is the only noninvertible matrix in the decomposition. Specifically, we may deduce the following theorem.

**Theorem 18.4** *If $J$ is the rank matrix of a matrix $A = M_1 J M_2$, qualitatively the system $AX = B$ is identical to $JY = C$, where $C = M_1^{-1} B$.*

**Proof:**

We simply observe that any solution $X$ of $AX = B$ corresponds to a solution $Y = M_2 X$ of $JY = C$, and vice versa:

$$AX = B \iff (M_1 J M_2)X = B \iff M_1(J(M_2 X)) = B$$
$$\iff M_1(JY) = B \iff M_1^{-1} M_1(JY) = M_1^{-1}B$$
$$\iff JY = C$$

Q.E.D.

This theorem may be used to easily explain many of the results in Lesson 16. For example, if $A$ is a $4 \times 3$ matrix with $rank(A) = 3$, then Theorem 18.4 implies that, qualitatively, $AX = B$ has the same type of solution set as $JY = C$, which may be described by the augmented matrix:

$$\begin{bmatrix} 1 & 0 & 0 & * \\ 0 & 1 & 0 & * \\ 0 & 0 & 1 & * \\ 0 & 0 & 0 & * \end{bmatrix}$$

Here we have written $*$'s in place of $C$ to indicate that they may take on any values. From this, it is clear that for *most* values of $C$, $JY = C$ will be overdetermined (compare Theorems 16.2a and 16.4b). On the other hand, for those particular values of $C$ for which $JX = C$ *has* a solution (i.e., when the last entry of $C$ is 0), there is *exactly* one solution. In particular, for every value of $C$, there is at *most* one solution (see Theorems 16.2b and 16.4a).

By Theorem 18.4, we may draw *exactly* the same conclusions for $AX = B$. Specifically, we may conclude that $AX = B$ always has at most one solution for any given $B$, and that for *most* values of $B$ there will be *no* solution.

### *Advanced Exercise 18.9*

Use Theorem 18.4 to explain why three equations in four unknowns will always have infinitely many solutions for every possible right-hand side, if the matrix of the left-hand side has rank 3.

## 18.6  The Geometry of Arbitrary Matrices

We conclude this lesson with a final observation. By expanding the rank decomposition even further, one may obtain a complete decomposition of any given matrix $A$ into easily described *geometric* transformations. Specifically, since $M_1$ and $M_2$ are invertible, we may use the observation of Section 18.4 to decompose $M_1$ and $M_2$ into products of elementary matrices. Thus, we may decompose an *arbitrary* matrix into elementary matrices and a rank matrix.

For example, we may decompose:

$$A = \begin{bmatrix} 0 & 3 \\ 2 & 2 \end{bmatrix}$$

into

$$A = \begin{bmatrix} 0 & 1 \\ 1 & 0 \end{bmatrix} \begin{bmatrix} 2 & 0 \\ 0 & 1 \end{bmatrix} \begin{bmatrix} 1 & 0 \\ 0 & 3 \end{bmatrix} \begin{bmatrix} 1 & 0 \\ 0 & 1 \end{bmatrix} \begin{bmatrix} 1 & 1 \\ 0 & 1 \end{bmatrix} \begin{bmatrix} 1 & 0 \\ 0 & 1 \end{bmatrix}$$

### Exercise 18.10

• Describe geometrically each of these elementary transformations. *Hint*: You have seen all of these types of matrices before in Lesson 9.

In this way, we have a complete, concrete description of $A$ as a sequence of simple, geometric transformations.

## 18.7   Summary

We have now achieved all of our objectives. We may now be confident that Gaussian elimination will not change the solution set of a system. We may use a decomposition of $A$ into $A = P^T LU$, where $P$ is a permutation matrix, $L$ is lower-triangular, and $U$ is row echelon, to solve equations such as $AX = B$ by repeated backsubstitution. Using this decomposition repeatedly, we may decompose $A$ *entirely* into elementary matrices and a "rank matrix." Thus, from one point of view, we have the complete story on matrices, in that we know how they are constructed from the basic building blocks of elementary matrices.

Of course this is an oversimplification! This corresponds to saying that because we know the periodic table, we know all about cellular biology, psychology, and sociology! Now that we have this fundamental understanding of matrix algebra and systems of equations, we will proceed to yet a higher-level view of matrices, as "linear transformations." But first, in Lesson 19 we will look at the definition of the determinant function and how to use the $P^T LU$ decomposition to compute determinants.

# Discovering Determinants

*Mathematica* Notebook

## 19.1   Introduction

We introduced the determinant in Lesson 16 as a real-valued function on square matrices that:

> – is denoted in *Mathematica* as Det[],
>
> – preserves products, and
>
> – takes invertible matrices to nonzero numbers.

One may deduce many important properties of the determinant function from just these two facts. In this lesson, we will:

> – discover the general definition of the determinant of a matrix,
>
> – learn some short-cuts in the case of $2 \times 2$ and $3 \times 3$ matrices,
>
> – discover the geometric significance of the determinant,
>
> – deduce some basic properties of the determinant function, including how it is affected by elementary row operations and transposition,
>
> – discover how to use Gaussian elimination to compute determinants,
>
> – derive the basic, inductive formula for determinants, and
>
> – prove a formula directly relating inverses and determinants.

For the most part, we will *discover* the properties of determinants through experimentation and simple reasoning, exploiting the fact that determinants preserve products.

We will not be able to explain *why* the determinant function preserves products. Unfortunately, the best answer to this question requires a foray into tensor algebra and is beyond the scope of this text. In abstract mathematical terms, there is a natural definition for the action of a matrix on a tensor, and because the space of $n$-dimensional alternating tensors is one-dimensional, this corresponds to multiplication by a single real number, which is the determinant of the matrix; because this definition is given in terms of an action, it *automatically* preserves products. Even without tensor algebra, we *could* give a formal proof of this result; however, this would take a long time and would not leave you any more convinced. Thus, we will take this result as an empirical fact, and *use* it to investigate the remaining properties of the determinant. This is in line with our principle of using the most *appropriate* proof technique, and so we will emphasize here the importance of this crucial result. On the other hand, we *will* be able to prove that a matrix is invertible precisely when its determinant is nonzero. This will follow directly from the inverse formula given in Theorem 19.5.

## 19.2  Discovering the Definition

In this section, we will discover the general definition of the determinant. We begin, much in the same way it was (probably) originally discovered, by performing general symbolic computations with small examples. We will then attempt to discover patterns in the resulting formulas, which will allow us to generalize them to higher dimensions.

### 19.2.1  Experiments with Small Matrices

We begin by experimenting with $2 \times 2$ and $3 \times 3$ examples. Our approach will also emphasize the single most important property of the determinant function:

**Theorem 19.1** *An $n \times n$ matrix $A$ is invertible iff its determinant $|A|$ is nonzero.*

As mentioned in Section 19.1, this result will follow immediately from the inverse formula given in Theorem 19.5.

We begin with a general $2 \times 2$ matrix $A$. The following *Mathematica* command will define a a $2 \times 2$ matrix with generic entries:

- ```
  A = Array[a,{2,2}]; MatrixForm[A]
  ```

To make our calculations more readable, we use the Format[] command, which instructs *Mathematica* to print out the entries of $A$ in subscript notatation:

- ```
  Format[a[i_,j_]] := Subscripted[a[i,j]]; MatrixForm[A]
  ```

Examining the general formula for the inverse of $A$, it becomes clear what formula must be nonzero so that the inverse is defined:

- ```
  Ainverse = Inverse[A]; MatrixForm[Ainverse]
  ```

### Exercise 19.1

- Deduce from this general formula for $A^{-1}$ a reasonable definition for the determinant of a general $2 \times 2$ matrix, so that Theorem 19.1 holds. You may verify your hypothesis by evaluating:

- ```
  Det[A]
  ```

## Exercise 19.2

Use your formula from Ex. 19.1 to compute the determinant of:

$$B = \begin{bmatrix} a & b \\ c & d \end{bmatrix}$$

We may now repeat our experiment on a general $3 \times 3$ matrix by evaluating:

- ```
A=Array[a,{3,3}]; Print["A  ==  ", MatrixForm[A]];
Ainverse = Inverse[A];
```

Because of its size, we will avoid printing the entire matrix for $A^{-1}$. It is more convenient to use the following commands to examine each entry individually. For example, to view the $(1, 1)$ entry of the inverse, we would evaluate:

- ```
i = 1; j = 1;
Print["(A", Superscript[-1], ")", Subscript[i,",",j],
      "  ==  ", Ainverse[[i,j]]];
```

## Exercise 19.3

- Examine a number of different entries of the inverse to determine the general formula for the determinant of a $3 \times 3$ matrix. Verify your answer, as in Ex. 19.1.

## Exercise 19.4

Pick one term in your determinant formula from Ex. 19.3 and determine the position in $A$ of each factor. Repeat this process with another term. Form a general observation regarding the factors of *any* term that appear in a determinant formula. Contrast the terms that are in this formula with expressions in the entries of $A$ that do *not* appear in the formula. Verify that your conjecture accurately describes each term in the formulas from Ex. 19.1 and Ex. 19.3.

## Exercise 19.5

Based on on your conjecture from Ex. 19.4, write down a possible term from the formula for the determinant of a general $4 \times 4$ matrix. Determine the total number of terms in this formula. How many terms would be in the formula for the determinant of a $5 \times 5$ matrix? A $6 \times 6$ matrix?

Now let $A$ be a $1 \times 1$ matrix, say $A = [a]$.

## Exercise 19.6

By hand, determine $A^{-1}$ and hypothesize a formula for Det[$A$]. Verify that your formula conforms to the general patterns you observed in Ex. 19.4.

## 19.2.2   Tricks for Small Matrices

For the moment, we focus on our formulas for the determinant of a $2 \times 2$ or $3 \times 3$ matrix and demonstrate two simple tricks for remembering these formulas that are often presented in a basic algebra text. These tricks are suggested by the following pictures:

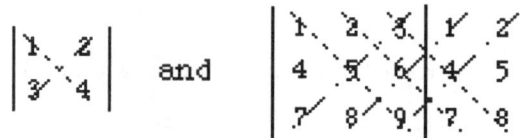

To compute the determinant, in either case, one should:

1. multiply along diagonal lines,

2. add the terms corresponding to the going down and to the right (with the smaller dashes), and subtract the terms for the lines going up and to the right (with the larger dashes).

It is a common mistake for students to assume that these rules for $2 \times 2$ and $3 \times 3$ matrices will generalize to higher dimensions. However, any such simple-minded scheme will not yield enough terms to obtain the correct determinant formula. Although these rules seem similar at first, they are really quite different and do not conform to any general pattern. For example, the second rule requires us to recopy the first two columns of the matrix, while the first does not. Because they do not work in general, these rules are only "tricks."

### *Drill Exercise 19.7*

For practice, record two $2 \times 2$ matrices and two more $3 \times 3$ matrices. Use these tricks to compute the determinant of each matrix, showing all work. Verify each of your answers with *Mathematica*.

## 19.2.3   The Determinant of a Permutation

In general, Det[$A$] is the sum of a number of products of the entries of A, where each term has sign +1 or $-1$. Although we have only examined two- and three-dimensional matrices, the patterns we have observed regarding the terms which appear in any determinant formula suggest a clear generalization to higher dimensions. Once we deduce the formula for $|P|$ for any permuation matrix $P$, we will be able to describe the general determinant formula in detail. We will infer the determinant of an arbitrary permutation from the fact that the determinant function "preserves" products. This fundamental property may be expressed as "the determinant of the product is the product of the determinants."

We first examine the simplest permutation, namely, the identity matrix $I$. Since $I = II$, we may take determinants of both sides to obtain:

$$|I| = |II| = |I||I|$$

The second equality follows from our fundamental property. This implies that $|I|$ must be a real number whose square is itself, that is, a solution of the equation:

$$x^2 = x \qquad \Longleftrightarrow \qquad 0 = x^2 - x = x(x-1)$$

Thus, $|I|$ is either 1 or 0. Since $I$ is clearly invertible (it is its own inverse!), we must have $|I| = 1$.

A similar argument applies if we replace $I$ by the next simplest type of permutation, namely a transposition $T$ (see Definition 17.1). In this case we have $TT = I$, so that $|T||T| = 1$, and $|T| = 1$ or $-1$.

### Exercise 19.8

Determine the $2 \times 2$ matrix that interchanges the two rows and compute its determinant. Now compute the determinant of the $3 \times 3$ matrix that interchanges the second and third rows. Form a conjecture regarding the determinant of *any* transposition.

Assuming that our conjecture is correct, we may now infer the determinant of an *arbitrary* permutation.

### Exercise 19.9

Consider an arbitrary permutation matrix $P$, given as a product of transpositions:

$$P = T_1 T_2 \cdots T_{n-1} T_n$$

Compute the determinant of $P$, as a formula in $n$. *Hint*: Use the fact that the determinant preserves products, and the result of Ex. 19.8.

Notice how surprising it is that we may even obtain such a formula for $|P|$. Any given permutation may be written a product of transpositions is a number of different ways, thus $n$ in your formula is not a well-defined quantity. For example:

- 
```
P  = {{0,0,1},{1,0,0},{0,1,0}};
T1 = {{0,1,0},{1,0,0},{0,0,1}};
T2 = {{1,0,0},{0,0,1},{0,1,0}};
T3 = {{0,0,1},{0,1,0},{1,0,0}};
P == T1.T2 == T2.T3 == T2.T1.T3.T2
```

It may be shown, however, that the number of transpositions in any decomposition must always *differ* by an *even* number. If the number of transpositions used to express $P$ is even, we say that $P$ is an "even" permutation, while if this number is odd, we say that $P$ is "odd." This property of a permutation is known as its "parity" and is indicated by the determinant. If $P$ is even then $|P| > 0$, while if $P$ is odd then $|P| < 0$.

## 19.2.4   The General Definition

We may now explain how the signs in the general determinant formula arise, as well as describing the general determinant formula. Re-examining the formula for a $3 \times 3$ matrix, we notice that the fourth term, for example, comes with a negative sign:

$$-(a_{1,3}\ a_{2,2}\ a_{3,1}) + a_{1,2}\ a_{2,3}\ a_{3,1} + a_{1,3}\ a_{2,1}\ a_{3,2} -$$

$$a_{1,1}\ a_{2,3}\ a_{3,2} - a_{1,2}\ a_{2,1}\ a_{3,3} + a_{1,1}\ a_{2,2}\ a_{3,3}$$

We may assume, for the moment, that $A$ is the matrix:

$$\begin{bmatrix} a_{1,1} & 0 & 0 \\ 0 & 0 & a_{2,3} \\ 0 & a_{3,2} & 0 \end{bmatrix}$$

so that all but the fourth term drops out of our formula, leaving:

```
Det[A] == -(a     a     a   )
             1,1   2,3   3,2
```

On the other hand, we may express $A$ as the product:

$$\begin{bmatrix} a_{1,1} & 0 & 0 \\ 0 & a_{2,3} & 0 \\ 0 & 0 & a_{3,2} \end{bmatrix} \begin{bmatrix} 1 & 0 & 0 \\ 0 & 0 & 1 \\ 0 & 1 & 0 \end{bmatrix}$$

Since we know that the determinant of the permutation on the right is $-1$, this calculation suggests two facts.

**Observation:**

(1) The sign of any given term in a determinant formula is the same as the determinant of the corresponding permutation matrix.

(2) The determinant of a diagonal matrix is the product of its nonzero entries.

Now we may finally give an operational definition for the determinant of an arbitrary, square matrix.

**Definition 19.1** The *determinant* of an $n \times n$ matrix $A$ may be computed as follows:

(a) Construct all possible terms formed by multiplying n entries of A chosen from distinct rows and columns,

(b) Multiply each term by the sign of the determinant of the corresponding permutation, and

(c) Add all the resulting terms together.

## Exercise 19.10

If an $n \times n$ matrix $A$ has a row or column of all 0's, explain why Det[$A$] must be 0, by appealing to Definition 19.1.

## 19.3    The Geometry of Determinants

There are a number of applications of the determinant function. In this section, we will discover that the *magnitude* of the determinant of a $2 \times 2$ matrix gives the *area* of the parallelogram bounded by the columns of the matrix, or equivalently, the factor by which the matrix scales any given unit area. We will also observe that the *sign* of the determinant distinguishes the natural *orientation* of the columns as an ordered set of vectors. Although we will only examine pairs of two-dimensional column vectors, these results generalize directly to triples of three-dimensional vectors, and so on. The results of this section will not be needed in the remainder of the text, so this material may be omitted.

### 19.3.1    The Magnitude of the Determinant

Consider the following $2 \times 2$ matrix:

- ```
  M = {{-1,1},{1,1}}; MatrixForm[M]
  ```

This transforms the unit square in the $x$-$y$ plane to a parallelogram in the $u$-$v$ plane:

- ```
  CoordinatePlot[M.{x,y},{x,0,1,1},{y,0,1,1}];
  ```

Notice how the two columns of $M$, call them $v_1$ and $v_2$, determine the result in the $u$-$v$ plane (compare Section 9.5), and vice versa. *Note*: This is the last time we will remind the reader to be sure to start *Mathematica* by opening "Start.ma."

- ```
  V = {v1, v2} = Transpose[M];
  ```

### Exercise 19.11

Compute the area of this parallelogram by analytic geometry, explaining your reasoning. Compare your result with the determinant of $M$.

This example was relatively easy, because $v_1$ and $v_2$ were perpendicular. We may repeat our observations with a more interesting set of vectors, and their associated transformation.

### Exercise 19.12

Repeat Ex. 19.11 with the set:

- ```
  W = {w1, w2} = {{5,3},{2,8}}; L = Transpose[W];
  CoordinatePlot[L.{x,y},{x,0,1,1},{y,0,1,1}];
  ```

Specifically, draw a picture of these two vectors, determine the coordinates of the remaining corner of the corresponding parallelogram, and compute the area of this figure by analytic geometry and trigonometry. Compare this area with the determinant of $L$.

Notice that, since the area of the unit square in the $x$-$y$ plane is 1, we may consider the magnitude of the determinant as representing the factor by which the area in the $x$-$y$ plane is multiplied to obtain the area in the $u$-$v$ plane.

### *Advanced Exercise 19.13*

Prove that the magnitude of determinant of the matrix transformation determined by two vectors in the plane will always equal the area of the parallelogram with those two vectors as sides. *Hint*: You may wish to use the dot product formula Eq. 27.2 from Lesson 27 and the Law of Cosines from trigonometry.

Although the same results will hold in higher dimensions, a proof using basic, analytic geometry would be quite difficult. However, once we discuss Gram-Schmidt orthonormalization in Section 29.5, we will provide a direct, geometric proof of this result.

## 19.3.2    The Sign of the Determinant

We have seen that the *magnitude* of the determinant of a matrix $M$ may be interpreted by considering the associated transformation of the plane. Specifically, Det[$M$] represents the factor by which the transformation scales any area as it is mapped by $M$ from the $x$-$y$ to the $u$-$v$ plane. It remains to discuss the meaning of the *sign* of Det[$M$]. This is related to the idea of "orientation," usually presented informally in physics in terms of the so-called right-hand rule.

For example, the pair of vectors

- ```
  S = {v1, v2} = {{1,1},{0,1}}; Show[Vector[S]];
  ```

in the plane is said to have a "positive," or "right-handed," orientation. One may explain this terminology in two ways. Geometrically, these two vectors are in the same relative position as the positive $x$- and $y$-axes. That is, the second is rotated in a counterclockwise direction from the first. Or, to put it another way, the standard basis vectors in the plane, $\{1, 0\}$ and $\{0, 1\}$, lying on the positive $x$- and $y$-axes, respectively, may be continuously "deformed" to obtain $v_1$ and $v_2$, respectively.

This geometric definition may be hard to verify in practice, which is why the determinant is useful here. Algebraically, $S$ is positively oriented if the associated matrix tranformation, which takes the standard basis vectors to $v_1$ and $v_2$, respectively, has a positive determinant:

- ```
  M = Transpose[S];Det[M]
  ```

In contrast, we would say that $S$ has a "negative" or "left-handed" orientation if this determinant were negative. If a set of vectors is negatively oriented, it is difficult to prove that they may *not* be continuously deformed onto the positive axes of the standard coordinate system. This result is a theorem from topology, an advanced branch of mathematics, which is beyond the scope of this text. That is why the determinantal characterization of orientation is so convenient.

For example, consider the three vectors in space:

- ```
  U = {u1, u2, u3} = {{1,0,0},{0,-1,0},{0,0,1}};Show[Vector[U]];
  ```

it would be difficult to show the positive $x$-, $y$-, and $z$-axes may not be continuously deformed in space until they coincide with $u_1$, $u_2$, and $u_3$, respectively. However, it is easy to verify that the determinant of the associated transformation is negative:
  ```

● `M = Transpose[U];Det[M]`

It may be proven that "positive" and "negative" sets of vectors differ only by a reflection. For example, one may observe that $U$ is the result of reflecting the standard basis vectors $\{1, 0, 0\}$, $\{0, 1, 0\}$, and $\{0, 0, 1\}$ through the $x$-$z$ plane.

These ideas are often applied to rectangular coordinate systems, by associating the positive axes of a coordinate system with column vectors. For example, the coordinate system on the left is associated with the vectors $\left\{ \begin{bmatrix} -1 \\ 0 \end{bmatrix}, \begin{bmatrix} 0 \\ 1 \end{bmatrix} \right\}$:

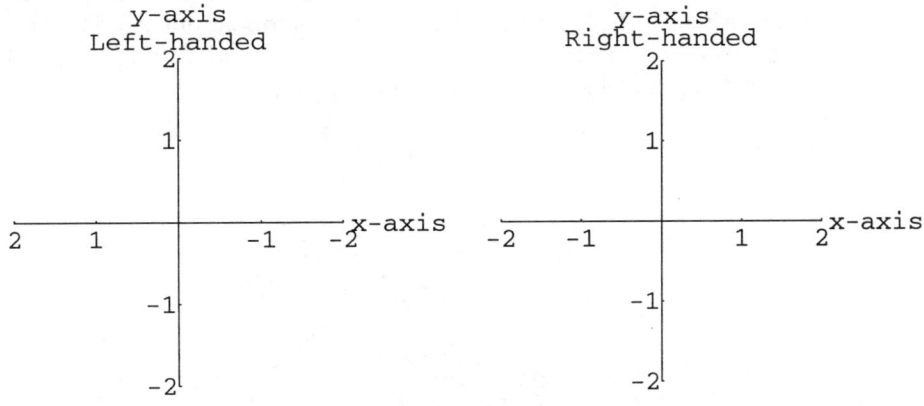

These two coordinate systems may not be superimposed via rotation, because the one is the reflection through the $y$-axis of the other and thus they have opposite orientation.

Likewise, the coordinate system:

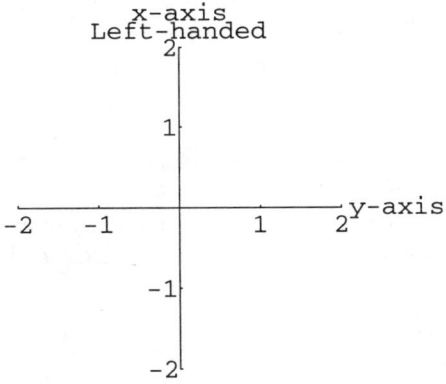

is associated with the set of vectors:

● `T = {{0,1},{1,0}};`

and the matrix transformation:

●     `M = Transpose[T]; MatrixForm[T]`

We have seen this matrix before in Lesson 9. It corresponds to reflection through the $y = x$ line. Thus, this coordinate system is "left-handed," as well. This would suggest that it *may* be rotated to obtain the previous left-handed coordinate system. One may easily verify that a 90° rotation will transform the one coordinate system to the other.

## 19.3.3   An Application to Calculus

One of the most common uses of the determinant occurs when performing a change of variables in a multiple integral. For example, suppose one wants to compute the area of the region $R$ bounded by the ellipses, $u^2 + 2v^2 = 4$ and $u^2 + 2v^2 = 9$, and the hyperbolas, $4u^2 - 8v^2 = 1$ and $4u^2 - 8v^2 = 9$, in the $u$-$v$ plane:

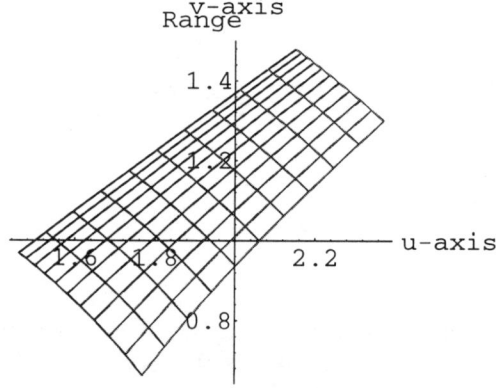

A rough visual estimate would say that this is approximately a rectangle that is 0.8 long and 0.4 tall with an area of about 0.32. The double integral

$$\iint_R dx\,dy$$

provides an exact expression for this area. However, in the $u$-$v$ plane, the limits of integration are rather difficult to calculate.

    With an appropriate change of variables, however, this region becomes easier to manage. For this we take $\{u, v\} = f[x, y]$, where $f$ is the transformation:

●     
```
Clear[x,y,u,v];
f[x_,y_] := {Sqrt[(y^2 + x^2)/2], Sqrt[x^2 - y^2]/2};
{u, v} = f[x, y];
```

With this change of variables, this complicated region in the $u$-$v$ plane is the image of a *rectangular* region in the $x$-$y$ plane

●     `CoordinatePlot[f[x,y],{x,2,3,.1},{y,.5,1.5,.1}];`

and we would express this as the integral:

$$\int_{.5}^{1.5} \int_{2}^{3} dx\,dy$$

Unfortunately, this integral would not yield the correct area, because we have not taken into account how this coordinate transformation distorts areas. Specifically, this integral sums infinitely many, infinitely small $dx$-by-$dy$ squares, called "area elements," with infinitesimal area $dA = dx\,dy$. As one may observe from the CoordinatePlot[], this change of variables distorts squares in the $x$-$y$ plane into parallelogram-like shapes in the $u$-$v$ plane. In fact, we saw in Lesson 9 that, for such infinitely small squares, this distortion at any given point is given by the matrix transformation $D(f)$.

- ```
  df[x_,y_] = {{D[u,x], D[u,y]},{D[v,x], D[v,y]}};
  MatrixForm[df[x,y]]
  ```

Together with our observations on determinants, we know that the area of the image of the area elements in the $x$-$y$ plane will have area equal to the absolute value of the determinant of $D(f)$ times $dx\,dy$. This multiplier is called the "Jacobian" of $f$ and written $J(f)$.

- ```
  jacobian = Abs[Det[df[x,y]]]
  ```

Combining terms by hand and integrating gives the actual area:

- ```
  NIntegrate[(x y)/Sqrt[2(x^4 - y^4)],{x,2,3},{y,.5,1.5}]
  ```

There are a number of other applications of the determinant. For example, it is a convenient tool for expressing many formulas from analytic geometry. We will encounter the determinant again when trying to solve differential equations. In particular, we will use it to compute "eigenvalues" (compare Section 18.2 and Lessons 32–33).

## 19.4  Computing Determinants

So far we have discovered the definition of the determinant and what it represents. Unfortunately, this definition is not very helpful for practical computations. For example, the determinant formula for a $10 \times 10$ matrix will have 3,628,800 distinct terms! Thus, we must prove some theorems to speed our calculations.

### 19.4.1  The Determinant of a Triangular Matrix and Elimination

One simple theorem concerns lower-triangular matrices (see Ex. 15.17).

#### Exercise 19.14

Create a $3 \times 3$ lower-triangular matrix with the maximum number of nonzero entries and compute its determinant. Conjecture a rule for computing the determinant of *any* lower- or upper-triangular matrix and explain why this should work in general, by appealing to Definition 19.1.
  ```

Using Ex. 19.8, we may compute the determinant of any transposition. Since they are lower-triangular, Ex. 19.14 shows how to compute the determinant of the remaining two types of elementary matrices.

The familar process of Gaussian elimination provides the most efficient technique for computing the determinant of a matrix $A$. We will first analyze the effect of each type of elementary matrix on the determinant of $A$. Specifically, we know that any step of Gaussian elimination may be accomplished via multiplication by an elementary matrix $E$ to obtain a new matrix $U = EA$. In the next few exercises, we analyze the relationship between $|A|$ and $|U|$ for each of the three types of row operations.

For example, consider the following matrices:

- 
```
A = {{3,6},{3,4}}; e = {{1/3,0},{0,1}}; U = e.A;
Print["A = ",MatrixForm[A],"    E = ",MatrixForm[e],"    U = ",MatrixForm[U]];
```

Here $E$ represents the "divide row 1 by 3" operation. Notice how the determinant of $A$ is related to the determinant of $U$:

- 
```
Print["|A| = ",Det[A]," and  |U| = ", Det[U]]
```

## Exercise 19.15

In general, for any choice of $A$, if $E$ represents any "divide" operation, prove that $|A| = (\text{divisor})|U|$.

## Exercise 19.16

Repeat Ex. 19.15 to determine corresponding formulas relating $|A|$ and $|U|$ for each of the other two types of elementary row operations.

Eventually, Gaussian elimination will obtain an upper-triangular matrix $U$, whose determinant is easy to compute, and we may use our observations in the previous exercises to work backwards and infer $|A|$.

## 19.4.2  The Determinant of an Arbitrary Matrix

In Section 19.4.1, we discussed the relationship between Gaussian elimination and determinants. By employing the $P^T LU$ decomposition and the fact that the determinant preserves products, we may give a general formula for the determinant, which is more computationally accessible than the original definition. Because this decomposition formula contains a transpose operation, we must pause to examine the relationship between deteminants and the transpose operator.

## Exercise 19.17

Create an arbitrary $3 \times 3$ lower-triangular matrix. Compare its determinant with that of its transpose. Form and prove a conjecture about the determinant of an arbitrary lower-triangular matrix and that of its transpose. Repeat this exercise with upper-triangular matrices.

## Exercise 19.18

Consider an arbitrary permutation matrix $P$

$$P = T_1 T_2 \cdots T_{n-1} T_n$$

expressed as a product of transpositions. Show that the determinant of the transpose of $P$ equals the determinant of $P$.   *Hint*: Use the fact that both determinants and transposes preserve products.

Using the results of the previous exercise and our earlier observations on permuations and triangular matrices, we have the following theorem:

**Theorem 19.2** *If a square matrix may be decomposed as $A = P^T LU$, where $P$ is a permutation, $L$ is lower-triangular, and $U$ is upper-triangular, then $|A| = |P||L||U|$, where:*

(a) $|P| = (-1)^{(number\ of\ row\ interchanges)}$,

(b) $|L| = product\ of\ divisors\ used\ during\ Gaussian\ elimination,\ and$

(c) $|U| = 1\ or\ 0,\ depending\ on\ whether\ or\ not\ U\ has\ 0's\ on\ its\ diagonal.$

## Exercise 19.19

- Generate a "nice" example of a $3 \times 3$ matrix, using the MakeNiceMatrix[] command that was introduced in Lesson 17:

- ```
  A = MakeNiceMatrix[3,3]; MatrixForm[A]
  ```

Compute the decomposition of $A$ and use it to compute $|A|$. Verify your answer by calculating $|A|$ another way.

Our theorem prompts two brief observations. First, it becomes clear from this formula that $A$ is invertible iff $|A|$ is not 0. The formula implies that $|A|$ is not zero precisely when its row echelon form $U$ has no zero rows, a condition that is equivalent to invertibility. It is further apparent that the *entire* decomposition is unnecessary to compute $|A|$. One only must record the *number* of row interchanges used, the divisors used to obtain each leading 1, and the rank of $A$.

Theorem 19.2 also allows us to prove a general relationship between $|A|$ and $|A^T|$.

## Exercise 19.20

Take an arbitrary $2 \times 2$ matrix and compare its determinant with the determinant of its transpose. Prove that what you observe in this case is a totally general phenomena.   *Hint*: Use the $P^T LU$ decomposition formula.

# 19.5   Three Theoretical Formulas

In this section, we prove three well-used formulas involving determinants. This section does *not* appear in the Notebook version of this lesson. These formulas are not of any practical, computational value, but do have many *theoretical* uses. We will not need them for the remainder of the text, so one may skip this section without consequence.
  ```

## 19.5.1   The Cofactor Expansion Formula

As it stands, Definition 19.1 produces a very messy formula that is not very useful in practice. However, by grouping terms in this formula, we may derive a commonly used inductive formula for the determinant of a matrix $A$ in terms of determinants of its submatrices, called "minors." This formula is known as a "Laplace expansion" or a "cofactor expansion."

Given an $n \times n$ matrix, we define its $k \times k$ minors (for $k < n$) as the determinants of all possible submatrices obtainable through removing all but $k$ of its rows and columns. The $(n-1) \times (n-1)$ minors, achieved by removing a *single* row and column, are the most commonly used and appear in the Laplace expansion formula. Notice that each such minor may be associated with the entry that is in both the row and column being removed. We may represent such a minor pictorially by crossing out the respective row and column from the original matrix. For example to obtain the $(2,3)$ minor of the following matrix, we draw a cross over the $(2,3)$ entry and compute the determinant of the remaining matrix, as follows:

$$\begin{bmatrix} 1 & 2 & 3 \\ 4 & 5 & 6 \\ 7 & 8 & 9 \end{bmatrix} \longrightarrow M_{2,3} = \begin{vmatrix} 1 & 2 \\ 7 & 8 \end{vmatrix} = 8 - 14 = -6$$

Observe how we refer to the $(2,3)$ minor as $M_{2,3}$. We will continue to use this notational convention for $(n-1) \times (n-1)$ minors throughout the text.

### Exercise 19.21

Write down the general formula for the determinant of a $3 \times 3$ matrix. For each of the entries in the first row:

(a) Group the terms with that entry as a factor, and factor out that entry.

(b) Identify the terms with which that entry is multiplied as a minor times 1 or $-1$. Describe the pattern of signs and minors that results.

We may reason that this type of grouping in the formula may be accomplished for an arbitrary $n \times n$ matrix $A$ with entries $a_{i,j}$. We know that every term in the determinant formula arises by multiplying entries from distinct rows and columns. Thus, each term has a factor $a_{1,j}$ from the first row, for some $j$. One may then group together all terms that contain $a_{1,1}$. From the remainder, one could then group those containing $a_{1,2}$, etc., until every term has been collected into some grouping. From the first grouping, we would factor out $a_{1,1}$, from the second $a_{1,2}$, etc., until we have the factorization $a_{1,1}(\cdots) + \cdots a_{1,n}(\cdots)$ computed in Ex. 19.21a.

If we now examine each part of this factorization more carefully, we will discover the minors $M_{1,j}$. Although it would be difficult to prove that all of the signs work out correctly, it is relatively simple to reason that we obtain all of the correct terms. Look, for example, at the second part, which is of the form $a_{1,2}(\cdots)$. Definition 19.1 stipulates that the factors of any term in the determinant formula must come from *distinct* rows and columns. In particular, none of the remaining factors in $(\cdots)$ may be from the first row or the second column. If we multiply out $a_{1,2}(\cdots)$, we must obtain all possible choices of entries from distinct rows and columns that contain $a_{1,2}(\cdots)$. Therefore, $(\cdots)$ will contain all possible choices of entries with distinct rows and columns that are *not* from the first row or the second column. These are precisely the terms in $M_{1,2}$.

With a little bit of care we could convince ourselves that the signs work out as well. However, at this point, a formal proof would not be much more convincing than the example we have already examined. The pattern from Ex. 19.21 suggests the general formula:

$$|A| = \sum_{i=1}^{n} (-1)^{i+1} a_{1,i} M_{1,i}$$

This formula is called a "cofactor expansion along the first row." We refer to a minor determinant $M_{i,j}$ with its associated sign $(-1)^{i+j}$ as a "cofactor" and denote it by $C_{i,j}$. With this notation, we may express our cofactor expansion more simply as:

$$|A| = \sum_{i=1}^{n} a_{1,i} C_{1,i}$$

If we place the minors of $A$ in a matrix $M$ and the cofactors of $A$ in a matrix $C$, it becomes easier to visualize the relationship between the two concepts. $C$ is the result of multiplying each entry of $M$ by the corresponding sign which appears in the following "checkerboard" pattern:

$$
\begin{array}{cccc}
+ & - & + & - \\
- & + & - & + \\
+ & - & + & - \\
- & + & - & +
\end{array}
$$

## Exercise 19.22

Verify that the $(i, j)$ entry of this "sign matrix" equals $(-1)^{i+j}$, by selecting three different entries and comparing each sign with that given by the formula.

It is clear that Ex. 19.21 could be repeated with any row or column. Alternatively, we have seen in Ex. 19.20 that that determinants are "invariant" ("in" = "not," "vary" = "change," i.e., "unchanging") under transposition, so we may transpose this formula to give a similar cofactor expansion for the first column. Moreover, by interchanging rows, we obtain expansions along *any* row (or column), provided we introduce the proper sign changes. Either argument leads to the general "cofactor expansion" formulas.

**Theorem 19.3** *If $A$ is any square matrix, $|A| = \sum_{j=1}^{n} a_{i,j} C_{i,j}$ and $|A| = \sum_{i=1}^{n} a_{i,j} C_{i,j}$. The first equation is known as a* column *expansion, while the second is a* row *expansion.*

### *Drill Exercise 19.23*

Create a $3 \times 3$ matrix and compute a cofactor expansion along two different rows and columns. Compare your answers with the actual determinant.

## 19.5.2   The Determinant Formula for an Inverse

So far, we have only examined the formula for $A^{-1}$ to identify the determinant as its common denominator. Using the cofactors matrix, we may now identify the numerators of $A^{-1}$, as well.

### Exercise 19.24

Write down the general formula for the inverse of a $2 \times 2$ matrix from Ex. 19.1, factor out the common denominator (namely the determinant) from all the entries, and write this as a scalar product of the reciprocal. Describe what is left in terms of the matrix of cofactors of $A$.

### Exercise 19.25

• Repeat Ex. 19.24 with a $3 \times 3$ matrix. You need not record every term, if you can identify the correct pattern from one or two terms. You may find it helpful to use *Mathematica* to compute the matrix of $2 \times 2$ minors of $A$ with the command:

• `Reverse[Map[Reverse, Minors[A,2]]]`

*Note*: Unfortunately, the Minors[] command does not produce the entries in the correct order, which is why we need to use the Reverse[] command to reorder them.

### Exercise 19.26

Give a general desciption of your results from Ex. 19.24 – 19.25 that will encompass both the $2 \times 2$ and $3 \times 3$ cases.

Using the cofactor expansion formula, we may show that the pattern of Ex. 19.26 generalizes directly to higher dimensions. This will also lead to an explicit formula for the inverse of a matrix in Theorem 14.2. To prove this, we will need a simple lemma. We use the term "lemma" to refer to a theorem when we want to emphasize that it is mainly used to prove another, more useful theorem.

**Lemma 19.4** *Assume that $A$ is an $n \times n$ matrix.*

(a) *If $A$ has a row of 0's, then $|A| = 0$.*

(b) *If $A$ has two rows that are equal, then $|A| = 0$.*

**Proof:**

In Ex. 19.10, we deduced Lemma 19.4a directly from Definition 19.1. It also follows easily from the $P^T LU$ decomposition of $A$. The final matrix $U$ in Gaussian elimination must also have a row of 0's. This means that it will be upper-triangular with a 0 on the diagonal, and so its determinant is 0. This implies that $|A| = |P||L||U| = |P||L|0 = 0$. Similarly, for Lemma 19.4b, if $E$ represents the row operation "subtract one row from the other" using the two identical rows, the result of this, call it $A' \equiv EA$, will then have a row of 0's. From Lemma 19.4a, we know that $|A'| = 0$. But we also have $|A'| = |E||A| = |A|$, since $E$ will be lower-triangular with 1's down the diagonal. Thus, $|A| = 0$. Q.E.D.

Now we define the "adjoint" of $A$ as the transpose of the matrix of cofactors, and denote it as $adj(A)$. Formally, we make the following definition:

**Definition 19.2** If $C$ is the matrix of cofactors of $A$, then we define the *adjoint* of $A$, $adj(A)$, as $adj(A)_{ij} = C_{ji}$.

With this notation, the equation from Theorem 19.3 becomes:

$$|A| = \sum_{i=1}^{n} a_{i,j} \, adj(A)_{j,i} \tag{19.1}$$

We may now prove the following:

**Theorem 19.5** *For any square matrix $A$, we have $adj(A)\,A = |A|I = A\,adj(A)$. In particular, $A$ is invertible iff $|A| \neq 0$ with $A^{-1} = \dfrac{adj(A)}{|A|}$.*

**Proof:**

We begin by computing $A\,adj(A)$. We must show that $(A\,adj(A))_{ii} = (|A|I)_{ii} = |A|$ and that $(A\,adj(A))_{ij} = (|A|I)_{ij} = 0$, when $i \neq j$. The first case follows immediately from the definition of matrix multiplication and Eq. 19.1:

$$(A\,adj(A))_{ii} = \sum_{j=1}^{n} a_{i,j}\,adj(A)_{j,i} = |A|$$

The other case, when $i \neq j$, is a bit trickier. We reason indirectly by considering the matrix $A'$ formed by replacing row $j$ of $A$ by row $i$. It is clear that $adj(A')_{k,j} = adj(A)_{k,j}$ for all $k$, since they are computed from exactly the same minors. In particular, a cofactor expansion along row $j$ of $A'$ yields:

$$|A'| = \sum_{k=1}^{n} a'_{j,k}\,adj(A')_{k,j} = \sum_{k=1}^{n} a_{i,k}\,adj(A)_{k,j} = (A\,adj(A))_{ij}$$

However, since rows $i$ and $j$ are equal, Lemma 19.4b says that $A'$ is 0, which is what we needed to show.

We have now shown that $A\,adj(A) = |A|I$. We could simply repeat the previous argument, using column expansions instead of row expansions to show the reversed equality $|A|I = adj(A)\,A$. When $|A| \neq 0$, we may divide by $|A|$ to give $A\,\frac{adj(A)}{|A|} = I$, i.e., $\frac{adj(A)}{|A|}$ is a right-inverse of $A$. By the Fredholm Alternative (Theorem 16.5), we have that $A^{-1} = \frac{adj(A)}{|A|}$. Q.E.D.

## Exercise 19.27

For a given matrix $A$, we have discussed four associated matrices: the matrix of minor determinants $M$, the cofactor matrix $C$, the adjoint matrix $adj(A)$, and the inverse of $A$. Summarize the results of this section by describing the relationship between these four matrices, both in words and in formulas.

### *Drill Exercise 19.28*

Create a $3 \times 3$ matrix $A$ and compute the associated matrices $M$, $C$, and $adj(A)$. Verify that $A\,adj(A) = det(A)I$ and $adj(A)\,A = det(A)I$. Is $A$ invertible? If so, what is $A^{-1}$?

## 19.6   Summary

In this lesson we have discussed how:

 – the determinant operator preserves products,

 – it takes invertible matrices to nonzero numbers,

 – subtracting a multiple of a row from another row or taking the transpose does not affect the determinant of a matrix,

 – interchanging rows of a matrix multiplies its determinant by $-1$,

 – multiplying a row of a matrix by a number multiplies its determinant by the same number,

 – there are two general formulas for the determinant, namely the cofactor expansion formula and the definition, and some handy tricks for small matrices, but the *best* way to compute determinants (and inverses) is by *Gaussian elimination*, and

 – we may give an explicit formula for the inverse in terms of determinants.

In general, it is easier to remember the procedure for Gaussian elimination than to memorize the two determinant formulas. One may also count the length of time (i.e., number of arithmetic operations) necessary to compute determinants and inverses by elimination and by any general formula. Such an exercise will verify that Gaussian elimination is *much faster*, as well.

Now with a good deal of computational experience and some basic definitions under our belts, we move on to the heart of linear algebra and discuss the notions of "vector space," "subspace," and "linear transformation." We will encounter yet another (geometric) view on the problem of solving systems of linear equations and prepare to discover the "real" story behind matrices and matrix multiplication.

# Systems of Equations from a Geometric Viewpoint

## Accompanied by Notebook

So far we have used matrices to solve systems of equations. We have also learned to view matrices both algebraically, as generalized numbers, and geometrically, as linear transformations. This suggests that we examine the process of solving systems of equations transformationally, that is, from a geometric point of view. This approach will lead to new insights regarding the existence, uniqueness, and structure of solutions to linear systems, as well as a greater understanding of matrix invertibility. Because "transformation" is simply another word for "function," we must approach the familiar notion of a function from a new geometric point of view. Although to this point we have only visualized the geometry of $2 \times 2$ matrices, after this lesson, we will be able to imagine the effect of a general $n \times m$ matrix.

Specifically, in this lesson we will:

- learn a new geometric language for discussing functions,

- learn to view functions and matrices from a novel geometric perspective,

- observe that a linear function is particularly well-behaved, in that it is largely determined by two sets, known as its "kernel" and "image,"

- describe the process of solving systems of equations and computing inverses from this new perspective, and

- discover a new geometric interpretation for the rank of a matrix.

In order to make the notions of kernel and image concrete, we will also explore the connection between the kernel and image of a linear transformation and the notion of linear combinations from Lesson 13. This will provide a convenient algebraic characterization of these ideas to supplement our geometric understanding. Our philosophy throughout this text is that any useful concept can and should be understood from many different points of view.

In the long run, this lesson will equip us with greater intuition regarding matrices and prepare us for a general discussion of linear transformations, bases, and subspaces.

In the short run, it will suggest a geometric definition of rank that will lead to an intuitive, geometric proof that $rank(A) = rank(A^T)$ in Lesson 28. In Lesson 16, we saw that many results on matrix inverses depend on this equation. From a geometric perspective, this collection of results on matrix inverses will become clear. In fact, many of the results on inverses actually have little to do with "linear" algebra but are very general results about *any* system of equations. Once we examine the process of equation solving from a functional point of view and understand more fully what a function is, these results will become quite intuitive.

## 20.1   Equation Solving and Functions

We begin by examining the process of solving a single equation in a single unknown. For example, consider the equation,

$$\frac{2}{5}x^2 - 5 = 1.4$$

This has the form $f(x) = y$, where $f(x) = \frac{2}{5}x^2 - 5$ and $y = 1.4$. Traditionally, we would visualize the solutions of this equation as the intersection of the horizontal line $y = 1.4$ with the graph $y = f(x)$:

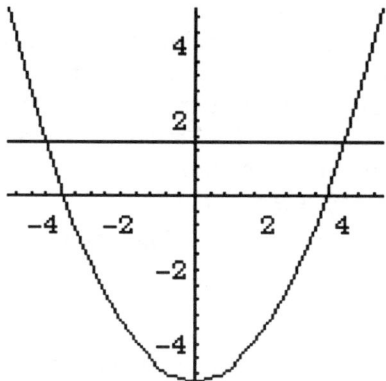

- *Note*: You may construct this picture in the accompanying *Mathematica* Notebook. From this picture one may determine the solutions $x = -4$ and $x = 4$. One may also predict the qualitative nature of solutions as we vary the right-hand side of the equation. If $y > -5$, the equation $f(x) = y$ will have two solutions, while if $y < -5$, there are no solutions. In the special case that $y = -5$, there will be exactly one solution.

  Instead of viewing $f(x)$ statically as a graph, we may instead visualize this function as a transformation of the real line onto itself. Traditionally, we relate the $x$- and $y$-values of a function by taking any given $x$-value, traveling vertically until we reach the graph of the function, and then proceeding horizontally until we hit the $y$-axis. An ani-
- mation of this familiar process is provided in the accompanying Notebook. In this way, the graph of a function associates each point in the domain (i.e., $x$-axis) with a unique point in the range (i.e., $y$-axis).

  Although it is most common to visualize a function in this way, acting one point at a
- time, *Mathematica* allows us to view this as a *global* process on the *entire* real line. There

is another animation in the accompanying Notebook that illustrates this point of view. View that animation now to see how $f(x)$ transforms the entire real line into itself.

Looking at this process, notice how it consists of two separate stages. At first, the $x$-axis is deformed vertically until it coincides with the graph of $f(x)$. This will usually cause some stretching of the line at some points and shrinking at others. Then, the graph is compressed horizontally until it coincides with a subset of the $y$-axis. This will eventually cause some points to be "collapsed" together. Notice how each point, $x$, on the original line eventually is transformed to its corresponding value, $f(x)$. For example, 1 ends up on $f(1) = \frac{2}{5}(1)^2 - 5 = -4.6$, 3 stops at $f(3) = \frac{2}{5}(3)^2 - 5 = -1.4$, and 5 is placed on $f(5) = 5$.

From this transformational point of view, we will ultimately forget that the process occurred in two dimensions and consider that we have simply mapped the $x$-axis onto itself. We refer to the original $x$-axis as the "domain" of $f(x)$ and the second copy as the "codomain" or "range" of $f(x)$. Notice how this is the same terminology we used in Lesson 9 in connection with matrix transformations. We provide a plot from this

• perspective, as well. Notice again how pairs of points in the domain are collapsed together into single points in the range. From this point of view, the function $f(x) = \frac{2}{5}(x)^2 - 5$ folds the domain in half, distorting it a bit in the process, mapping the result onto the range so that it covers every point greater than $-5$. Specifically, the folding process collapses points together that have the same value under $f(x)$. Just as before, this illustrates how each point in the range greater than $-5$ comes from precisely two points in the domain.

For example, the point $-1.4$ is mapped onto by 3 and $-3$ (i.e., $f(3) = f(-3) = -1.4$). Geometrically, we say that $-1.4$ is the "image" of 3 and $-3$ (see Lesson 9), or conversely that the set $\{3, -3\}$ is the "pre-image" of $-1.4$ under $f(x)$. *Note*: This is just another way of saying that $\{3, -3\}$ is the solution set of $f(x) = \frac{2}{5}(x)^2 - 5 = -1.4$. Likewise, the point $-5$ is mapped onto exactly once by $f(x)$, while no point less than $-5$ is "hit" by the function.

## Exercise 20.1

What is the image of 6 under $f(x)$? What is the pre-image of 9.4? What is the pre-image of $-5.4$? *Hint*: First translate each question into an equation.

We refer to the set of all values achieved by $f(x)$ in the range as the "image of $f(x)$" and denote it by $Image(f)$. Notice how we use the term "image" in two different senses, depending on the context. The image of a *single point* in the domain is a *single point* in the range, while the image of the entire *function* is a *set of points* in the range. For example, since $f(-1) = -4.6$, we would say that "$-4.6$ is the image of $-1$ under $f(x)$" or "$f(x)$ maps $-1$ to $-4.6$" or "$-1$ maps to $-4.6$ under $f(x)$" or "$-1$ hits $-4.6$ via $f(x)$." *Note*: When referring to a function, you may refer to it as $f(x)$ or simply $f$. In contrast, the image of the function $f$ is the set of *all* possible images of individual points under $f$. In the previous example, $Image(f) = \{x \mid -5 \le x\}$. The notion of the image of a function should not be entirely new to you, in that it is discussed in most algebra II or precalculus texts.

### Exercise 20.2

- If $g(x) = -2|x - 1| + 5$, describe geometrically, as we did for $f(x)$, the effect of $g$ on the real line. What is $Image(g)$? *Note*: The command to plot this in *Mathematica* is already entered for you in the accompanying Notebook.

We may now describe the process of solving the equation $f(x) = y$ in an entirely different way. For example, if we have a point $y$ in the range of $f(x)$, we solve $f(x) = y$ for the *pre-image* of $y$. Likewise, we may describe the effect of any function $f(x)$ as in two stages, as collapsing each pre-image to a single point and then mapping the results in a 1–1 fashion onto the image of $f(x)$.

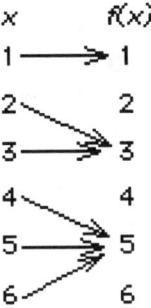

Figure 20.1: Geometry of a Function of a Single Variable

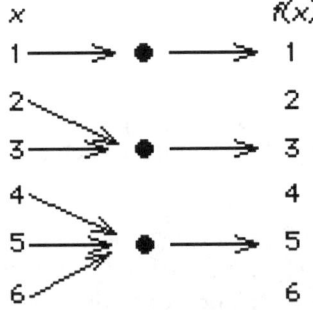

Figure 20.2: A Function in Two Steps

For example, consider the function indicated by Fig. 20.1. We may view this as consisting of two steps, as in Fig. 20.2. In the first step, each pre-image, $\{1\}$, $\{2, 3\}$, and $\{4, 5, 6\}$, collapses to a single point. In the second, the resulting three points map in a 1–1 fashion onto $Image(f) = \{1, 3, 5\}$ as a subset of the range. In Section 20.2, we will continue to explore this point of view for functions of more than one variable. We will specifically examine *linear* functions and informally discover a number of properties of the solution sets of linear systems. Much of the remainder of this text will revolve around providing precise proofs of the insights this geometric picture suggests.

## 20.2  A More General Description

The pictures that we have used so far to visualize the action of a function have been limited to one dimension, because the functions we have examined take in a single real number and map that to a single real number. We denote such a function, $f(x)$, by:

$$f : R \rightarrow R$$

This notation is designed to indicate the domain and range of $f$. Notice that this notation does not show specifically *how* $f(x)$ transforms its inputs to outputs. A detailed description of the mapping must also be given separately.

To solve a higher-dimensional system of equations, such as

$$
\begin{aligned}
x^2 + y^2 + z^2 &= 25 \\
3x - 4y + 5z &= -1
\end{aligned}
$$

we should describe this in a manner similar to the one-dimensional example from Section 20.1. However, now our function would be described as:

$$f : R^3 \rightarrow R^2$$

That is, in this example $f$ takes in three real numbers and gives out two real values. Specifically, $f$ takes $(x, y, z)$ to $f(x, y, z) = (x^2 + y^2 + z^2, 3x - 4y + 5z)$. We may use geometric language to describe this problem Section 20.1. We wish to solve for all choice of inputs $(x, y, z)$ so that $f(x, y, z) = (25, -1)$, that is, all of the points in the pre-image of $(25, -1)$. Although the concept is exactly the same, *visualizing* this process is slightly more difficult than in our previous example because the domain and range involve so many dimensions.

In order to discuss systems of equations in many variables geometrically, we must supply a new picture that will effectively describe a transformation of *any* dimension. No matter their dimension, the domain and range of a function are fundamentally *sets* of numbers. A generic set is usually pictured as a big "blob" so that a general function may be described pictorially as in Fig. 20.3.

More precisely, we may indicate how each pre-image collapses to a single point in the image of $f(x)$, as in Fig. 20.4. Just as in Fig. 20.2, the pre-images of a function may come in various shapes and sizes, sometimes consisting of just a single point. Although we have not drawn them all, every point in the domain is in *some* pre-image. Thus, one should consider the domain as completely partitioned into pre-image sets. The function has the effect of collapsing each pre-image set into a single point and then mapping that set of results into the range. Unlike the domain, many points in the range may be "missed" by the function. That is, the image of a function is usually a proper subset of the range.

When the function comes from a system of *linear* equations, the picture is particularly

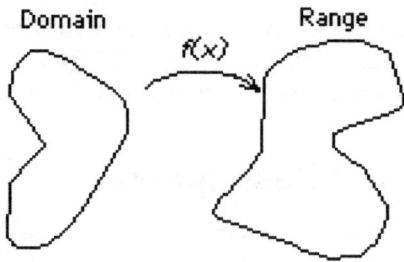

Figure 20.3: Geometry of a General Function

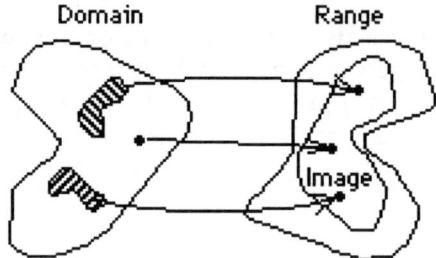

Figure 20.4: Detailed Geometry of a General Function

simple. For example, consider the system:

$$
\begin{aligned}
4x + 6y &= 2 \\
6x + 9y &= 3
\end{aligned}
\tag{20.1}
$$

This may be written as $f(x, y) = (2, 3)$ for $f : R^2 \rightarrow R^2$, $f(x, y) = (4x + 6y, 6x + 9y)$. *Aside*: In Lesson 21, we will use $R^3$ to refer to three-dimensional column vectors. It is common also to use this notation, as we do in this section, to refer to the set of 3-tuples of real numbers, and to refer to this set as "three-dimensional Euclidean space." It represents our usual notion of "space" and is pictured using the standard $x$-$y$-$z$ coordinate axes. Although $R^2$ represents the usual $x$-$y$ coordinate plane, it is still referred to as "two-dimensional Euclidean *space*." In this way, we use a consistent terminology, no matter what dimension we are discussing.

Because $f$ is a *linear* function, it may be written in matrix notation:

$$
f(x, y) = \begin{bmatrix} 4 & 6 \\ 6 & 9 \end{bmatrix} \begin{bmatrix} x \\ y \end{bmatrix}
$$

That is, a linear function is given by matrix multiplication. We have already examined the effect multiplication by a 2 × 2 matrix has on a two-dimensional picture. Now we re-examine this situation more generally, in terms of our new functional picture. We will see precisely how pre-image sets from the domain collapse into single points in the range. The *Mathematica* command MatrixPicture[] generates a picture similar to Fig. 20.4 for a 2 × 2 matrix. Use this command now to view the effect of $\begin{bmatrix} 4 & 6 \\ 6 & 9 \end{bmatrix}$. It is already typed into the accompanying Notebook. Notice how it plots each pre-image and its corresponding image point in the same color.

## 20.3   Observations on Linear Functions

Unlike non-linear functions, all (nonempty) pre-image sets for a matrix *look very similar!* In the example from Section 20.2, all pre-images are straight lines with slope $-\frac{2}{3}$. In particular, they are all *parallel*, so that every pre-image is simply a translated version of any other. This implies that, if we solve one system of equations involving $f$, we may obtain the general solution of any other system by determining a *single* solution $X$ and "translating" the solution to the previous system so that it contains $X$.

We usually focus on the pre-image through the origin in the domain. This pre-image through the origin must map to the origin in the range. That is because $A \cdot 0 = 0$ for any matrix $A$. Thus, this particular pre-image is the solution to:

$$\begin{aligned} 4x + 6y &= 0 \\ 6x + 9y &= 0 \end{aligned} \qquad (20.2)$$

This is easily solved to give $x = -\frac{3}{2}y$ or $y = -\frac{2}{3}x$, that is, the line through the origin with slope $-\frac{2}{3}$, as we have already observed. Since this particular solution set to a large extent determines all of the others, it is called the "kernel" of $f$ and denoted $Kernel(f)$. This line can also be expressed as $(x, y) = (-\frac{3}{2}y, y) = y(-\frac{3}{2}, 1)$. Such an equation is known as a "parametric equation," in that it traces out the entire line as the "parameter" $y$ varies over all possible values.

We may solve for the image of $f$, as well. In this case it is also a straight line with slope $\frac{3}{2}$ through the origin. Since this line contains the point $(2, 3)$, we can conclude that Eq. 20.1 has a solution. In fact, it is easily verified that $(x, y) = (\frac{1}{2}, 0)$ is a particular solution to Eq. 20.1. Moreover, because we know that the pre-image of each point is a line, we know that there is a one-dimensional "infinity" of solutions. That is, if we perform Gaussian elimination, we would expect the general solution to have one free variable. We may almost give the general solution from the picture alone. In that the solution set (i.e., the pre-image) is a line with slope $-\frac{2}{3}$, the general solution can be written as $y = -\frac{2}{3}x +$ ___ or $x = -\frac{3}{2}y +$ ___, depending on which variable one wishes to consider as free. Since we know that $(\frac{1}{2}, 0)$ is a particular solution, we may plug-in to determine that $x = -\frac{3}{2}y + \frac{1}{2}$. Parametrically, the general point in the pre-image looks like $(x, y) = (-\frac{3}{2}y + \frac{1}{2}, y) = (-\frac{3}{2}y, y) + (\frac{1}{2}, 0) = y(-\frac{3}{2}, 1) + (\frac{1}{2}, 0)$.

### Exercise 20.3

- Use Gaussian elimination to verify that this is the general solution to Eq. 20.1. Compare this solution set to $Kernel(f)$ both geometrically and algebraically.

### Exercise 20.4

Reason as above to determine the general solution to $\begin{aligned} 4x + 6y &= -12 \\ 6x + 9y &= -18 \end{aligned}$ *without* performing Gaussian elimination.

### Exercise 20.5

Does $\begin{matrix} 4x + 6y & = & 8 \\ 6x + 9y & = & 8 \end{matrix}$ have any solutions? Explain your answer geometrically by appealing to the picture for $f$ as a transformation from its domain to its range. If we picked the numbers on the right-hand side of our equations at random, would you tend to expect there to be solutions or not? Explain.

Notice how the kernel and the image of a linear function $f$ allow us to characterize the solutions to any system of equations involving $f$. In particular, we suggest answers to our two fundamental questions on existence and uniqueness of solutions to a system of equations:

**Observation:**

(1) A system of $n$ equations in $m$ unknowns has at least one solution if and only if the right-hand side of the equations, viewed as a point in $R^n$, lies in the image of the associated linear function $f : R^m \to R^n$.

(2) Such a system will have unique solutions if and only if the kernel of $f$ contains exactly one point, that is, the only point that maps to the origin in the range is the origin in the domain. In this case, we say that the kernel is "trivial."

## 20.4   The Kernel, Image, and Linear Combinations

To this point, we have pictured the kernel and image of a linear function from the plane into itself, that is, corresponding to two equations in two unknowns. Although it becomes more difficult to picture more equations in more unknowns, the geometric intuition that we have gained from the previous example will not mislead us. If we learn to manipulate points in higher dimensions algebraically, we may reason about solutions sets in precisely the same ways as in Section 20.3.

For example, consider the system of equations:

$$\begin{matrix} -2x + y + z & = & 1 \\ x - 2y + z & = & 1 \\ x + y - 2z & = & -2 \end{matrix} \qquad (20.3)$$

This corresponds to a function $h : R^3 \to R^3$, that is, a function of 3-space to itself.

### Exercise 20.6

Give a formula for $h(x, y, z)$. Show that the kernel of $h$ may be expressed parametrically as $(x, y, z) = (z, z, z)$, where $z$ is a free parameter. *Hint*: The kernel is the pre-image of the origin $(0, 0, 0)$, so you may solve the corresponding system of equations. Now express the general solution to Eq. 20.3 in parametric form. This is the pre-image of the point $(1, 1, -2)$. How do these two pre-images compare algebraically? How is this similar to our $2 \times 2$ example?

Notice that even though the solution sets for our two- and three-dimensional examples reside in different dimensions, they are similar in that they both contain infinitely many points. Moreover, in each case there is only one free variable, so they must be one-dimensional.

## Exercise 20.7

• What do you think a linear one-dimensional set of points looks like? How do you think the two pre-images from Ex. 20.6 should compare *geometrically*? The accompanying Notebook allows you to use *Mathematica* to verify your intuition.

## Exercise 20.8

From our geometric description of a linear function, the image is a copy of the domain after collapsing down each pre-image to a point. Since each pre-image of $h$ is identical to the kernel and hence one-dimensional, how many dimensions will the image of $h$ have? How do you think it will look?

Algebraically, the points in the image of $h$ are obtained by multiplying points $(x, y, z)$ from the domain, written as a column vectors $\begin{bmatrix} x \\ y \\ z \end{bmatrix}$ by the matrix $\begin{bmatrix} -2 & 1 & 1 \\ 1 & -2 & 1 \\ 1 & 1 & -2 \end{bmatrix}$.

We may descibe this in terms of linear combinations, as in Lesson 13. Specifically, every point in the image is of the form:

$$\begin{bmatrix} -2 & 1 & 1 \\ 1 & -2 & 1 \\ 1 & 1 & -2 \end{bmatrix} \begin{bmatrix} x \\ y \\ z \end{bmatrix} = x \begin{bmatrix} -2 \\ 1 \\ 1 \end{bmatrix} + y \begin{bmatrix} 1 \\ -2 \\ 1 \end{bmatrix} + z \begin{bmatrix} 1 \\ 1 \\ -2 \end{bmatrix}$$

That is,

**Observation:**

Each point in the image of $h$ may be produced as a linear combination of the columns of the matrix for $h$.

Put another way, the image is the set of points generated in a linear fashion from the set of points $S = \left\{ \begin{bmatrix} -2 \\ 1 \\ 1 \end{bmatrix}, \begin{bmatrix} 1 \\ -2 \\ 1 \end{bmatrix}, \begin{bmatrix} 1 \\ 1 \\ -2 \end{bmatrix} \right\}$. Geometrically, we may consider

$Image(h)$ as the set of all points that can be reached from the points in $S$ by forming linear combinations (see Ex. 13.5). We say that $S$ "spans" $Image(h)$ or that $Image(h)$ is the "span" of $S$.

For reference, denote the vectors in $S$ by $v_1$, $v_2$, and $v_3$, respectively. Now notice that

$$(-1)v_1 + (-1)v_2 = (-1) \begin{bmatrix} -2 \\ 1 \\ 1 \end{bmatrix} + (-1) \begin{bmatrix} 1 \\ -2 \\ 1 \end{bmatrix} = \begin{bmatrix} 1 \\ 1 \\ -2 \end{bmatrix} = v_3. \text{ This implies that } v_3$$

may be dropped from the set $S$ *without decreasing the span*, since one may immediately recover $v_3$ from $v_1$ and $v_2$. We say that $v_3$ "depends linearly" on $v_1$ and $v_2$, or that $v_3$ is "linearly dependent" on $v_1$ and $v_2$. Geometrically, this implies that one may picture $Image(h)$ as all possible linear combinations of $v_1$ and $v_2$:

$$Image(h) = S = \left\{ x \begin{bmatrix} -2 \\ 1 \\ 1 \end{bmatrix} + y \begin{bmatrix} 1 \\ -2 \\ 1 \end{bmatrix} \;\middle|\; x \text{ and } y \text{ are any real numbers} \right\}$$

In particular, since this set of points depends on two arbitrary real numbers, it should be two-dimensional. We may use *Mathematica* to visualize this set of points using the SpanningPicture[] command.

### Exercise 20.9

•   Use the SpanningPicture[] command to see a selection of points in the span of $v_1$ and $v_2$. What would we get if we took *all* possible linear combinations of $v_1$ and $v_2$? Does this agree with your answer to Ex. 20.8? Explain.

We should also observe that the kernel of $h$ may be described using the language of linear combinations as well. We know that a point $(x, y, z)$ is in the kernel of $h$ if it maps to the origin. This means that:

$$\begin{bmatrix} 0 \\ 0 \\ 0 \end{bmatrix} = \begin{bmatrix} -2 & 1 & 1 \\ 1 & -2 & 1 \\ 1 & 1 & -2 \end{bmatrix} \begin{bmatrix} x \\ y \\ z \end{bmatrix} = x \begin{bmatrix} -2 \\ 1 \\ 1 \end{bmatrix} + y \begin{bmatrix} 1 \\ -2 \\ 1 \end{bmatrix} + z \begin{bmatrix} 1 \\ 1 \\ -2 \end{bmatrix}$$

that is, we may combine the vectors in $S$, using the coordinates of our point as coefficients, to obtain the 0 vector.

### Advanced Exercise 20.10

There is a direct connection between the fact that the vectors in $S$ are linearly dependent and that the kernel of $h$ is nontrivial (i.e., consists of more than just the 0 vector). Explain.

## 20.5   The Geometry of Inverses

The geometric perspective on functions given by Fig. 20.4 also may be related to the problem of computing inverses. For example, a left-inverse for a function $f : X \to Y$ is a function $g : Y \to X$, so that $g(f(x)) = x$. You should remember from algebra II that applying one function directly after another is called "composition" and "$g \circ f$" stands for the composition of $f$ and $g$. Thus, if we define the identity function by the formula $Id(x) \equiv x$, then the defining equation for a left-inverse becomes $g \circ f = Id$. With this notation, the notion of a left-inverse of a function strongly resembles the corresponding notion for matrices. That is because $Id(x)$ is linear *function* that corresponds to the identity *matrix*.

In layman's terms, $g$ "cancels out" or "undoes" the effect of $f$. Specifically, if $y = f(x)$, then $g(y) = g(f(x)) = x$. Using the language of this lesson, we observe that if $f(x)$ has a left-inverse, the pre-image of any point $y$ in the image of $f$ consists of a *single* point $g(y)$. In other words:

**Observation:**

A function $f$ has a left-inverse if and only if $f(x) = y$ has at most one solution for each $y$, that is, $f$ is injective.

Notice that we are using the same terminology that we introduced for matrices. We have already given a formal proof of this fact in Theorem 16.2b in the context of linear systems. Intuitively, if any pre-image of $f$ *did* consist of *multiple* points, $f$ would collapse them together. Just as one may not separate the whites from the yolk after one has scrambled an egg, there is no way to reverse the collapsing process. Thus, if $f$ has a left-inverse, such a situation could not happen, and each pre-image must contain at most one point.

Just as for matrices, we say that such a function is "injective." Since it cannot collapse any points in $X$, the function "injects" an "identical" copy of the domain $X$ into $Y$ (although $X$ may get a little *distorted* in the process). Geometrically, we picture this as in Fig. 20.5. This picture implies that all pre-images are single points. If $f$ is given by a matrix, this situation occurs precisely when $Kernel(f)$ is trivial.

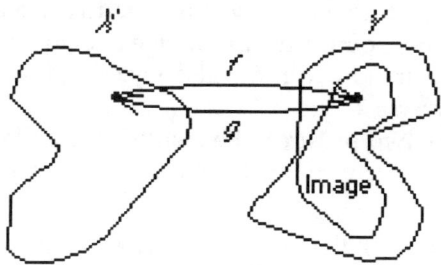

Figure 20.5: Geometry of a Left-Invertible Function

From the picture we can infer two things about the nature of any left-inverse of an injective function. On the one hand, there is only one possible way to define $g(y)$ for each $y$ in the image of $f$. On the other hand, we are free to define $g$ on the remainder of $Y$ however we like. This corresponds to our observation in Lesson 15 that left-inverses are not usually unique. However, if $Image(f) = Y$, then there is no ambiguity in the definition of $g$, $f$ is in fact a 1–1 correspondence, and $g$ is a two-sided inverse.

Similarly, a right-inverse for a function $f : X \to Y$ is a function $g : Y \to X$, so that $f \circ g = Id$. This implies that for each $y$ in the range of $f(x)$, $g(y)$ is a point in the pre-image of $y$ under $f$. Since $g$ must assign a value to *every* $y$ in $Y$, we cannot have any empty pre-images. In other words, $f(x) = y$ must have a solution for every $y$. In this case, the image of $f$ equals the entire range $Y$ and $f$ is "onto" or "surjective." Again, we use the same terminology as for matrices. The term "surjective" comes from the French, "sur" meaning "onto." We may picture this situation as in Fig. 20.6.

Theorem 16.2a expressed this connection between right-invertibility and surjectivity for matrices. In functional terms, this theorem says:

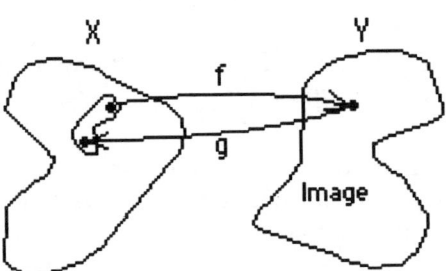

Figure 20.6: Geometry of a Right-Invertible Function

**Observation:**
A function $f$ has a right-inverse if and only if $f(x) = y$ has at least one solution for each $y$, that is, $f$ is surjective.

In other words $f$ has a right-inverse iff $Image(f) = Y$.

From this picture, one may conclude that we are free to define $g$ at any particular point $y$ to be *any* point in the pre-image of $y$ under $f$. Thus, left-inverses are not usually unique either. If, however, $f$ is also injective, then $f$ is a 1–1 correspondence and $g$ is a two-sided inverse.

From these two pictures, the results of Corollary 16.3 become intuitively clear. An $n \times m$ matrix $A$ corresponds to a linear function $f : R^m \to R^n$, and there are two cases to consider.

- If $m > n$, there must be some collapsing to map from a higher-dimensional space to a lower-dimensional one; thus, the kernel cannot be trivial, $f$ cannot be injective, and $f$ cannot have a left-inverse.

- If $m < n$, we are mapping a from a lower-dimensional space to a higher one; thus, $f$ cannot map the domain onto the entire range; that is, $f$ cannot be surjective and $f$ cannot have a right-inverse.

In the same way, Corollary 16.5, the so-called Fredholm Alternative, becomes clear. This assumes that $m = n$, so that both the domain and range are the same dimension.

- If a matrix $A$ is injective, there is no collapsing and so it must map the domain onto the entire range; that is, $A$ is also surjective.

- If $A$ is surjective, there could not have been any collapsing and $A$ must be injective as well.

- In either case, $A$ is a 1–1 correspondence, and both left- and right-inverses are unique and equal to the unique two-sided inverse.

## 20.6  The Geometric Significance of Rank

By examining a number of two-dimensional examples, one may easily determine a geometric definition for the rank of a matrix. A number of sample matrices are already typed in *Mathematica*.

### Exercise 20.11

• Experiment with MatrixPicture[] to complete the following chart:

| Rank | Dimension of Image | Dimension of Kernel |
|------|--------------------|--------------------|
| 0    |                    |                    |
| 1    |                    |                    |
| 2    |                    |                    |

Describe the three geometrically possible shapes for an image or kernel of a 2 × 2 matrix.

### Exercise 20.12

Give a rule relating the dimensions of the image and the kernel of a matrix to its dimensions and rank. Try to generalize to the case of a non-square matrix, say a 2 × 3 matrix. *Hint*: Think about our geometric description of a matrix.

### *Advanced Exercise 20.13*

Use your rule from Ex. 20.12 and the observations from this lesson to explain the results of Theorem 16.4.

## 20.7   Summary

In this lesson, one should have learned:

- how we may visualize any function in terms of collapsing groups of points, called pre-images, in the domain to single points and then mapping the results in a 1–1 fashion into the range,

- the definitions of the kernel and image of a matrix $A$,

- how all the pre-images of a matrix $A$ should look and how they are related, as well as how the image of $A$ should look,

- how $Kernel(A) = \{ X \mid AX = 0 \}$ corresponds to the uniqueness of solutions to any system of equations $AX = B$, as well as left-inverses of $A$,

- how $Image(A) = \{ Y \mid AX = Y \text{ for some } X \}$ corresponds to the existence of solutions to $AX = B$, as well as left-inverses of $A$,

- how $Kernel(A)$ and $Image(A)$ may be described in terms of linear combinations of the columns of $A$, and

– how the rank of $A$ is related to the dimensions of $Kernel(A)$ and $Image(A)$.

This again shows how linear functions are so much nicer and more orderly than general non-linear functions. This is why matrices are used so often in applications in science, engineering, and economics.

The kernel and image of a matrix are two important examples of "linear subspaces." In a certain sense, they are the only examples of subspaces. Although these two examples are very simple, they provide very good insight into the nature of linear subspaces. We have already seen how these subspaces relate to the existence and uniqueness of solutions to linear systems of equations. In Lesson 28, we will see how knowledge of the image will also suggest how to construct *approximate* solutions even for overdetermined systems (using projections, or the so-called Method of Least Squares). This in itself would be a sufficient payoff for our effort to gain a geometric understanding of systems of equations, but we have obtained so much more. Now we have an easy way to remember the connection between rank and solvability. This picture also provides us with a new perspective on left- and right-inverses.

Intutively, based on the two examples of this lesson, we should realize that subspaces look like points, lines, planes, etc. through the origin. In order to provide a precise definition of a linear subspace, we must supply a formal definition of a "vector space" in Lesson 21. The abstract concept of a vector space will allow us to greatly widen the applicability of linear algebra. We will see how linear algebra may be applied in a great many different settings, because, despite superficial dissimilarities, many applications are all fundamentally the same.

# PART FIVE

# Abstract Vector Spaces and Bases

# Vectors and Vector Spaces

In previous lessons, we have given an *intuitive* introduction to the notions of "linear subspaces" and "linear transformations." We saw in Lesson 20 that geometrically low-dimensional linear subspaces are lines and planes through the origin. In Lesson 9, we examined concrete examples of geometric transformations, such as rotations and reflections in the plane. Later, in Lessons 18 and 20, we observed that any linear transformation may be viewed essentially as a collapse followed by a 1–1 correspondence, followed by an inclusion. In order to provide rigorous, mathematical definitions of a linear subspace and a linear transformation, we must first define what we mean by a "linear space" or "vector space."

Simply put, a vector space is a "complete" collection of vectors. In this lesson, we will discuss three examples of vectors that may seem, on the surface, to be rather different. With experience, one notices that all of these examples have an underlying similarity. We will discuss how to describe such similarities as a collection of properties or "axioms." These axioms are together known as the "vector space axioms," and *any* example that shares these properties may be legally referred to as a "vector space." This is the "abstract," mathematical definition of a vector space.

Much of the impetus for such an abstract definition came from physics. Physicists, such as Einstein in his theory of relativity, have realized that there is no "universal" system of measuring time and space. Therefore, this theory requires one to represent physical quantities in a "coordinate-free" manner and/or a clear description of how measurements of physical quantities depend explicitly on a *choice* of coordinate system. This abstract definition of a vector space provides just such a tool for coordinate-free descriptions, as well as for a careful analysis of coordinate systems. Ultimately, it also widens the applicability of the concepts of linear algebra.

## 21.1 Three Examples of Vectors

The notion of a "vector" includes a number of very different-looking examples. This means that the techniques and theorems of linear algebra may be applied to many different disciplines from statistics to differential equations, from analyzing experimental data to building better car-suspension systems! However, in all of these examples, a vector has the same basic properties. Fundamentally, anything that one may add or multiply in a reasonable way may be referred to as a vector, and a vector space is a "complete" collection of such objects. For the remainder of this section, this will serve as our working definition of a vector space. The three most important examples of vectors are:

1. column vectors,

2. geometric vectors, and

3. functions.

This section is designed to introduce these three examples by allowing you discover the rules for addition and multiplication in each example.

## 21.1.1  Column Vectors

Our first example is that of "column vectors". We denote the set of $n$-dimensional column vectors by $R^n$, since they consist of $n$ real numbers. We have already seen these in Lesson 13. These are the easiest type of vector with which to work, since they are simply matrices with one column. We may define row vectors in a similar fashion, but we will work mainly with column vectors. We may add, subtract, and multiply column vectors by scalars, just as we do with matrices. Since this is already familiar, we will not belabor this example right now. However, bear in mind that it is the most important example, in that it is so concrete and convenient for computations.

Although we have presented matrices first and column vectors second, logically, column vectors and their rules for arithmetic come first. We will see in Lesson 30 that a matrix represents a collection of column vectors. This implies that the arithmetic for matrices follows naturally as a consequence of that for column vectors. However, even the rules for column vectors were originally suggested by looking at something else, namely, geometric vectors, which we discuss in our next example.

## 21.1.2  Geometric Vectors

Most trigonometry, calculus, or physics texts introduce the notion of a "geometric vector." This often gives the impression that these are the *only* example of vectors. One of the goals of this lesson is to dispel this mistaken belief. Moreover, it is designed to demonstrate how the arithmetic for geometric vectors follows naturally from their geometric and physical interpretation, and are not simply made-up arbitrarily. We will distinguish a geometric vector from a column vector in the text (as opposed to graphics) by employing the traditional arrow notation. That is, while we might refer to a column vector as $v$, we will tend to refer to a geometric vector as $\vec{v}$.

Historically, geometric vectors were the first type of vectors to be studied. They arise very naturally in physics to represent displacement, velocity, acceleration, forces, and so on. They are pictured as directed line segments or "arrows," such as:

or

In fact, the word "vector" comes from the German word for "arrow," motivated by this example. Notice that a geometric vector has a specific length and direction. The length and direction are used to distinguish one vector from another, and that *alone*! In other words, even if two arrows start at different points, if they have the same length and direction, we will think of them as the *same* vector. Consider for example, the arrows $\vec{v}$ and $\vec{w}$:

Because they both have the same length and point in the same direction, we will consider them as representing the same vector, much in the same way that we consider $\frac{1}{2}$ and $\frac{2}{4}$ as representing the same fraction.

These previous examples of geometric vectors were meant to lie in the plane. We may picture three-dimensional geometric vectors, that is in space, as well. To make this clear, we usually draw in $x$-, $y$-, and $z$-axes as a frame of reference:

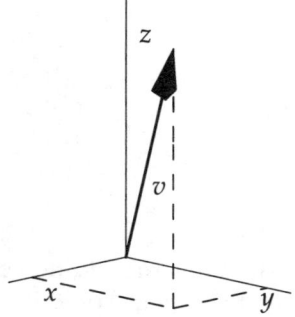

We will be able to illustrate most concepts with vectors in the plane, employing three dimensions only when necessary. *Aside*: Although we may *conceive* of vectors in four or more dimensions, it is difficult to *visualize* them. We will denote the set of geometric vectors in $n$-dimensional space by $E^n$. We use the letter $E$ after "Euclid," the Greek mathematician who is called the "father of geometry."

For such arrows to be honest-to-goodness vectors, we must be able to add and multiply them. There is only one possible set of definitions for addition and multiplication that is consistent with the physical interpretations of displacement, velocity, and so on. The next three exercises suggest how to add and multiply geometric vectors.

Consider three points in the plane $A$, $B$, and $C$. Let $\vec{v}$ represent the displacement of traveling from $A$ to $B$. Similarly, let the geometric vector $\vec{w}$ represent moving from $B$

to $C$. We may visualize this situation as:

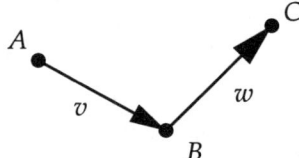

Clearly, the sum $\vec{v} + \vec{w}$ should be the geometric vector of the total displacement from $A$ to $C$.

## Exercise 21.1

On a sheet of paper, draw and clearly label $\vec{v}$, $\vec{w}$, and your conjecture for the sum $\vec{v} + \vec{w}$. Describe the addition process in words.

## Exercise 21.2

In the windy city of Chicago, if one is walking at a given speed in a given direction, represented by the vector $\vec{v}$, and a strong gust of wind blows across one's path, indicated by the vector $\vec{w}$, one will be deflected off to the right. What direction would one actually travel? Carefully sketch a copy of this situation on a separate piece of paper and draw in your resulting velocity, given by the sum $\vec{v} + \vec{w}$.

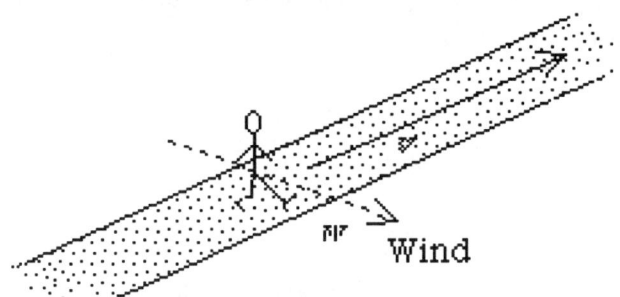

Revise your description of the addition process from Ex. 21.1 so that it applies to this more general situation. *Hint*: Would it help to reposition the vectors?

## Exercise 21.3

One would expect the difference $\vec{u} = \vec{v} - \vec{w}$ to be a geometric vector $\vec{u}$ so that $\vec{u} + \vec{w} = \vec{v}$. For example, consider the vectors:

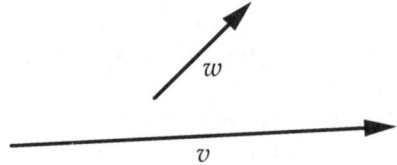

On a separate piece of paper, draw what you think the vector $\vec{u} = \vec{v} - \vec{w}$ should be and clearly label it. *Hint*: Formulate a conjecture for $\vec{u}$ and verify whether or not $\vec{u} + \vec{w} = \vec{v}$. Describe the subtraction process in words.

The next exercise concerns "scalar multiplication." If Frank has velocity $\vec{v}$, and Barbara is going twice as fast, her velocity should be $2 \cdot \vec{v}$.

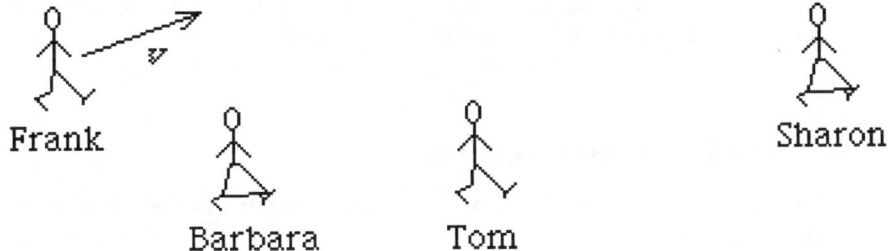

### Exercise 21.4

Make a copy of $\vec{v}$ on a separate piece of paper and carefully sketch in Barbara's velocity vector and label it. *Hint*: If our multiplication is reasonable, we should have $2 \cdot \vec{v} = \vec{v} + \vec{v}$. Repeat for Tom, who is going $\frac{1}{2}$ as fast with velocity $\frac{1}{2} \cdot \vec{v}$. Sketch in Sharon's velocity vector, if her velocity is $(-1) \cdot \vec{v}$. Describe the effect of scalar multiplication on geometric vectors, in words.

## 21.1.3 Polynomial Functions

Our final commonly used example is that of polynomial functions. Given two polynomials such as $P(x) = x^2 + 3x - 1$ and $Q(x) = 2x^2 - x + 4$, basic algebra determines how to compute their sum $P + Q$ and the product $2 \cdot P$. According to our working definition on p. 200, we may rightfully refer to these polynomials as "vectors," since we have a good definition of addition and scalar multiplication. We denote the set of polynomials of degree less than or equal to $(n-1)$ by $P_n$. This means that $P(x) = x^2 + 3x - 1$ is in $P_3$. Although this may seem strange at first, in all of our other examples (i.e., $R^n$ and $E^n$) the superscript corresponded to the dimension of the example, and so we maintain that convention.

### Exercise 21.5

Explain why a polynomial such as $Q(x) = 2x^2 - x + 4$ might be considered "three-dimensional."

## Exercise 21.6

Compare the operations of addition and scalar multiplication of polynomials with that of column vectors.

*Note*: For simplicity, we will consider *polynomial* functions almost exclusively. However, the reader should be aware that this example may be extended to many different collections of real-valued functions. This is a very important example for the study of differential equations and in numerical analysis. For the remainder of the text, we will denote the set of all real-valued functions with domain $X$ by $F(X)$.

## 21.2   The Vector Space Axioms

To this point, we have been quite informal about the definition of a vector. We have said that a vector is "anything that may be added and multiplied in a reasonable way." The key phrase here is "in a reasonable way." In mathematics, this means that addition and scalar multiplication conform to a list of rules or axioms. These axioms are agreed upon among mathematicians by considering:

1. what we wish to be true, and

2. what is actually true in the important examples.

A "complete" collection or set of vectors of a particular type form a "vector space." *Note*: Although we use the word "space," this does not mean that vectors must be geometric objects. We often will assign a letter, such as $V$, to such a collection, so that we may refer to the entire vector space more easily. We have just seen three important examples of vector spaces:

1. the set of all column vectors of a given length,

2. the set of all geometric vectors of a given dimension, and

3. the set of all polynomials with degree less than a given natural number, with real numbers as coefficients.

The addition and scalar multiplication in these examples all satisfy the same collection of familiar-sounding axioms. Instead of simply listing them all, the following exercises will use all three of these examples to investigate some of the vector space axioms.

One of the most basic vector space axioms is that every vector space $V$ should contain a particular vector that we usually call "0," and that vector should *act* like zero. In other words, adding 0 to something should have no effect:

$$v + 0 = v, \quad \text{for any vector } v \in V$$

## Exercise 21.7

Which column vector should be called 0? Why?

Similarly, for every vector $v$, there should be another vector $w$, so that $v+w = 0$. This vector $w$ is called the "additive inverse" of $v$ and is usually denoted by $-v$ to indicate its dependence on $v$.

## Exercise 21.8

What polynomial $Q(x)$ would be the additive inverse of $P(x) = x^2 + 3x - 1$? Verify your answer.

In practice, we often abbreviate $v + -w$ as $v - w$, just as in basic algebra. The next exercise shows why this practice is justified.

## Exercise 21.9

Consider the geometric vectors from Ex. 21.3:

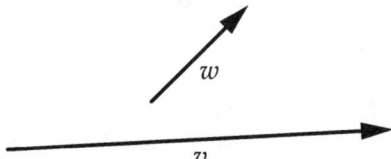

Make an accurate copy of $\vec{w}$ and $\vec{v}$ on a separate piece of paper and carefully sketch the geometric vectors $(-1)\cdot\vec{w}$ and $\vec{v}+(-1)\cdot\vec{w}$. Compare your answer for $\vec{v}+(-1)\cdot\vec{w}$ with your answer for $\vec{u} = \vec{v} - \vec{w}$ from Ex. 21.3. Is this what you would expect? Explain.

The process described in Ex. 21.9 may be carried out in every vector space. That is because of the following theorem:

**Theorem 21.1** *In any vector space $V$, $(-1) \cdot w = -w$, for any vector $w \in V$.*

In other words, scalar multiplication by $-1$ must yield the additive inverse. Although this may seem obvious, that is only because Theorem 21.1 may be proven in a number of commonly used algebraic systems. It is not an axiom; it is a theorem, because it may be *proven* from the more basic axioms, which we will supply in Definition 21.1.

Just as addition by 0 should have no effect, scalar multiplication by 1 should have no effect. That is:

$$1 \cdot v = v, \qquad \text{for any vector } v \in V.$$

## Exercise 21.10

Pick an arbitrary column vector and verify that this property of scalar multiplication by 1 is satisfied.

So far these properties may all seem pretty obvious. It is the fact that so many significant and useful results follow from such simple and intuitive axioms that makes linear
algebra of such tremendous value. As a final example, we will consider a simple axiom
that does *not* look quite as obvious in the world of geometric vectors, namely that of
"associativity." We take for granted the associative property of ordinary addition:

$$u + (v + w) = (u + v) + w \tag{21.1}$$

If one examines this equation for the following three geometric vectors

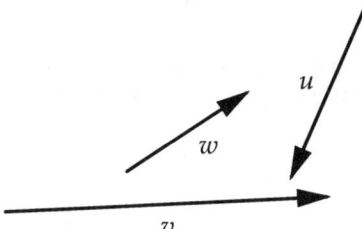

Eq. 21.1 no longer seems quite so obvious.

## Exercise 21.11

Copy $\vec{w}$, $\vec{u}$, and $\vec{v}$ onto a separate piece of paper and carefully sketch the geometric
vectors $\vec{v} + \vec{w}$ and $\vec{u} + (\vec{v} + \vec{w})$. Then sketch $\vec{u} + \vec{v}$ and $(\vec{u} + \vec{v}) + \vec{w}$. Compare
your two answers. *Note*: For this to be a useful exercise, you should try to make as
accurate a sketch as possible.

## *Advanced Exercise 21.12*

Explain geometrically why associativity will always hold for vectors in $E^2$.

We could continue to motivate all of the vector space axioms in this manner. Instead
we simpy state the complete, rigororous definition of a vector space, listing all of the
necessary axioms for future reference:

**Definition 21.1** A set $V$ is called a *vector space* (and its elements are called *vectors*) if it
is equipped with well-defined operations of:

(a) addition, $+ : V \times V \to V$ and

(b) scalar multiplication, $\cdot : R \times V \to V$

which satisfy the following axioms:

    I.   Properties of addition:

        (1)  For every $v$ and $w \in V$, $v + w = w + v$.

        (2)  For every $v$, $w$ and $u \in V$, $(v + w) + u = v + (w + u)$.

(3) There is an element $0 \in V$ so that $v + 0 = v$, for every $v \in V$.

(4) For every $v \in V$, there is an element $-v \in V$ so that $v + (-v) = 0$.

II. Properties of scalar multiplication:

(1) For every $v \in V$ and real numbers $r$ and $s$, $r \cdot (s \cdot v) = (rs) \cdot v$.

(2) For every $v \in V$, $1 \cdot v = v$.

III. Combined properties:

(1) For every $v \in V$ and real numbers $r$ and $s$, $(r + s) \cdot v = r \cdot v + s \cdot v$.

(2) For every $v$ and $w \in V$ and real number $r$, $r \cdot (v + w) = r \cdot v + r \cdot w$.

## Exercise 21.13

Determine those axioms which we have not examined in the previous exercises. Pick three such axioms and verify that they are satisfied, using any one of our three examples of a vector space. You should use each type of example at least once.

## Exercise 21.14

Match the following verbal descriptions with the corresponding axiom:

- The multiplicative identity is the natural choice.

- There are additive inverses.

- Addition is associative.

- Scalar multiplication distributes over vector addition.

- Scalar multiplication distributes over scalar addition.

- Addition is commutative.

- Scalar multiplication is associative.

- There is an additive identity.

*Hint*: The terms "associative," "commutative," and "distributive" mean the same as in high school algebra.

## *Advanced Exercise 21.15*

Give a formal proof of Theorem 21.1.

## 21.3  Summary

Although the first vectors commonly used were geometric, mathematicians observed other objects in many different areas of mathematics behaving in a similar manner. By distilling the essential similarities, they discovered the collection of axioms that define a modern vector space. Now we refer to any object that behaves as a vector should (i.e., that is part of a larger collection satisfying the axioms of a vector space), as a vector. This abstract point of view has the advantage of widening the applicability of any result that we discover. Once we prove a result in an *abstract* vector space, we may apply that result to *any* example of a vector space.

In this lesson, we have explored three main examples $R^n$, $E^n$, and $P_n$. We have discussed how to perform the fundamental operations of addition and scalar multiplication in each of these examples and have become familiar with the vector space axioms. Although we eventually wish to consider the notions of linear subspaces and linear transformations in greater detail, we will put off their formal definitions until Lesson 26. Our next major goal will be to discover how an abstract vector space may be described in concrete terms through the use of coordinates. Along the way, we will discover why the rules for addition and scalar multiplication are so similar in all of our examples. Specifically, we will learn that they are all, in a certain sense, the *same*!

# Coordinate Systems
# and Bases

*Mathematica* Notebook

## 22.1   Introduction

In the last lesson, we learned the precise definition of a vector space. We learned that a vector space is defined by its operations of addition and scalar multiplication. We also noticed that the operations of addition and scalar multiplication were suspiciously similar in all of our basic examples. This is no accident! In this lesson, we want to discuss the notions of "linear independence" of vectors, a "spanning" set of vectors, and a "basis." Once we understand these ideas, we will be able to:

- provide a more concrete description of *any* vector space,

- formulate a precise definition of the concept of dimension, and

- explain the basic similarity of *all* (finite-dimensional) vector spaces.

These are some of the most fundamental concepts of linear algebra, making this one of the most important lessons of this text. In a more traditional text, these concepts often seem difficult and abstract, even though they are really very intuitive and familiar ideas. That is because they are usually presented originally in terms of formal definitions. This text, however, will begin by relating these concepts to the familiar notions of coordinate systems and plotting points. Only after we have laid the geometric foundation will we provide the corresponding formal definitions. In the process, we will discuss what we mean by a "coordinate system" and explore some nonstandard coordinate systems for the plane.

## 22.2   Vectors in *Mathematica*

We begin by discussing how *Mathematica* manipulates each of our three examples of vectors. We will also demonstrate a few more *Mathematica* tricks.

## 22.2.1   Column Vectors

We said in Lesson 21 that column vectors are the same as matrices with a single column. Although notationally they look the same, there is actually an important conceptual distinction between these two types of objects. An $n$-dimensional column vector is an *element* of $R^n$, while an $n \times 1$ matrix represents a linear *transformation* (from $R^1$ to $R^n$). *Mathematica* maintains this distinction. For example

$$\begin{bmatrix} 1 \\ 2 \\ 3 \end{bmatrix}$$

may be represented in *Mathematica* as the column vector

-   `v = {1, 2, 3}; MatrixForm[v]`

or as:

-   `M = {{1}, {2}, {3}}; MatrixForm[M]`

Notice that the MatrixForm[] command works equally well on column vectors as with matrices. Although, we will not *write* them any differently, one should keep in mind this distinction.

     *Mathematica* will perform algebra with column vectors, just as with matrices. For example, to compute

$$3 \cdot \begin{bmatrix} 1 \\ -4 \\ 2 \end{bmatrix} + \begin{bmatrix} 1 \\ 2 \\ 3 \end{bmatrix}$$

one would enter:

-   `3 {1,-4,2} + {1,2,3}`

### Exercise 22.1

-   Evaluate this cell and verify the computer's work by hand.

## 22.2.2   Geometric Vectors

We have provided the MyArrow[] and Vector[] commands in the "GVects.m" package in order to construct geometric vectors. For example, MyArrow[$\{-2, 1\}, \{3, 2\}$] will return a *Mathematica* graphics object representing the geometric vector starting at the point $(-2, 1)$ and ending at the point $(3, 2)$. To view this object, we must use the Show[] command:

-   `Show[MyArrow[{-2,1},{3,2}],`
         `PlotRange->{{-4,4},{-4,4}}];`

Notice that all of the commands in this section work equally well for three-dimensional vectors, with the obvious modifications.

### Exercise 22.2

- Determine the correct command to plot a three-dimensional geometric vector.

   Because vectors were originally invented to be independent of any given coordinate system, it makes sense to use the Axes->None option in the Show[] command to view a geometric vector without explicit reference to a coordinate system:

- ```
  Show[MyArrow[{-2,1},{3,2}],
       PlotRange->{{-4,4},{-4,4}}, Axes->None];
  ```

   Although the same vector may be drawn anywhere in the plane, if we have imposed a coordinate system on the plane, we will often draw the vector in "standard position," that is, starting at the origin. For example:

- ```
  Show[Vector[{3,2}]];
  ```

   The Vector[] command will also take a set of vectors and attempt to scale each vector appropriately to give an aesthetically pleasing result, such as:

- ```
  Show[Vector[{{3,2},{-1,1}}], PlotRange->{{-3,3},{0,4}}];
  ```

   If one wants finer control of the output, one may use the TailWidth and HeadScale options, such as:

- ```
  u = Vector[{3,2},TailWidth->.01,HeadScale->.2];
  v = Vector[{-1,1},TailWidth->.01,HeadScale->.15];
  Show[u, v, PlotRange->{{-3,3},{0,4}},
          Axes->True, AspectRatio->Automatic];
  ```

### Exercise 22.3

Accurately copy this picture onto a sheet of paper and draw in the sum $\vec{x} = \vec{u} + \vec{v}$, clearly labeling all vectors. What are the coordinates of the endpoint of $\vec{x}$ ? Formulate a general algebraic procedure for adding two geometric vectors in standard position in the plane.

### Exercise 22.4

Include $\vec{w} = 2\vec{v}$ (in standard position) into your picture from Ex. 22.3. What are the coordinates of the endpoint of $\vec{w}$ ? Formulate a general algebraic procedure for performing scalar multiplication with geometric vectors in standard position.

### Exercise 22.5

Compare the algebra of geometric vectors in standard position with that of column vectors and polynomials, extending your comparison between column vectors and polynomials from Ex. 21.6.

*Important*: Because of this convenient correspondence between points and geometric vectors (in standard postion), we will often blur the distinction between the two types of objects.

## 22.2.3  Polynomial Functions

*Mathematica* is well suited to manipulating polynomials. One enters a polynomial just as one would write it, except one must use the Shift-6 key (^) to indicate an exponent and square brackets to indicate a function. For example, the polynomial

$$P(x) = x^2 + 3x - 1$$

would be entered as:

- `P[x] = x^2+3x-1`

### Exercise 22.6

- Evaluate this cell and compare the output with the original input.

Notice that *Mathematica* converts all input into its own "standard" form, for its own convenience, but it will not change the data in an essential way.

In *Mathematica* the "vector" operations on polynomials are performed as one would expect. For example, if

- `Q[x] = 2x^2-x+4`

we may then compute:

- `3 P[x] + Q[x]`

### Exercise 22.7

- Evaluate these two cells and describe the results.

One may always refer to the output of the *last* evaluated cell with the percent sign (%). This is a *Mathematica* abbreviation for the corresponding Out[] command. Thus, we may simplify the linear combination of polynomials, 3P + Q, with the command:

- `Simplify[%]`

### Exercise 22.8

- Evaluate this command and verify the computer's work by hand, showing all intermediate steps.

## 22.3  Linear Combinations Revisited

We have already introduced the concept of a linear combination, in the context of column vectors, in Lesson 13. Now that we know how to perform the linear operations of addition and scalar multiplication in the context of geometric vectors or polynomial functions, as well, we may *compute* linear combinations in any of our three main vector spaces, without understanding what they mean. In Lesson 20, we saw how linear combinations of column vectors arise in a natural, meaningful way, in the context of linear systems. In this section, we discuss how linear combinations with *geometric* vectors may be understood in terms of coordinate systems.

We are all familiar with the usual Cartesian coordinate system in the plane (named after the French mathematician and philosopher, Renés Descartes), given by the $x$- and $y$-axes:

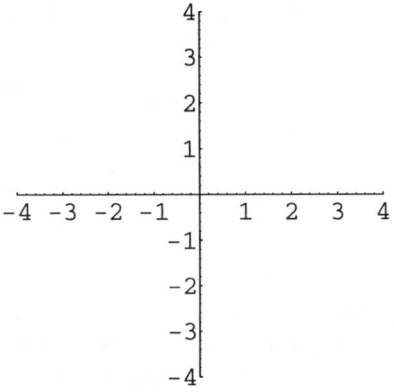

*Note*: As usual, in our mind's eye, we extend this picture infinitely far in both directions.

In order to construct this coordinate system, we must first choose where to put our two axes, and the units. Once we choose an origin, all of this information is summarized by two geometric vectors, usually called $\vec{i}$ and $\vec{j}$.

Geometricallly, the concept of a linear combination of vectors is very closely related to the familiar process of plotting points. For example, to plot the point (3,2), we "go over three and up two":

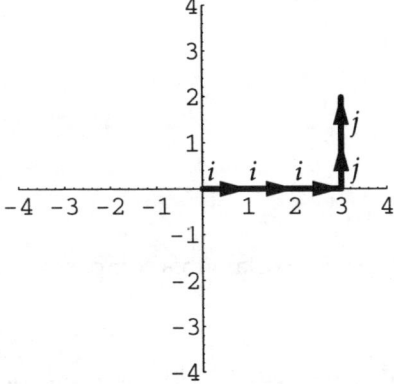

We see that this picture represents a sum of geometric vectors. Algebraically, this corresponds to the linear combination:

$$\vec{v} = 3\vec{i} + 2\vec{j}$$

We may graph this equation as:

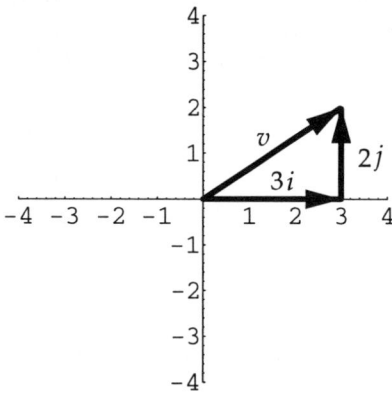

Here we say that the vector $\vec{v}$ is a "linear combination" of the vectors $\vec{i}$ and $\vec{j}$, because we have combined the two vectors using only the linear operations of addition and scalar multiplication that define the vector space. The scalars 2 and 3 are called the "coefficients" of this linear combination. We have discussed this type of equation in connection with *column* vectors in Lesson 13, but now we see that we may consider such equations in *any* vector space.

*Note*: If this notation is familiar (from trigonometry, calculus, or physics), focus on mastering the associated terminology. These terms are central to the language of linear algebra and are *crucial* for an understanding of the remainder of the text.

## Exercise 22.9

Neatly sketch and label the following linear combinations precisely as before. *Note*: It may be convenient to print a copy of the coordinate axes (using the Print Selection command from the File menu).

(a) $\vec{w} = 2\vec{i} + 3\vec{j}$

(b) $\vec{r} = 3\vec{j} + 2\vec{i}$

(c) $\vec{s} = -2\vec{i} + -3\vec{j}$

## Exercise 22.10

Compare $\vec{w}$ and $\vec{r}$. Explain this comparison in algebraic terms. Repeat this exercise for $\vec{w}$ and $\vec{s}$.

Notice that this process of graphing linear combinations of vectors is quite similar to using ordinary graph paper, such as:

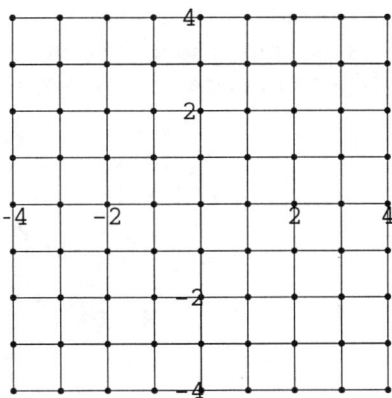

### Exercise 22.11

On a copy of this grid, sketch the following vectors. *Only show the final vector*, not its components in the $\vec{i}$ and $\vec{j}$ directions:

(a) $\vec{t} = 3\vec{i} + 2\vec{j}$

(b) $\vec{u} = 3\vec{i} - 4\vec{j}$

(c) $\vec{a} = -\frac{5}{2}\vec{i} + \frac{2}{3}\vec{j}$

*Note*: We will refer back to these linear combinations in the next section as we examine different nonstandard coordinate systems.

## 22.4   Nonstandard Coordinate Systems

### 22.4.1   Introduction

We emphasized in the last lesson how important it is to be able to look at many different examples (such as column vectors and polynomials) and *think* of them in the same manner, as vectors. As we will continue to see throughout the text, this ability to change our point of view is crucial in the study of linear algebra and is often the key step in solving a given problem. A different type of change that is equally important is a change of coordinate system. Physicists often use the term "frame of reference" for a mathematical coordinate system. For example, in trigonometry, calculus, or physics one often encounters polar coordinates (in two dimensions) and spherical or cylindrical coordinates (in three dimensions), and learns how these coordinate systems simplify any problem with a rotational symmetry.

So far in this lesson, we have emphasized:

1. how Cartesian coordinates are determined by two vectors, $\vec{i}$ and $\vec{j}$,

2. how to visualize linear combinations of these vectors, using graph paper, and

3. how ordinary graph paper allows us to visualize the standard, Cartesian coordinate system.

Now we move on to examine some other types of coordinate systems.

First, we discuss the graph paper associated with any given coordinate system. For example, graph paper for polar coordinates would look like this:

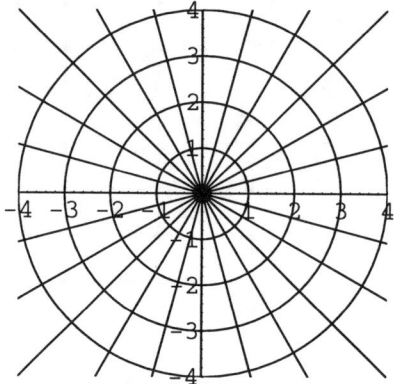

Remember that the polar coordinates of a point specify the angle $\theta$, which describes the ray on which the given point lies, and the radius $r$, which specifies the circle on which it lies. Together, these two numbers uniquely identify a point in the plane, as the intersection of a circle and a ray. The graph paper then supplies a visual definition of the coordinate system.

In general, graph paper for any coordinate system is constructed by tracing a number of different curves. Each curve is the locus of points where one coordinate is constant. By constructing the curves with a variety of equally constant values, one obtains a sequence of curves associated with one coordinate. Repeating this process with every coordinate creates a "grid" of curves that may be used to uniquely specify any given point. That is,

    – every point in the plane can be expressed by some set of coordinates, and

    – there is only one set of coordinates for each point.

## Exercise 22.12

•    Polar coordinates form a "nonlinear" coordinate system, as opposed to Cartesian coordinates, which are linear. Explain why we might use this terminology.

*Note*: In fact, polar coordinates *do not work* (!) at the origin, because there are many possible coordinates for the same point.

Linear algebra is only concerned with *linear* coordinate systems. From a linear algebraic point of view, graph paper is a picture of all possible linear combinations (with integer coefficients) of two given vectors $\vec{\imath}$ and $\vec{\jmath}$. In the remainder of this lesson, we examine *different* kinds of linear coordinate systems (besides the familiar Cartesian coordinates), using graph paper.

## 22.4.2  Constructing Graph Paper

Before proceeding, it is instructive to consider how one would construct the graph paper associated with a set of vectors. For example, given a pair of vectors, $\vec{\imath}$ and $\vec{\jmath}$, we would construct its associated graph paper by:

1. extending each vector into both directions to form a pair of axes, using each vector as a unit length;

2. plotting the endpoints of all possible linear combinations of $\vec{\imath}$ and $\vec{\jmath}$ with integer coefficients; and

3. translating each axis to each point along the other axis.

To see this process carried out, "animate" the following group of graphics. Remember how we do this, by centering one frame on the screen, selecting the entire group, and hitting the Command-y key (or simply double-clicking on a single frame).

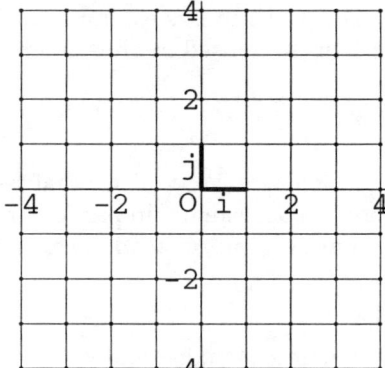

*Note*: In the Notebook version of this lesson, this appears as an animation.

Notice how the entire coordinate system is generated by our choice of $\vec{\imath}$ and $\vec{\jmath}$.

*Important*: Make sure to examine all of the examples of this section in order, because they are desiged to build on one another.

## 22.4.3  Example 1

One of the most common ways to modify of a coordinate system is to choose a different scale on one or both of our axes. For example, we may shrink the scale on the $y$-axis to half its original size. This corresponds to choosing $\vec{\jmath}$ half as long as $\vec{\imath}$ and gives the following coordinate system:

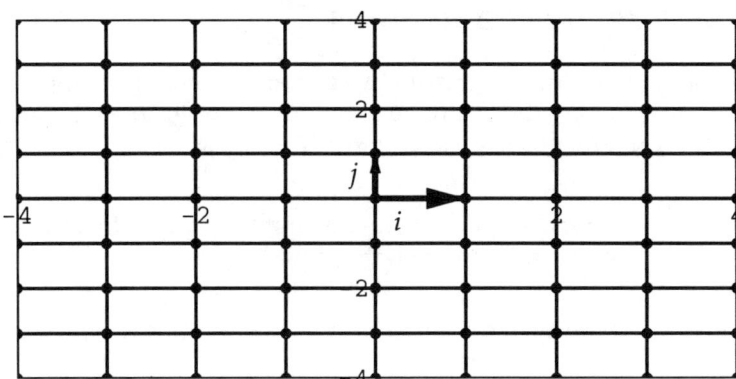

## Exercise 22.13

On a copy of this grid, sketch $\vec{b} = 2\vec{\imath} + 3\vec{\jmath}$. Geometrically, compare this with $\vec{w}$ from Ex. 22.9.

*Warning*: Although we use the same names for the two basic vectors to emphasize the analogy, the vectors $\vec{\imath}$ and $\vec{\jmath}$ in this example (and the following ones) are *not* the same as those in Ex. 22.9, or those used in physics, etc.

## 22.4.4  Example 2

Now Pandora's box is open! We will see that we have a great deal of freedom in designing usable coordinate systems. In particular, there is no need to require $\vec{\imath}$ and $\vec{\jmath}$ to be perpendicular. For example, instead of shrinking $\vec{\jmath}$, we may shear it to the right.

## Exercise 22.14

Sketch the coordinate system generated by:

On your grid, sketch the following linear combinations:

(a) $\vec{c} = 3\vec{\imath} + 2\vec{\jmath}$

(b) $\vec{d} = 3\vec{\imath} - 4\vec{\jmath}$

(c) $\vec{e} = -\frac{5}{2}\vec{\imath} + \frac{2}{3}\vec{\jmath}$

Notice that even though these are the same linear combinations as before, since $\vec{\imath}$ and $\vec{\jmath}$ are different we do not obtain the same results. Geometrically, compare $\vec{\imath}$, $\vec{u}$, and $\vec{a}$ from Ex. 22.11 with $\vec{c}$, $\vec{d}$, and $\vec{e}$.

*Important*: Because of this connection between linear combinations and coordinate systems, we will sometimes refer to the coefficients of a linear combination as the "coordinates" of the resulting vector. Thus, for example, we would say: "The coordinates of $\vec{c}$ with respect to the vectors $\vec{i}$ and $\vec{j}$ are $(3, 2)$." We have seen that the same coordinates may give rise to different vectors, if we use different vectors in our linear combination. Thus, it is very important to include the modifier "with respect to the vectors . . ." and clearly specify the correct set of vectors.

### 22.4.5   Example 3

Likewise, we may "flip" $\vec{j}$ over to obtain a coordinate system with a nonstandard orientation (see Section 19.3.2) to obtain the coordinate system. We have provided the BasisPicture[] from the "MPicts.m" package to make pictures of such nonstandard coordinate systems. To create a picture of this coordinate system, we must specify $\vec{i}$ and $\vec{j}$ as column vectors:

- ```
  i = {1,0}; j = {1,-1};
  picture = BasisPicture[{i,j},4, HeadScale->.45,TailWidth->.004];
  Show[picture,Axes->None];
  ```

*Note*: One will need to resize this picture to make $\vec{i}$ and $\vec{j}$ the same length as before.

### Exercise 22.15

On a copy of this grid, sketch the following linear combinations:

(a) $\vec{f} = 3\vec{i} + 2\vec{j}$

(b) $\vec{g} = 3\vec{i} - 4\vec{j}$

(c) $\vec{h} = -\frac{5}{2}\vec{i} + \frac{2}{3}\vec{j}$

Geometrically, compare $\vec{f}$, $\vec{g}$, and $\vec{h}$ with the corresponding vectors from Ex. 22.11 and Ex. 22.14. *Hint*: First compare the respective $\vec{i}$ and $\vec{j}$ vectors.

### 22.4.6   Example 4

It begins to seem that there are no restrictions on how $\vec{i}$ and $\vec{j}$ can be chosen. We know that there are two important restrictions on a coordinate system, given at the beginning of this section. However, within those restrictions, there still remains a great deal of flexibility. To this point, we have kept $\vec{i}$ as the traditional unit vector in the $x$-direction. Our final example demonstrates how this is unnecessary, as well.

If we shear the standard coordinate system up and to the right by equal amounts, the result appears stretched along the $y = x$ line:

- ```
  i = {1,2}; j = {2,1};
  picture = BasisPicture[{i,j},4, HeadScale->.3,TailWidth->.004];
  Show[picture,Axes->None];
  ```

### Exercise 22.16

On a copy of this grid, sketch the following linear combinations:

(a) $\vec{l} = 3\vec{\imath} + 2\vec{\jmath}$

(b) $\vec{m} = 3\vec{\imath} - 4\vec{\jmath}$

(c) $\vec{n} = -\frac{5}{2}\vec{\imath} + \frac{2}{3}\vec{\jmath}$

Geometrically, compare $\vec{l}$, $\vec{m}$, and $\vec{n}$ with the corresponding vectors from Ex. 22.11–22.15. *Hint*: For each pair of coordinate systems, try to describe a single geometric transformation of the plane that takes the given set of vectors under one coordinate system to the corresponding set of vectors in another.

*Important*: From the previous exercises, one notices that whatever occurred geometrically to the coordinate vectors occurred to every linear combination as well. We will discuss this phenomena formally in Lesson 30. In the language of linear algebra, this illustrates the principle of Section 9.5:

**Observation:**

The effect of a linear transformation is determined by its effect on a basis.

## 22.5   "Good" *vs.* "Bad" Coordinate Systems

We have now seen quite a variety of coordinate systems. We have already mentioned that every good coordinate system has two important properties. This implies that there are certain choices of vectors that will *not* generate a usable coordinate system. For a linear coordinate system, these two crucial properties correspond directly to two important linear algebra concepts. Specifically, for a collection of vectors to generate a good coordinate system, they must be

- linearly independent and

- spanning.

We have already briefly discussed these concepts in the case of column vectors in Lesson 20. We ultimately wish to supply a formal definition of these terms that would make sense in an arbitrary vector space and thus could be applied to *any* of our three examples. However, we will *start* by examining their significance in the context of geometric vectors and *then* state their formal definitions. By combining our previous experience with column vectors and our natural intuition regarding geometric vectors, we can make our definitions seem much more natural than if we were to state them immediately.

Because all of our coordinate systems, to this point, have been generated by two vectors, it should be instructive to change the number of vectors involved and examine the results. Notice that the instructions that we gave before on constructing graph paper work equally well for *any* set of vectors. Thus, we may equally well examine sets of one or three vectors.

## 22.5.1    Graph Paper from One Vector

For example, if we try to only employ a single vector $\vec{\imath}$ to generate a coordinate system, we will obtain the following:

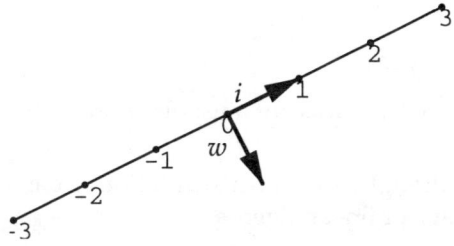

Notice that, in this case, linear combinations of $\vec{\imath}$, such as $2\vec{\imath} + (.5)\vec{\imath} - \vec{\imath} = (1.5)\vec{\imath}$, will only yield vectors of the form $r\vec{\imath}$, where $r$ is any real number.

### Exercise 22.17

Would this form a good coordinate system for the entire *plane*? Why or why not? *Hint*: If $\vec{w}$ is obtained by rotating $\vec{\imath}$ 90° clockwise, as shown, what would the coordinates for $\vec{w}$ be with respect to the coordinate system given by $\vec{\imath}$ alone? Which of the two properties of a coordinate system does this fail to satisfy? To which of the two properties of vectors would this seem to correspond?

## 22.5.2    Graph Paper from Three Vectors

On the other hand, if we were to use*three* vectors in the plane $\vec{\imath}$, $\vec{\jmath}$, and $\vec{k}$, such as:

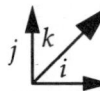

to generate a coordinate system, we would obtain:

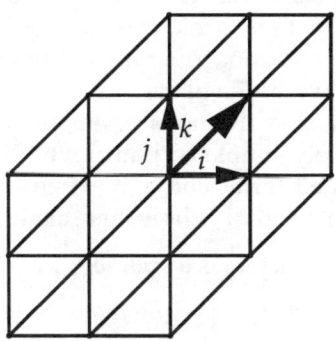

### Exercise 22.18

On a copy of this grid, sketch the following linear combinations:

(a) $\vec{p} = -\vec{\imath} - 2\vec{\jmath} + \vec{k}$

(b) $\vec{q} = -\vec{\imath} + \vec{k}$

(c) $\vec{r} = \vec{\imath} + \vec{\jmath} - \vec{k}$

Intuitively, why is this not a usable coordinate system for the plane?

With a little thought and experimentation, we may refine our intuition and express it in the language of linear algebra.

### Exercise 22.19

Re-examine the three linear combinations from Ex. 22.18. Use those results to determine which of the two properties of a coordinate system this example fails to satisfy. To which property of vectors should this correspond?

## 22.6　Now for the Definitions

In the last section, we illustrated the concepts of independent and spanning sets of vectors, using *negative* examples. That is, we examined sets of geometric vectors that were linearly dependent or *not* spanning. In this section, we provide formal definitions for these terms and explain their grammatical usage. We stress the grammar because in order to *think* about linear algebra, one must learn the language of linear algebra. In fact, the next three lessons will not require any essentially new algorithms. We will simply learn to rephrase and reinterpret what we already know, using the language of linear algebra.

## 22.6.1　Spanning Sets

The first negative example illustrates the notion of a "spanning set" of vectors. A single vector *cannot* form a spanning (adjective) set for the plane. That is because the points that have coordinates with respect to that set of vectors (namely, the set of all linear combinations of that set), which we call the "span" (noun) of that set of vectors, is only a line. Since a line cannot cover the entire plane, we say that the given set of vectors does not "span" (verb) the plane.

　　Since the idea of spanning is essentially the same as "covering" or "reaching," the concept is relatively simple and intuitive. The greater challenge is to learn to use this collection of related terms correctly when writing and speaking. To illustrate proper usage, it is helpful to observe how there are four different ways of saying the same thing:

- "a set of vectors is a *spanning set* for a given vector space,"

- "a set of vectors is *spanning* with respect to a given space,"

- "a set of vectors *spans* a given space," or

– "the *span of a set* of vectors equals the entire space."

It is important to practice using this terminology as much as possible, even when thinking on one's own about any given situation.

Having established a solid intuitive foundation, we now supply the formal definition.

**Definition 22.1** A set of vectors $X$ from a given vector space $V$ is said *to span* (verb) $S$ iff every vector in $S$ may be expressed as a linear combination of the vectors in $X$. Likewise, the *span* (noun) of $X$ is the set $S$ of all possible linear combinations of the vectors in $X$.

### Exercise 22.20

Is the second example of the three vectors $\left\{ \vec{\imath}, \vec{\jmath}, \vec{k} \right\}$ from the Section 22.5 a spanning set for the plane? Why or why not?

*Note*: We have already introduced this concept in the case of column vectors in Lesson 20. The span of a set of column vectors is precisely the same as the image of the linear function defined by the matrix with the given vectors as columns. In fact, once we learn the general definition of a linear function (i.e., transformation), we will see that the concepts of span and image are equivalent notions in *any* vector space.

## 22.6.2   Linear Independence

The second example of Section 22.5 illustrated a set of vectors that were *not* linearly independent. Three vectors in the plane cannot form a "linearly independent" (adjective) set. That is because one of the vectors will necessarily be redundant. We say it is "linearly dependent on" (adjective) the others, because it will be expressible as a linear combination of the other vectors. In our example, we had $\vec{\jmath} = -\vec{\imath} + \vec{k}$. Thus, we may remove $\vec{\jmath}$ from the set without changing the span of the set (i.e., $\vec{\jmath}$ is unnecessary). This type of equation is referred to as a "linear dependence relationship," or "dependence relation," for short. Specifically, we say that this is a dependence relation *between* $\vec{\jmath}$ and the set $\{\vec{\imath}, \vec{k}\}$.

Since this equation may be rewritten in any number of ways, such as

$$\vec{\imath} = -\vec{\jmath} + \vec{k} \qquad \text{or} \qquad \vec{k} = \vec{\imath} + \vec{\jmath}$$

it is not specifically $\vec{\jmath}$'s "fault," so we also say that the entire set is "linearly dependent" (without the word "on"). Algebraically, this corresponds to writing the previous equation as:

$$0 = \vec{\imath} + \vec{\jmath} - \vec{k}$$

This last form of the equation is referred to as a "dependence relationship within S."

In this context, we again see that there are many equivalent ways of saying the same thing:

– "a set of vectors is *linearly dependent*,"

- "one vector in a set is *linearly dependent on* the others,"

- "one vector in a set *depends linearly on* the others," or

- "there is a *linear dependence* relationship within a set of vectors."

The formal definitions of these terms are straightforward generalizations of this example.

**Definition 22.2** If $v$ is a single vector and $S$ is a subset of a given vector space $V$, a linear *dependence relation between $v$ and $S$* is an equation expressing $v$ as a linear combination of the vectors in $S$. If there exists a dependence relation between $v$ and $S$, we say that $v$ is *linearly dependent on $S$*.

**Definition 22.3** A *linear dependence relation* within a set of vectors $S$ is an equation expressing the $0$ vector as a linear combination of the vectors in $S$. We say that $S$ is *linearly dependent* iff there exists a *nontrivial* dependence relation among the vectors in $S$.

*Note*: Since we may always express $0$ as $0v + 0w + \cdots$ for any set of vectors, we must put in the caveat that the linear combination is nontrivial; that is, not all of the coefficients equal $0$.

   Although, in principle, this also provides a precise definition for a set of vectors to *not* be linearly dependent, phrasing this can be tricky. We say:

- "a set of vectors is *linearly independent*,"

- "every vector in the set is *linearly independent of* the others,"

- "*no* vector in the set *depends linearly on* the others," or

- "there are *no linear dependencies* among the vectors."

One formal definition of this concept is obtained by simply negating Definition 22.3:

**Definition 22.4** A set of vectors $S$ is said to be *linearly independent* iff we can *only* express $0$ as a linear combination of the vectors in $S$ in the trivial way, that is, with all $0$ coefficients.

This definition is often rephrased equivalently in terms of a test:

**Definition 22.5** A set of vectors, $S = \{v_1, \ldots, v_n\}$, is said to be *linearly independent* iff

$$a_1 v_1 + \cdots + a_n v_n = 0 \quad \Rightarrow \quad a_1 = 0, \ldots, a_n = 0$$

That is, if we have a dependence relation that *might* show that the set is linearly dependent, we can prove that the relation is in fact the trivial relation, and thus of no consequence. This formulation of the definition provides a prescription for verifying whether or not a set of vectors is linearly independent. We will examine this definition in greater detail in the next lesson.

Finally, we should observe that this concept is very closely related to the notion of the kernel of a linear transformation that we introduced in Lesson 20. That is because we may consider a linear equation either as a statement about column vectors or as a corresponding statement regarding a matrix product, which in turn may be regarded as a description of a linear function. For example, a set of column vectors is linearly independent if and only if the matrix with those vectors as columns has a trivial kernel.

### 22.6.3  Bases

Now for the final definition. In the previous section, the first example failed to provide a good coordinate system because a single vector alone could not span the plane (although it was independent; a single nonzero vector always forms an independent set) and so we could not obtain coordinates for every point. On the other hand, the second example was not a good coordinate system because the set of vectors was not independent (although it was spanning) and so it could not specify unique coordinates for each point. Thus, to be a good coordinate system a set of vectors must be both linearly independent and spanning. We call such a set of vectors a "basis."

**Definition 22.6** A set of vectors $S$ is said to form a *basis* for a vector space $V$ iff it is linearly independent and spans $V$. *Note*: The plural of "basis" is "bases" (pronounced "bay-sees").

To this point, we have seen that the two fundamental properties of a good coordinate system, when expressed in the language of linear algebra, correspond precisely to the two defining properties of a basis. Thus, linear coordinate systems and bases are logically equivalent concepts. You are probably very familiar with the concept of a basis, although you may not have used this language before.

### Exercise 22.21

- In all our examples, you should have observed that the only good coordinate systems for the plane were generated by precisely two vectors. Why is this intuitively reasonable?

## 22.7  Summary

In this lesson, we have discussed the notions of:

- a spanning set of vectors for a vector space, which corresponded to the *existence* of coordinates for any given vector (with respect to that set),

- the independence of a set of vectors, which corresponded to the *uniqueness* of coordinates (with respect to that set), and

- a basis of a vector space, which is simply an independent set of vectors that spans the vector space.

Now we may also define the "dimension" of a vector space to be the number of vectors in a basis (*Aside*: We have not yet proven that this is a *good* definition, because we have no reason to believe, at this point, that every basis will always have the same number of vectors in it! We will prove this important result later.). Intuitively, we have seen, in the case of geometric vectors, that a basis corresponds to a (good) linear coordinate system, and the linear algebra notion of dimension corresponds to our usual notion of dimension.

The example of geometric vectors provides us with solid intuition for these concepts. However, you should learn their formal definitions as well, because in the next lesson, we will examine these concepts in concrete situations other than as geometric vectors (namely, column vectors and polynomials). In these situations, we must rely on the definitions more heavily. We will show how a basis provides a more concrete description of a linear subspace and why all (finite-dimensional) vectors spaces are essentially the same!

# Independent and Spanning Sets

### Accompanied by Notebook

## 23.1 Introduction

In the last lesson, we dicussed the definitions of linearly independent and spanning subsets of a vector space, and saw their interpretations in two-dimensional Euclidean space (i.e., geometric vectors in the plane). In this lesson, we will explore these definitions in the vector spaces of column vectors and polynomials. In both cases, we will examine linear combinations and linear dependencies, which naturally lead to systems of linear equations. Because Gaussian elimination preserves linear combinations, we will be able to quickly discover dependencies by inspection of an associated reduced row echelon matrix.

From a more theoretical point of view, we will provide alternative descriptions of the kernel and image of a system of equations, highlighting the connection between the adjectives "independent" and "spanning" and the properties of uniqueness and existence. In this way, we will tie together these abstract notions of a vector space and our concrete starting point of linear systems.

## 23.2 Gaussian Elimination and Dependence Relations

Both Definition 22.1 and 22.5 involve linear dependence relations among a given subset $S$ of a vector space $V$. One may view Definition 22.5 as saying that there is a *unique* dependence relation between 0 and $S$ (namely the "trivial" relation with all 0 coefficients). Likewise, Definition 22.1 requires one to show that there *exists* a dependence relation between any vector $v$ in $V$ and $S$. In either case, we must solve for a dependence relation between a given vector $v$ and a set of vectors:

$$S = \{v_1, v_2, \ldots, v_n\}$$

In other words, we must solve the vector equation:

$$v = a_1 v_1 + a_2 v_2 + \cdots + a_n v_n \tag{23.1}$$

as an equation in the coefficients $a_1, a_2, \ldots, a_n$. In any concrete example, this will naturally lead to a system of equations in the $a_i$'s, which we may then solve explicitly by Gaussian elimination.

In this section, we re-interpret the process of solving a system of equations as searching for dependence relations among the columns of the system. Most calculations in linear algebra may be phrased in this way, especially calculations with bases. We will observe that Gaussian elimination does not change the *dependence relations* among the columns.

First, observe that, if $S$ is a set of column vectors in $R^n$, the set of *dependence relations* within $S$ is the same as the set of *solutions* to $AX = 0$ (that is, the kernel of $A$; see Definitions 22.3 and the discussion in Section 20.3), where $A = \begin{bmatrix} v_1 & \cdots & v_n \end{bmatrix}$ and $X = \begin{bmatrix} a_1 \\ \vdots \\ a_n \end{bmatrix}$. In particular, we know that:

### Observation:

Solutions to a system of linear equations correspond directly to dependence relations among the columns of a matrix.

That is because this set of dependence relations corresponds to the solutions of Eq. 23.1 with $v = 0$. Using the Principle of Partitioned Matrices, this equation may be rewritten as the matrix equation:

$$0 = a_1 v_1 + a_2 v_2 + \cdots + a_n v_n = \begin{bmatrix} v_1 & \cdots & v_n \end{bmatrix} \begin{bmatrix} a_1 \\ \vdots \\ a_n \end{bmatrix} = AX$$

Because row operations do not change the solutions to a system of equations, they will not change the dependencies among the columns. In particular, we can conclude:

### Observation:

All dependence relations among the columns of a matrix may be determined in the reduced row echelon form of the matrix.

Because a reduced row echelon matrix is much simpler than a general matrix, dependencies among the columns are much easier to determine by inspection.

Although we did not use this terminology earlier, most of the computations so far in this text have been based on these two observations. We have stated every question as solving a corresponding system of equations, and we have used its reduced row echelon form to determine its general solution. In recording the general solution, we were simply deducing the obvious dependence relations among the columns of the reduce row echelon form of the system, and thus of the original system.

For example, consider the system of equations:

$$\begin{aligned} 3x + 6y &= 12 \\ 2x + 8y &= 20 \end{aligned} \tag{23.2}$$

Instead of viewing this system as rows, consider it as a statement about columns:

**Question:**

What linear combination $x \begin{bmatrix} 3 \\ 2 \end{bmatrix} + y \begin{bmatrix} 6 \\ 8 \end{bmatrix}$ with coefficients $(x, y)$ yields $\begin{bmatrix} 12 \\ 20 \end{bmatrix}$?

- Using *Mathematica*, one computes the reduced row echelon form of the augmented matrix to be:

$$\begin{bmatrix} 1 & 0 & -2 \\ 0 & 1 & 3 \end{bmatrix}$$

we may conclude that the third column, $w_3 = \begin{bmatrix} -2 \\ 3 \end{bmatrix}$, is equal to $-2$ times the first, $w_1 = \begin{bmatrix} 1 \\ 0 \end{bmatrix}$, plus 3 times the second, $w_2 = \begin{bmatrix} 0 \\ 1 \end{bmatrix}$. In other words, these three columns $w_1$, $w_2$, and $w_3$, satisfy the vector equation:

$$-2w_1 + 3w_2 = w_3 \tag{23.3}$$

## Exercise 23.1

Verify that the *original* three columns of $A$, $v_1 = \begin{bmatrix} 3 \\ 2 \end{bmatrix}$, $v_2 = \begin{bmatrix} 6 \\ 8 \end{bmatrix}$, and $v_3 = \begin{bmatrix} 12 \\ 20 \end{bmatrix}$, *also* satisfy Eq. 23.3, with the $v_i$'s in place of the $w_i$'s.

In terms of the original system, this dependence relation implies that $(-2, 3)$ is a solution to Eq. 23.2.

As a more general example, we now consider the general solution to our system from Lesson 4:

$$\begin{array}{rrrrrrrrrr} 2x & + & 4y & - & 2z & + & 4w & = & 8 \\ -x & - & 2y & + & 4z & + & 10w & = & 11 \\ 3x & + & 5y & - & 6z & + & 5w & = & 7 \\ 2x & + & 5y & - & 3z & - & 11w & = & -7 \end{array}$$

and observe how it corresponds to linear dependencies among the columns of the associated augmented matrix. Using *Mathematica*, one determines that the columns of the reduced row echelon form of the corresponding augmented matrix are:

$$w_1 = \begin{bmatrix} 1 \\ 0 \\ 0 \\ 0 \end{bmatrix}, w_2 = \begin{bmatrix} 0 \\ 1 \\ 0 \\ 0 \end{bmatrix}, w_3 = \begin{bmatrix} 0 \\ 0 \\ 1 \\ 0 \end{bmatrix}, w_4 = \begin{bmatrix} 28 \\ -11 \\ 4 \\ 0 \end{bmatrix}, \text{ and } w_5 = \begin{bmatrix} 29 \\ -10 \\ 5 \\ 0 \end{bmatrix}$$

Clearly, the final column $w_5$ may be written in terms of first four, $w_1, \ldots, w_4$; in fact, setting the coefficient of $w_4$ to 0, $w_5$ may easily be written in terms of $w_1, \ldots, w_3$:

$$\begin{bmatrix} 29 \\ -10 \\ 5 \\ 0 \end{bmatrix} = (29) \begin{bmatrix} 1 \\ 0 \\ 0 \\ 0 \end{bmatrix} + (-10) \begin{bmatrix} 0 \\ 1 \\ 0 \\ 0 \end{bmatrix} + (5) \begin{bmatrix} 0 \\ 0 \\ 1 \\ 0 \end{bmatrix} + (0) \begin{bmatrix} 28 \\ -11 \\ 4 \\ 0 \end{bmatrix}$$

giving the dependence relation:

$$w_5 = 29w_1 + (-10)w_2 + 5w_3 \tag{23.4}$$

In terms of a solution to the original system, this relation corresponds to the solution $x = 29, y = -10, z = 5, w = 0$.

## Exercise 23.2

Verify that Eq. 23.4 holds among the original columns

$$v_1 = \begin{bmatrix} 2 \\ -1 \\ 3 \\ 2 \end{bmatrix}, v_2 = \begin{bmatrix} 4 \\ -2 \\ 5 \\ 5 \end{bmatrix}, v_3 = \begin{bmatrix} -2 \\ 4 \\ -6 \\ -3 \end{bmatrix}, v_4 = \begin{bmatrix} 4 \\ 10 \\ 5 \\ -11 \end{bmatrix}, \text{ and } v_5 = \begin{bmatrix} 8 \\ 11 \\ 7 \\ -7 \end{bmatrix}$$

as well.

To determine the general relationship between *all* the columns, including $w_4$, consider changing the coefficient $w$ to a nonzero value. For example, letting $w = 1$ leads to the equation:

$$\begin{bmatrix} 29 \\ -10 \\ 5 \\ 0 \end{bmatrix} = - \begin{bmatrix} 1 \\ 0 \\ 0 \\ 0 \end{bmatrix} + - \begin{bmatrix} 0 \\ 1 \\ 0 \\ 0 \end{bmatrix} + - \begin{bmatrix} 0 \\ 0 \\ 1 \\ 0 \end{bmatrix} + 1 \begin{bmatrix} 28 \\ -11 \\ 4 \\ 0 \end{bmatrix} \tag{23.5}$$

Focusing on the top row, since we have added 28 into this row, we must decrease the coefficient of $w_1$ to $29 - 28 = 1$ to obtain the same total as before, namely $w_5$.

## Exercise 23.3

Determine the remaining coefficients of Eq. 23.5. Verify your answer. Verify that the same dependence relation holds for the original columns of the system $v_1 - v_5$.

## Exercise 23.4

Repeat Ex. 23.3 with $w = 2$.

Generalizing, we may determine *all possible* dependencies among the columns.

## Exercise 23.5

Determine a general pattern for the coefficients in terms of $w$. *Hint*: Think about the parametric form of the general solution to the system.

*Note*: We will discuss this exercise again more fully in Section 24.2

## 23.3 Searching for Dependence Relations

In Section 23.2, we saw how dependence relations among a set of column vectors $S$ correspond to solutions of a system of equations, where the vectors of $S$ correspond to the columns of the augmented matrix. In this section, we focus on determining dependence relations between any given collection of vectors as our primary goal. We will extend our procedures to include the vector space of polynomials $P_n$, as well as $R^n$.

### 23.3.1 Column Vectors

For example, let:

$$S = \{v_1, v_2, v_3, v_4\} \left\{ \begin{bmatrix} 1 \\ 0 \\ 0 \end{bmatrix}, \begin{bmatrix} 2 \\ 2 \\ 0 \end{bmatrix}, \begin{bmatrix} 1 \\ -1 \\ 1 \end{bmatrix}, \begin{bmatrix} 2 \\ 0 \\ -1 \end{bmatrix} \right\} \quad \text{and} \quad v = \begin{bmatrix} -1 \\ 1 \\ 2 \end{bmatrix}$$

- and consider the problem of determining a dependence relation between $v$ and $S$. By our observations in Section 23.2, we may deduce any such dependence relation from the reduced row echelon form of $\begin{bmatrix} 1 & 2 & 1 & 2 & -1 \\ 0 & 2 & -1 & 0 & 1 \\ 0 & 0 & 1 & -1 & 2 \end{bmatrix}$. Using *Mathematica*, we quickly solve this with the RowReduce[] command to obtain the following reduced row echelon matrix:

$$\begin{bmatrix} 1 & 0 & 0 & 4 & -6 \\ 0 & 1 & 0 & -\frac{1}{2} & \frac{3}{2} \\ 0 & 0 & 1 & -1 & 2 \end{bmatrix}$$

*Note*: The necessary command is pretyped in the accompanying Notebook. *Warning*: In *Mathematica* a set of column vectors, such as $S$, *looks* just like a matrix, such as $A$. However, one should observe that these are *not* the same. Here we must place the vectors in $S$ as *columns* of $A$, while *Mathematica* would interpret $S$ as a matrix with the vectors as *rows*.

From this it is easy to see that the final column may be written as a linear combination of the first three. This suggests the following dependence relation among the *original* columns:

$$v = -6 \cdot v_1 + \frac{3}{2} \cdot v_2 + 2 \cdot v_3$$

Notice that as a system of equations, the coefficient of the fourth column vector is a free variable. Setting this equal to 1, for example, leads to a more convenient dependence relation with integer coefficients:

$$v = -10 \cdot v_1 + 2 \cdot v_2 + 3 \cdot v_3 + 1 \cdot v_4 \tag{23.6}$$

### Exercise 23.6

Verify that $v$ is given by this linear combination, by substituting by hand the values for each vector into Eq. 23.6 and simplifying.

### Exercise 23.7

Determine a third dependence relation, similar to Eq. 23.6. Verify your relation, as in Ex. 23.6.

## Constructing New Exercises

As we did for linear systems, we have provided the RandomVectors[] and Random-Combination[] commands, from the "Vects.m" package, to allow us to create supplementary exercises. To construct a new exercise, we must first generate a set $S$ of vectors. Although we may do so by hand, the command:

- `S = RandomVectors[2]`

will generate two random vectors. Although we may explicitly specify the desired ambient vector space $V$, the default option is for three-dimensional column vectors. *Remember*: Even though the output of this command looks like a *matrix*, in this case, we want to consider this a *set* of vectors. The RandomCombination[] command will then randomly generate a vector $v$ *in the span of* $S$.

- `v = RandomCombination[S]`

### Exercise 23.8

- Use these two commands to generate a set $S$ and a vector $v$ in its span. Determine the linear combination that generates $v$ from $S$. Verify your answer as in Ex. 23.6.

### *Drill Exercise 23.9*

- Use the following command to generate three more vectors in the span of $S$:

- `Do[Print[RandomCombination[S]], {i,1,3}]`

Determine the dependence relations for each vector and then verify each of your answers. *Hint*: Notice that this is another example of parallel systems of equations (see Section 15.2), so that you do not need to perform Gaussian elimination three separate times.

One may also use the built-in *Mathematica* command Random[] to generate each entry of a column vector separately. For example:

- `w = Table[Random[Integer, {-5,5}], {i,1,3}]`

will randomly generate a three-dimensional column vector (not necessarily in the span of $S$), with integer entries between 5 and −5.

### Exercise 23.10

- Randomly generate a vector $w$ and determine if it is in the span of $S$. That is, either supply an explicit dependence relation or explain why no such relation exists.

### Exercise 23.11

Explain why one should not generally expect $w$ to be in the span of $S$. *Hint*: Geometrically, what does the span of $S$ look like? How much room does it take up in three dimensions?

## 23.3.2   Polynomials

Now consider the vector space of polynomials. At first, it would seem that the techniques of Section 23.3.1 would not apply. However, if we write out the definitions, we will see otherwise. To determine if the polynomial $p(t) = -1 + t + 2t^2$ is in the span of $T = \{1, 2 + 2t, 1 - t + t^2, 2 - t^2\}$, we may proceed in *essentially* the same manner as with column vectors. In this context, Eq. 23.1 becomes:

$$-1 + t + 2t^2 = a_1 \cdot 1 + a_2 \left(2 + 2t\right) + a_3 \left(1 - t + t^2\right) + a_4 \left(2 - t^2\right)$$

Multiplying out and collecting like terms yields:

$$
\begin{aligned}
-1 + t + 2t^2 &= a_1 \cdot 1 + a_2 \left(2 + 2t\right) + a_3 \left(1 - t + t^2\right) + a_4 \left(2 - t^2\right) \\
&= a_1 + 2a_2 + 2a_2 t + a_3 - a_3 t + a_3 t^2 + 2a_4 - a_4 t^2 \\
&= a_1 + 2a_2 + a_3 + 2a_4 + 2a_2 t - a_3 t + a_3 t^2 - a_4 t^2 \\
&= \left(a_1 + 2a_2 + a_3 + 2a_4\right) + \left(2a_2 - a_3\right) t + \left(a_3 - a_4\right) t^2
\end{aligned}
$$

Because two polynomials are equal iff all their coefficients are equal, this last equation is equivalent to the system of equations:

$$
\begin{array}{rcrcrcrcr}
a_1 & + & 2a_2 & + & a_3 & + & 2a_4 & = & -1 \\
    &   & 2a_2 & - & a_3 &   &      & = & 1 \\
    &   &      &   & a_3 & - & a_4  & = & 2
\end{array}
$$

which in matrix form is *just the same* as that in Section 23.3.1.

### Exercise 23.12

•    Demonstrate that $p$ is in the span of $T$ by supplying an explicit dependence relation. Verify your answer as in Ex. 23.6.

### Exercise 23.13

Explain why the resulting matrices were so similar. *Hint*: Compare $S$ and $v$ with $T$ and $p$. Describe, in general, how one could quickly write down the correct system of equations for any given set of polynomials *without* performing the intermediary algebra.

## Constructing New Exercises

Notice that the RandomVectors[] and RandomCombination[] commands work equally well on polynomials. For example:

- ```
  Clear[t]; T = RandomVectors[2, Basis->{1,t,t^2}]
  ```

will generate a new set of "vectors" $T$ in the vector space of quadratic polynomials $P_3$. Likewise:

- ```
  p = RandomCombination[T]
  ```

will generate a new polynomial $p(t)$ in the span of $T$. However, in this case, *Mathematica* leaves the terms uncombined, thus showing the desired linear combination. Thus, to obtain a more useful result, one should immediately Expand[] the result:

- ```
  p(t) = Expand[RandomCombination[T]]
  ```

## Exercise 23.14

- Use these two commands to generate a set $T$ and a polynomial $p(t)$ dependent upon $T$. Determine the linear combination that generates $p(t)$ from $T$. Verify your answer as in Ex. 23.6.

### *Drill Exercise 23.15*

- Generate three more polynomials in the span of $T$, using the command:

- ```
  Do[Print[Expand[RandomCombination[T]]], {i,1,3}]
  ```

Determine the dependence relations for each polynomial and verify each answer. *Hint*: Compare this problem with Ex. 23.9.

Similar to the case of column vectors:

- ```
  q(t) = Sum[Random[Integer, {-5,5}]t^i, {i,0,2}]
  ```

will randomly generate a polynomial (not necessarily in the span of $T$).

## Exercise 23.16

- Generate a random polynomial, $q(t)$, in this way and determine whether or not it is in the span of $T$. That is, either supply an explicit dependence relation or explain why no such relation exists.

## Exercise 23.17

Explain why one should not generally expect $q(t)$ to be in the span of $T$. *Hint*: Although this is not an explicitly geometric example, appeal to geometric reasoning.
  ```

## 23.4  Determining If a Set Is Independent

According to Definition 22.5, a set

$$S = \{v_1, v_2, \ldots, v_n\}$$

is independent iff there are no nontrivial solutions to

$$0 = a_1 v_1 + a_2 v_2 + \cdots + a_n v_n$$

as an equation in the coefficients $a_1, a_2, \ldots, a_n$. In other words, "There are no nontrivial dependence relations within $S$." This should look very familiar; in fact, this is just a special case of Eq. 22.5, where $v$ is 0!

### 23.4.1  Column Vectors

For example, if we again take the set of column vectors:

$$S = \left\{ \begin{bmatrix} 1 \\ 0 \\ 0 \end{bmatrix}, \begin{bmatrix} 2 \\ 2 \\ 0 \end{bmatrix}, \begin{bmatrix} 1 \\ -1 \\ 1 \end{bmatrix}, \begin{bmatrix} 2 \\ 0 \\ -1 \end{bmatrix} \right\}$$

Eq. 22.5 becomes:

$$\begin{bmatrix} 0 \\ 0 \\ 0 \end{bmatrix} = a_1 \begin{bmatrix} 1 \\ 0 \\ 0 \end{bmatrix} + a_2 \begin{bmatrix} 2 \\ 2 \\ 0 \end{bmatrix} + a_3 \begin{bmatrix} 1 \\ -1 \\ 1 \end{bmatrix} + a_4 \begin{bmatrix} 2 \\ 0 \\ -1 \end{bmatrix}$$

We may again employ Gaussian elimination to simplify these equations. Either we will see that:

- the solution of all 0's is the unique solution, in which case, we have proven that these are independent, or

- we will produce a nontrivial solution, which provides an explicit dependence relation among the vectors.

•    This time, we would use the RowReduce[] command to simplify:

$$\begin{bmatrix} 1 & 2 & 1 & 2 & 0 \\ 0 & 2 & -1 & 0 & 0 \\ 0 & 0 & 1 & -1 & 0 \end{bmatrix}$$

We will obtain the same result as before, except the final column will consist entirely of 0's:

$$\begin{bmatrix} 1 & 0 & 0 & 4 & 0 \\ 0 & 1 & 0 & -\frac{1}{2} & 0 \\ 0 & 0 & 1 & -1 & 0 \end{bmatrix}$$

From this, we may conclude that there *are* nontrivial dependence relations among the vectors of $S$, so that it *is* dependent. Since $a_4$ is a free variable, we assign it any value, say 2 (to cancel out the denominators), and solve to obtain the explicit dependence relation:

$$0 = -8 \cdot v_1 + v_2 + 2 \cdot v_3 + 2 \cdot v_4 \tag{23.7}$$

### Exercise 23.18

Verify that $S$ does in fact satisfy this dependence relation, by hand.

### Exercise 23.19

Determine a second dependence relation by choosing a different value for $a_4$. Is it essentially different than the first? Why or why not?

With a bit of practice, one learns to deduce such relations by inspection. For now, we will experiment a bit to develop our intuition regarding "linear independence." For instance, in this example one should notice that each vector in $S$ depends on the other three.

### Exercise 23.20

Express the third vector as a linear combination of the other three, and verify your answer.

### Exercise 23.21

Create a set of four three-dimensional column vectors, where the fourth vector is *independent* of (i.e., cannot be written as a linear combination of) the first three. *Hint*: This implies that the first three cannot be independent. What can you observe about the reduced row echelon form of the associated matrix?

### Exercise 23.22

- If one randomly chooses two vectors in the plane, would you expect them to be independent or not? How about three vectors? Use the following command to generate sets $S$ and $T$ of two and three vectors, respectively, in the plane and verify whether or not they are independent.

```
S = RandomVectors[2, Basis->{{1,0},{0,1}}]
T = RandomVectors[3, Basis->{{1,0},{0,1}}]
```

Explain your results geometrically.

## 23.4.2  Polynomials

The transition to polynomials is just as before. Say, for example, we wish to determine whether or not the set:
$$S = \left\{1, 2+2t, 1-t+t^2, 2-t^2\right\}$$
is linearly independent. This time, Eq. 23.1 is:
$$0 = a_1 \cdot 1 + a_2\left(2+2t\right) + a_3\left(1-t+t^2\right) + a_4\left(2-t^2\right)$$
Multiplying out, collecting terms, and identifying like terms gives:
$$
\begin{array}{rrrrrrrl}
a_1 & + & 2a_2 & + & a_3 & + & 2a_4 & = 0 \\
 & & 2a_2 & - & a_3 & & & = 0 \\
 & & & & a_3 & - & a_4 & = 0 \\
\end{array}
$$

### Exercise 23.23

Compare this system of equations with that of Section 23.4.1. Are these polynomials independent? Explain.

### Exercise 23.24

Express the first polynomial as a linear combination of the other three. Verify your answer by direct simplification.

### Exercise 23.25

Would you expect a set of four quadratic polynomials to be independent or not? Explain. *Hint*: Reason by analogy with Ex. 23.22.

### *Advanced Exercise 23.26*

If $S$ is a nonempty set of vectors, prove that $S$ is independent iff every vector in the span of $S$ may be expressed in exactly one way as a linear combination of vectors in $S$.

## 23.4.3   Verification with *Mathematica*

Because independence is a more theoretical property, it becomes more difficult to verify a claim of independence. Thus, we provided the IndependentQ[] from the "Bases.m" package, which will take a list of column vectors and return "True" if they are independent and "False" if they are dependent.

For example, if $S$ and $T$ are given by

$$S = \left\{ \begin{bmatrix} 1 \\ 2 \\ 3 \end{bmatrix}, \begin{bmatrix} 2 \\ 4 \\ 6 \end{bmatrix} \right\} \quad \text{and} \quad T = \left\{ \begin{bmatrix} 1 \\ 0 \\ 3 \end{bmatrix}, \begin{bmatrix} 0 \\ 1 \\ 2 \end{bmatrix} \right\}$$

$S$ is clearly a dependent set, while the vectors in $T$ are independent. Evaluate the following commands in the accompanying Notebook to see what happens:

- ```
  S = {{1,2,3},{2,4,6}};
  T = {{1,0,3},{0,1,2}};
  Print["Is S independent?  ", IndependentQ[S]];
  Print["Is T independent?  ", IndependentQ[T]];
  ```

One may use this command to confirm a determination of independence. On the other hand, if one believes a set of vectors to be dependent, producing an explicit dependence relation is far more convincing than any computerized result. In either case, the computer should only serve as a *check*; one should always be ready and able to explain one's reasoning on any given problem to another person in a clear and effective manner.
  ```

### 23.4.4 Dependence, the Kernel, and Injectivity

One should notice that when searching for dependencies we are looking at the *kernel* of a matrix. In the case of column vectors, the matrix is precisely the matrix with the original vectors as columns (with 0's for the right-hand side). Thus, we have another interpretation for the kernel of a matrix:

**Observation:**

$Kernel(A)$ is the collection of *dependence relations* among the columns of $A$.

Moreover, we again see the kernel as being connected with *uniqueness* of solutions to a given problem — whether it is a system of equations or the "coordinates" (i.e., coefficients of a linear combination) of a vector. We may summarize all of the relationships between the concepts from Lessons 8, 16, 20, and 22 by the following theorem.

**Theorem 23.1** *If $A = [\; v_1 \;\; \cdots \;\; v_q \;]$ is an $p \times q$ matrix and $B = [b_j]$ a $q \times 1$ column vector, then the following are equivalent:*

(a) *the solutions to $AX = B$ are unique for every $B$,*

(b) *$A$ is injective,*

(c) *$A$ has a left-inverse,*

(d) *$S = \{v_1, \ldots, v_q\}$ is linearly independent, and*

(e) *$Kernel(A) = 0$, the trivial subspace consisting of only the zero vector.*

## 23.5 Determining If a Set Is Spanning

Determining whether a set of vectors $S$ spans a given vector space $V$ is a bit more difficult than determining independence. Intuitively, Definition 22.1 says that *every* vector $w$ in the given space $V$ may be represented as a linear combination of the vectors in $S$, i.e., every possible $w$ is in the span of the vectors in $S$. We have already discussed how to determine whether *one* vector $w$ is linearly dependent on a group of vectors $S = \{v_1, v_2, \ldots, v_n\}$. By Definition 22.2, this means that we must solve:

$$w = a_1 v_1 + a_2 v_2 + \cdots + a_n v_n$$

The difficulty lies in that Definition 22.1 would seem to require us to do this for *every* vector in $V$. Because there are *infinitely* many vectors in most vector spaces, this would be practically impossible. However, if we *already know* that a set $T$ forms a spanning set, then we only must verify that the vectors of $T$ are in the span of $S$.

### 23.5.1 Column Vectors

For example, consider the example of $3 \times 1$ column vectors. It is easy to verify that the set

$$T = \left\{ \begin{bmatrix} 1 \\ 0 \\ 0 \end{bmatrix}, \begin{bmatrix} 0 \\ 1 \\ 0 \end{bmatrix}, \begin{bmatrix} 0 \\ 0 \\ 1 \end{bmatrix} \right\}$$

forms a spanning set (in fact, it forms a basis). For reference, we will refer to these as $e_1$, $e_2$, and $e_3$. Any column vector in $R^3$ may be written as

$$v = a_1 \begin{bmatrix} 1 \\ 0 \\ 0 \end{bmatrix} + a_2 \begin{bmatrix} 0 \\ 1 \\ 0 \end{bmatrix} + a_3 \begin{bmatrix} 0 \\ 0 \\ 1 \end{bmatrix} = a_1 e_1 + a_2 e_2 + a_3 e_3 \tag{23.8}$$

for some $a_i$'s, showing that $T$ spans $R^3$. To then verify that

$$S = \left\{ \begin{bmatrix} 1 \\ 0 \\ 0 \end{bmatrix}, \begin{bmatrix} 2 \\ 2 \\ 0 \end{bmatrix}, \begin{bmatrix} 1 \\ -1 \\ 1 \end{bmatrix}, \begin{bmatrix} 2 \\ 0 \\ -1 \end{bmatrix} \right\}$$

from Section 23.4.1 is spanning, we only need demonstrate that each vector in $T$ is a linear combination of the vectors in $S$.

For example, consider vector $e_2$ in $T$. To determine a dependence of $e_2$ on $S$, as before, we may RowReduce[]:

$$\begin{bmatrix} 1 & 2 & 1 & 2 & 0 \\ 0 & 2 & -1 & 0 & 1 \\ 0 & 0 & 1 & -1 & 0 \end{bmatrix}$$

- Do this now by executing the RowReduce[] command in the accompanying Notebook. For this result, we may identify the explicit dependence relation:

$$e_2 = -v_1 + \frac{1}{2} v_2 \tag{23.9}$$

## Exercise 23.27

- Verify Eq. 23.9, and then proceed to derive dependence relations for $e_1$ and $e_3$. *Hint*: Instead of performing Gaussian elimination three times, one may again observe that this is another example of parallel systems of equations (see Ex. 23.9).

## Exercise 23.28

Use the results of the Ex. 23.27 and Eq. 23.8 to derive coordinates for the vector $v = \begin{bmatrix} 2 \\ 3 \\ -1 \end{bmatrix}$ with respect to $S$. *Hint*: First determine what the coeficients $a_i$ in Eq. 23.8 would be for this particular $v$, then substitute the three equations from Ex. 23.27 into Eq. 23.8 to obtain an equation for $v$ in terms of the $v_i$ alone. Repeat this for the general vector $\begin{bmatrix} a_1 \\ a_2 \\ a_3 \end{bmatrix}$. *Note*: By this process you provide a formal proof that $S$ is spanning.

## Exercise 23.29

- Determine whether or not the given set of vectors spans the indicated vector space. You should supply explicit dependence relations whenever possible to justify your answers.

(a) Does $S = \left\{ \begin{bmatrix} 1 \\ -1 \\ 0 \end{bmatrix}, \begin{bmatrix} 1 \\ 4 \\ -2 \end{bmatrix}, \begin{bmatrix} 0 \\ 0 \\ 5 \end{bmatrix}, \begin{bmatrix} 2 \\ 3 \\ -4 \end{bmatrix} \right\}$ span the set of $3 \times 1$ column vectors? Explain.

(b) Does $S = \left\{ \begin{bmatrix} -1 \\ 3 \\ 5 \\ 2 \end{bmatrix}, \begin{bmatrix} 2 \\ -1 \\ 0 \\ 1 \end{bmatrix}, \begin{bmatrix} 1 \\ -8 \\ 5 \\ 3 \end{bmatrix} \right\}$ span the set of $4 \times 1$ column vectors?
Does this make intuitive, geometric sense? Explain.

(c) Does $S = \left\{ \begin{bmatrix} -3 \\ 0 \\ 3 \end{bmatrix}, \begin{bmatrix} 0 \\ -3 \\ -3 \end{bmatrix}, \begin{bmatrix} 3 \\ 3 \\ 0 \end{bmatrix} \right\}$ span the set of $3 \times 1$ column vectors
spanned by $T = \left\{ \begin{bmatrix} 1 \\ -2 \\ 1 \end{bmatrix}, \begin{bmatrix} -2 \\ 1 \\ 0 \end{bmatrix}, \begin{bmatrix} 1 \\ 1 \\ 1 \end{bmatrix} \right\}$? Explain.

## 23.5.2  Polynomials

To solve similar problems with polynomials, we may proceed as in Section 23.5.1 once we identify a spanning set of polynomials, $T$. As in the case of column vectors, there is a natural choice. For example, in the vector space of all polynomials of degree 3 or less, every vector looks like:

$$P(t) = a_1 + a_2 t + a_3 t^2 \tag{23.10}$$

Thus, the set:

$$T = \{1, t, t^2\}$$

is clearly a spanning set (and a basis). We may determine whether or not any given set of quadratic polynomials spans $P_3$ just as before. In fact, we may quickly determine the correct matrix to RowReduce[], if we simply replace each polynomial by its corresponding column vector. This is logically valid, because both $R^3$ and $P_3$ satisfy the same formal properties of a vector spaces, so that similar computations may be performed in a similar manner.

### Drill Exercise 23.30

- Verify whether or not the given set of vectors spans the indicated vector space. You should supply (and verify) explicit dependence relations whenever possible to justify your answers.

  (a) Does $S = \{2, 3 + t, 2 - t^2\}$ span the set of quadratic polynomials $P_3$? Explain.

  (b) Does $S = \{1, 2 + 2t, 1 - t + t^2, 2 - t^2\}$ span $P_3$? Explain.

  (c) Does the set $S = \{1 + t, t + t^2\}$ span $P_3$? Explain. Does this make intuitive sense? Why or why not? *Hint*: Think geometrically.

### 23.5.3  Verification with *Mathematica*

If one believes that one set of vectors $S$ spans the same space as another set of vectors $T$, one should be able to determine explicit dependence relations between each vector in $T$ and the set $S$, which may be explicitly verified. It is more difficult to provide an independent verification that $S$ does *not* span $T$. Thus, we provide the SpanningQ[], from the "Bases.m" package, which takes two lists of column vectors and returns "True" if the second set is in the span of the first, and "False" otherwise.

For example, if $S$ and $T$ are given by:

$$S = \left\{ \begin{bmatrix} 1 \\ 2 \\ 0 \end{bmatrix}, \begin{bmatrix} 3 \\ 2 \\ 0 \end{bmatrix} \right\} \quad \text{and} \quad T = \left\{ \begin{bmatrix} 1 \\ 2 \\ 3 \end{bmatrix}, \begin{bmatrix} 2 \\ 1 \\ 3 \end{bmatrix}, \begin{bmatrix} 1 \\ 0 \\ 1 \end{bmatrix} \right\}$$

- this command claims that $T$ is not in the span of $S$, since we cannot generate a vector with nonzero third entry from the vectors in $S$. Evaluate the SpanningQ[] command in the accompanying Notebook to see this.

- If one omits the second set of vectors, SpanningQ[] will assume that $T$ is the standard basis (i.e., the columns of the identity matrix). To see this, evaluate the following command:

- ```
  S = {{1,2,0},{3,2,0}};
  Print["Does S span 3-dimensional Euclidean space?  ", SpanningQ[S]];
  ```

Similar to IndependentQ[], this command provides a helpful verification that a set $S$ of vectors is *not* spanning. However, if one claims that $S$ *is* spanning, then one should determine and verify explicit dependence relations. As usual, one should be ready and able to explain one's reasoning on any given problem to another person in a clear and effective manner, by appealing to the results of an appropriately chosen RowReduce[] command.

### 23.5.4  The Span, the Image, and Surjectivity

As in Section 23.4.4, in this section we highlight the connections between the notion of a spanning set and the *image* of a matrix. Specifically, a column vector $w$ is in the image of a matrix $A$ precisely when $w$ is in the span of the columns of $A$. Thus, we have another interpretation for the *image* of a matrix:

**Observation:**

*Image*$(A)$ is the *span* of the columns of $A$.

Moreover, we again see the image as being connected with the *existence* of solutions to a given problem — whether it is a system of equations or the coordinates of a vector.

In particular, consider the problem of verifying whether or not a set $S = \{v_1, \ldots, v_q\}$ of column vectors span $R^q$. To solve for dependence relations for each of the vectors of a spanning set simultaneously, we would perform Gaussian elimination on a multiply
  ```

augmented matrix with the set of $q$-dimensional vectors

$$T = \left\{ \begin{bmatrix} 1 \\ \vdots \\ 0 \end{bmatrix}, \cdots, \begin{bmatrix} 0 \\ \vdots \\ 1 \end{bmatrix} \right\}$$

appended. One should notice that this procedure is *exactly* the same as for computing a right-inverse! In general, we may summarize all of the relationships between the concepts from Lessons 8, 16, 20, and 22 by the following theorem.

**Theorem 23.2** *If $A = \begin{bmatrix} v_1 & \cdots & v_q \end{bmatrix}$ is an $p \times q$ matrix and $B = [b_j]$ is a $q \times 1$ column vector, then the following are equivalent:*

(a) *$S$ is a spanning set,*

(b) *$AX = B$ has a solution for every $B$,*

(c) *$A$ has a right-inverse,*

(d) *$A$ is surjective,*

(e) *the span of the $S$ is the entire space, and*

(f) *the image of the $A$ is the entire range.*

One should compare Theorems 23.1 and 23.2 to see how they are similar.

## 23.6   Summary

In this lesson, we have discussed:

- how to verify the basic properties of linear dependence, independence, and "spanningness," by converting each property into a statement about dependence relations and deducing the necessary relations from the reduced row echelon form of the corresponding matrix;

- how the concepts of independent and spanning sets are intertwined with the earlier notions of uniqueness and existence of solutions, as well as left- and right-inverses;

- how computations in the vector spaces of column vectors and polymonials are intimately related; and

- how some vector spaces, such as $R^n$ and $P_n$, are defined in terms of a natural choice of basis (columns and polynomials), while others, such as $E^n$, require us to choose a basis.

We may now fill in some logical gaps that we have avoided so far in our presentation regarding:

- why the dimension of a vector space (number of vectors in a basis) is a well-defined number,

- why the rank of a matrix (number of leading 1's after Gaussian elimination) is a well-defined number,

     − why the rank of $A$ and its transpose are equal, and

     − why the $LU$ decomposition (i.e., without interchanges) is unique.

The first fact is a simple logical argument. The second fact follows from *redefining* the rank of $A$ as the dimension of the image of $A$ (which is clearly a well-defined number) and then proving that this is the same as our original definition. The third comes from looking at the span of the rows (the "row space") and the columns (the "column space") of $A$ and how this is affected by Gaussian elimination. The last fact requires a little bit of algebra, using the Principle of Partitioned Matrices.

     We will begin to provide details of each of these arguments in Lesson 25. Before we do so, however, we must provide the formal definition of a linear subspace and discuss how to use bases to give concrete descriptions of subspaces of Euclidean space (i.e., any vector space of column vectors), in Lesson 24. At this point, we have only have seen two concrete examples of bases (for polynomials and column vectors); thus we must also show how to *construct* other sets of basis vectors.

# Constructing Bases

*Mathematica* Notebook

## 24.1 Introduction

In Lesson 23, we learned to verify the two fundamental properties of a basis, namely, linear independence and "spanningness." So far we have only seen two concrete examples of bases (i.e., the standard bases for column vectors and polynomials). Although the geometric bases that we used (in Lesson 22) are not concrete in that they are not amenable to computations, they demonstrated, on an intuitive level, how a basis allows one to represent any vector by its coordinates (with respect to that basis). Coordinates are very concrete, in that they consist of ordinary numbers that may be easily manipulated. Thus, bases are used to convert abstract questions in a vector space into specific, numerical questions that are amenable to direct, computerized calculations.

In this lesson, we will demonstrate how to construct bases in various settings. Specifically, we will discuss how to:

- delete vectors from a spanning set to obtain a basis, and

- add vectors to a linearly independent set to obtain a basis.

Both of these techniques require us to *begin* with a spanning set for the vector space in question. Thus, we will also discuss how to:

- identify spanning sets for two important spaces, namely the image and the kernel of a matrix.

Although this may seem rather theoretical, the ability to construct an appropriate basis is often the first step in solving any practical problem. We begin to discuss this in greater detail in the next lesson. The theme of constructing appropriate bases will dominate the remainder of the text.

## 24.2 Spanning Sets

While a basis for any given subspace may be difficult to determine, spanning sets arise naturally in two important examples. We observed on p. 191 that the image of a matrix $A$ is the same as the span of its columns, or in other words, the columns of $A$ form a spanning set for $Image(A)$. Likewise, there is a natural spanning set for $Kernel(A)$. However, it takes a bit more effort to deduce its form.

Since we know how to determine the general solution to $AX = 0$, we know, in principle, how to determine the general form of a vector in $Kernel(A)$. For example, if we RowReduce[] $A$:

- 
```
A = Partition[{1,2,1,2,-3,3,6,4,-1,2,4,8,5,1,-1,-2,-4,-3,3,-5},5];
Print["The reduced row echelon form of A = ", MatrixForm[A],
      " is ", MatrixForm[RowReduce[A]]]
```

we may read off the general solution to $AX = 0$, as a system in $x$, $y$, $z$, $u$, and $w$:

$$
\begin{aligned}
x &= 0 - 2y - 9u + 14w \\
y &= \text{anything} \\
z &= 0 + 7u - 11w \\
u &= \text{anything} \\
w &= \text{anything}
\end{aligned}
$$

*Note*: We do not bother to add on the right–hand side of the equation to our matrix before computing, since it consists entirely of $0$'s and does not change. Writing this in matrix form yields:

$$
\begin{bmatrix} x \\ y \\ z \\ u \\ w \end{bmatrix} = y \begin{bmatrix} -2 \\ 1 \\ 0 \\ 0 \\ 0 \end{bmatrix} + u \begin{bmatrix} -9 \\ 0 \\ 7 \\ 1 \\ 0 \end{bmatrix} + w \begin{bmatrix} 14 \\ 0 \\ -11 \\ 0 \\ 1 \end{bmatrix}
$$

In particular,

$$
S = \left\{ \begin{bmatrix} -2 \\ 1 \\ 0 \\ 0 \\ 0 \end{bmatrix}, \begin{bmatrix} -9 \\ 0 \\ 7 \\ 1 \\ 0 \end{bmatrix}, \begin{bmatrix} 14 \\ 0 \\ -11 \\ 0 \\ 1 \end{bmatrix} \right\}
$$

forms a spanning set for $Kernel(A)$. This set is, in fact, a basis. With a little practice, one learns to construct this spanning set directly from the reduced row echelon form of $A$. *Hint*: Erasing the rows from the vectors in $S$ corresponding to the basic variables leaves an identity matrix; the remaining entries in each vector come directly from the free columns of the reduced matrix.

## Exercise 24.1

Show that these vectors are independent by appealing to Definition 22.5. *Hint*: Focus on the pattern of $0$'s and $1$'s.

NullSpace[] is a built-in *Mathematica* command that will compute a basis for the kernel of a matrix automatically. *Note*: The terms "nullspace" and "kernel" are interchangeable. For example:

- `NullSpace[A]`

Notice how this produces exactly the same spanning set $S$ as we obtained by hand. As usual, one should learn to determine a basis for $Kernel(A)$ directly from the reduced row echelon form of $A$, only using the computer as an additional verification.

Notice that the dimension of $Kernel(A)$ equals the number of free variables in $AX = 0$, which equals (number of columns $-$ $rank(A)$). Thus, to generate a matrix with a three-dimensional kernel, we would say:

- `A = MakeNiceMatrix[4,7, Rank->4];MatrixForm[A]`

### Exercise 24.2

- Compute a basis for $Kernel(A)$. *Note*: Although you may use the RowReduce[] command, you should only use NullSpace[] command to check your work. Verify directly that each of your basis vectors are in $Kernel(A)$ by multiplication.

### Exercise 24.3

Use the computer to create two more sample matrices of various sizes and ranks and repeat Ex. 24.2 on your examples.

## 24.3   Deleting from a Spanning Set

While a spanning set $S$ of a subspace $V$ is relatively easy to calculate, it is not quite as convenient as a basis for calculations. Although every vector in $V$ has *some* coordinates with respect to $S$, the coordinates may not be *unique*. It would be convenient if we could take a spanning set $S$ and somehow produce a basis $B$ from it. One simple-minded strategy would be to successively "throw out" dependent vectors, since they do not contribute anything essentially new to the span. The hope is that we will eventually be left with an independent set, which would *still* span $V$. This strategy works quite well. In fact, it is possible to determine which vectors to "throw away" all at once.

### 24.3.1   A Familiar Example

We have actually carried out this strategy before, in Lesson 20. There we considered the subspace $V = Image(M)$ where:

- `M = {{-2, 1, 1},{ 1,-2, 1},{ 1, 1,-2}}; MatrixForm[M]`

In that lesson, we observed that $Image(M) = Span(S)$, where $S$ is the set of columns of $M$. We may see this equality in practice if we define the set $S$ of columns of $M$ in *Mathematica*:

- `S = {v1, v2, v3} = {{-2,1,1}, {1,-2,1}, {1,1,-2}}`

Notice that although this looks just like a matrix, this represents the *set* of vectors given by the *columns* of $M$. On the one hand:

- ```
  v = M.{-1,3,2}; MatrixForm[v]
  ```

  is a vector in the image of $M$. If we consider $S$ as a spanning set for $V$, then $v$ may be considered as the vector with coordinates $(-1, 3, 2)$ (with respect to $S$):

- ```
  v = -1{-2,1,1} + 3{1,-2,1} + 2{1,1,-2}
  ```

  The equality of these two calculations follows, in general, by the Principle of Partitioned Matrices.

### Exercise 24.4

- Compute the vector in the span of $S$ with coordinates (with respect to $S$) which are $(3, -2, 1)$.

  *Note*: Because of this equality, $Image(M)$ is also referred to as the "column space" of $M$. In Ex. 20.9, we observed that:

- ```
  v3 == (-1) v1 + (-1) v2
  ```

  Given our experience from the previous lesson, this becomes clear from the reduced row echelon form of $M$:

- ```
  MatrixForm[RowReduce[M]]
  ```

  We then reasoned that the first two vectors of $S$ alone should form a spanning set for $Image(M) = Span(S)$. Moreover, because Gaussian elimination preserves columns, we may predict that, after dropping the third column vector, the first two columns will continue to be independent. This is easily verified:

- ```
  MatrixForm[RowReduce[{{-2, 1},{ 1,-2},{ 1, 1}}]]
  ```

  Thus, the first two columns alone are independent *and* span $V$; that is, they form a basis for $Image(M)$.

### Exercise 24.5

  For example, show how the vector $v$ from above is in the span of the first two columns by *only* using the dependence relation given above and the fact that $v$ has coordinates $(-1, 3, 2)$ with respect to $S$.

## 24.3.2   A More General Example

  In this section, we consider a more general example and provide a careful explanation of our answer. In general, one must be careful when dropping vectors from a spanning set, so that the span of the remaining vectors does not decrease. For example, consider the set of vectors:

- ```
  T = {v1,v2,v3,v4} = {{1,2,1},{1,0,-1},{1,-2,1},{1,4,3}}
  ```

  To prepare to use RowReduce[], we must first place these vectors as the columns of a matrix $A$.

- ```
  A = {{1,1,1,1},{2,0,-2,4},{1,-1,1,3}}; MatrixForm[A]
  ```
  ```

### Exercise 24.6

Describe the relationship between $T$ interpreted as a matrix and the partitioned matrix $A = \begin{bmatrix} v_1 & v_2 & v_3 & v_4 \end{bmatrix}$ in terms of matrix arithmetic. *Note*: This observation allows one to enter the data for any given problem more quickly, since one need only enter either the set *or* the matrix manually and *compute* the other by matrix arithmetic.

We may then determine the dependencies within $T$ by using RowReduce[]:

- `MatrixForm[RowReduce[A]]`

For example, we may observe that one cannot obtain $v_3$ from any of the other three vectors. Thus, we cannot throw out $v_3$ and still span the same subspace. On the other hand, each of the *other* three may be obtained from the remaining vectors so we may eliminate any one of these.

*Remember*: Any dependency we observe in the reduced row echelon form is exactly reflected in the original columns; thus, although we may argue by examining the columns of the reduced row echelon form, our conclusions are equally valid for the *original* set of vectors.

As with the previous example, we may even go on to predict that if we throw out any of the other vectors, the resulting set will be linearly independent. We do not need the computer to deduce that when we RowReduce[] again (after canceling out one of the other columns) we will obtain an indentity matrix. We may simply cancel out the column from the reduced row echelon form and reduce the remaining matrix in our heads. One should step through this process to convince oneself that it is valid.

### Exercise 24.7

Cancel out the second column from $A$ to obtain a matrix $A_1$. If we let the reduced row echelon form of $A$ be $U$, delete the second column from $U$ to obtain $U_1$. Explain why the reduced row echelon form of $U_1$ should be the identity matrix. Verify that $A_1$ and $U_1$ both row-reduce to $I$.

To summarize, we may conclude that $B = \{v_1, v_3, v_4\}$ is a basis for $Span(T)$, because:

- the corresponding columns in the row–reduced form of $A$ are independent, so $B$ is also linearly independent, and

- the second column in the row–reduced form of $A$ is equal to twice the first minus the fourth, so we know that $v_2 = 2v_1 - v_4$; in particular, $B$ spans the same space as $T$.

Although we have seen that relations among the columns are preserved under row operations, we may still want to double-check that our observed relation is valid:

- `v2 == 2v1 - v4`

*Note*: Although it may be tempting to put a given spanning set as rows of $A$ (since this would be easiest to type into *Mathematica*), Gaussian elimination *does not* preserve relations among the rows, so that would be a bad idea. Gaussian elimination does not even preserve the *order* of the rows!

## 24.3.3   Creating Supplementary Exercises

As before, we may use the RandomVectors[] command to create additional exercises. For example:

- ```
  U = RandomVectors[5]
  ```

produces a set of five three-dimensional vectors.

### Exercise 24.8

- Determine an appropriate subset of vectors of $U$ that serves as a basis for $Span(U)$. Explain why your choice is a basis, as in Section 24.3.2.

### *Drill Exercise 24.9*

- Repeat Ex. 24.8 for a set $P$ of five quadratic polynomials:

- ```
  Clear[t]; P = RandomVectors[5, Basis->{1,t,t^2}]
  ```

  *Hint*: Rewrite this problem in terms of column vectors.

**Definition 24.1** The *column space*, $Column(A)$, of an $n \times m$ matrix $A$ is the subspace of $R^n$ given as the span of the columns of $A$. Likewise, the *row space*, $Row(A)$, is the span of the rows of $A$ in $R^m$.

By our previous observations, $Column(A) = Image(A)$. This implies that $Row(A) = Image(A^T)$. In particular, solving problems involving $Row(A)$ and $Column(A)$ is the same as solving the problems we have already encountered involving spanning sets, except that here we are using a slightly different language.

### Exercise 24.10

- Use the following command to generate a matrix A:

- ```
  A = MakeNiceMatrix[3,4, Rank->2];MatrixForm[A]
  ```

  Determine a basis $B$ for the column space of $A$. Make sure to explain why your choice is independent and $span(B) = Column(A)$. Likewise, determine a basis $B'$ for the row space of $A$ and justify your answer. *Hint*: Write out the spanning sets for $Row(A)$ and $Column(A)$, and apply the techniques of this lesson.
  ```

## 24.3.4   Verification in *Mathematica*

We have seen (in Lesson 22) that there are many possible bases for the same space. Thus, one may obtain a correct answer that is different from others. Using the commands IndependentQ[] and SpanningQ[] that were introduced in Lesson 23, one may verify whether or not a set of vectors forms a basis, since a basis is simply a linearly independent set that spans the given space. For simplicity, we have combined these two commands into one command called BasisQ[], also in the "Bases.m" package. Similar to SpanningQ[], it will take two sets of vectors $S$ and $T$ and return "True" if $S$ is a basis for the subspace spanned by $T$, and "False" otherwise. If $T$ is omitted, it will determine if $S$ forms a basis for all of $R^n$ (of the proper dimension).

For example, to verify our calculations in Section 24.3.2, we would evaluate:

```
A = {{1,1,1,1},{2,0,-2,4},{1,-1,1,3}};
B = {{1, 2, 1}, {1, -2, 1}, {1, 4, 3}};
T = {{1, 2, 1}, {1, 0, -1}, {1, -2, 1}, {1, 4, 3}};
Print["Does B form a basis for the span of T?  ",
    BasisQ[B,T]]
```

In the same way, to verify that we have a basis for all three-dimensional column vectors, we would choose our spanning set $T$ as the standard basis (i.e., the columns of the $3 \times 3$ identity matrix). Omitting the second argument is equivalent to entering the standard basis. For example, to verify that the column space of $A$ is all of three-space, one would evaluate:

```
Print["Does B form a basis for 3-d Euclidean space?  ",
    BasisQ[B]]
```

### Exercise 24.11

Consider the set of vectors $V = \left\{ \begin{bmatrix} 1 \\ -1 \\ 2 \end{bmatrix}, \begin{bmatrix} -1 \\ 0 \\ 3 \end{bmatrix}, \begin{bmatrix} 0 \\ -1 \\ 5 \end{bmatrix}, \begin{bmatrix} 3 \\ -2 \\ 2 \end{bmatrix} \right\}$. Show that $V$ spans all of three-dimensional Euclidean space $R^3$, by showing that each of the vectors in the standard basis of Euclidean space $\left\{ \begin{bmatrix} 1 \\ 0 \\ 0 \end{bmatrix}, \begin{bmatrix} 0 \\ 1 \\ 0 \end{bmatrix}, \begin{bmatrix} 0 \\ 0 \\ 1 \end{bmatrix} \right\}$ is in the span of $V$. Find a subset of $V$, which forms another basis for three-dimensional space. As before, explain why your answer is correct. *Hint*: You can answer both questions by looking at the reduced row echelon form of a single matrix.

### *Drill Exercise 24.12*

Consider the set of vectors $Q = \{1, 2 + 2t, 1 - t + t^2, 2 - t^2\}$. Show that $Q$ spans the space of all polynomials of degree less than 3 (that is, $P_3$), by showing that each of the vectors in the standard basis of this space $\{1, t, t^2\}$ is in the span of $Q$. Find a subset of $Q$, which forms a basis for this space. Explain why your answer

- is correct. *Hint*: You can answer both questions by examining the reduced row echelon form of a single matrix.

## 24.4  Adding Vectors to a Linearly Independent Set

We may also obtain a basis by starting with a linearly independent set and *augmenting* it with independent vectors until we obtain a spanning set. Instead of doing this one vector at a time, we usually try to do so all at once. Actually, we usually first make it *too* large, by throwing in an *entire* spanning set, and then throwing out the *new* vectors that are unnecessary. The following exercises illustrate this technique.

### Exercise 24.13

- Expand the set $S = \left\{ \begin{bmatrix} 1 \\ 2 \\ 1 \\ 2 \end{bmatrix}, \begin{bmatrix} 0 \\ 2 \\ 1 \\ 2 \end{bmatrix} \right\}$ to a basis of $R^4$, by deleting vectors from:

$$S' = \left\{ \begin{bmatrix} 1 \\ 2 \\ 1 \\ 2 \end{bmatrix}, \begin{bmatrix} 0 \\ 2 \\ 1 \\ 2 \end{bmatrix}, \begin{bmatrix} 1 \\ 0 \\ 0 \\ 0 \end{bmatrix}, \begin{bmatrix} 0 \\ 1 \\ 0 \\ 0 \end{bmatrix}, \begin{bmatrix} 0 \\ 0 \\ 1 \\ 0 \end{bmatrix}, \begin{bmatrix} 0 \\ 0 \\ 0 \\ 1 \end{bmatrix} \right\}$$

### Exercise 24.14

- Expand the set $S = \left\{ \begin{bmatrix} 1 \\ 2 \\ 1 \\ 2 \end{bmatrix}, \begin{bmatrix} 0 \\ 2 \\ 1 \\ 2 \end{bmatrix} \right\}$ to a basis of the subspace spanned by $T =$

$\left\{ \begin{bmatrix} 3 \\ -2 \\ -1 \\ -2 \end{bmatrix}, \begin{bmatrix} 2 \\ 4 \\ 3 \\ 1 \end{bmatrix}, \begin{bmatrix} 1 \\ 4 \\ 2 \\ 4 \end{bmatrix} \right\}$, by deleting vectors from:

$$S' = S \cup T = \left\{ \begin{bmatrix} 1 \\ 2 \\ 1 \\ 2 \end{bmatrix}, \begin{bmatrix} 0 \\ 2 \\ 1 \\ 2 \end{bmatrix}, \begin{bmatrix} 3 \\ -2 \\ -1 \\ -2 \end{bmatrix}, \begin{bmatrix} 2 \\ 4 \\ 3 \\ 1 \end{bmatrix}, \begin{bmatrix} 1 \\ 4 \\ 2 \\ 4 \end{bmatrix} \right\}$$

## 24.5  Summary

Before we end this lesson, we should observe the fundamental fact that:

**Observation:**

Every vector space has a basis.

Since we may construct a spanning set for *any* vector space — simply take *every vector in the space* — we may then proceed to throw out dependent vectors to obtain a basis. Because in this case $S$ is an infinite set, the formal proof is a bit tricky, but this is the essential idea behind the proof. In particular, *every vector space has a basis*. However, there is no guarantee that that basis will only have a *finite* number of vectors in it! In this text, we will *only* consider vector spaces with finite bases, so that all of our vector spaces will be of finite dimension.

### Advanced Exercise 24.15

What would the coordinates of a vector in an infinite-dimensional vector space look like? Give three examples of infinite-dimensional vector spaces.

In this lesson, one should have learned how to:

- identify spanning sets for the image and the kernel of a matrix,

- delete vectors from a spanning set to obtain a basis, and

- add vectors to a linearly independent set to obtain a basis.

Now that we can *construct* bases, in the next lesson we will begin to explore how to *use* bases to discover:

- how bases give concrete descriptions of subspaces of Euclidean space,

- why the dimension of a vector space (i.e., number of vectors in a basis) is a well-defined number,

- why the rank of a matrix is also well-defined, and

- why the reduced row echelon form of a matrix is unique.

# The Theory of Bases

## Accompanied by Notebook

In Lesson 24 we discussed various ways to construct bases, ususally by considering the image and kernel of an appropriate matrix. In this lesson, we begin to examine the uses for bases. We will see that a basis $B$ leads to a concrete description of an abstract vector space $V$ by determining a coordinate isomorphism $C_B$ of $V$ with $R^n$, where $n$ is the dimension of $V$. In particular, we will show:

– how a basis provides a concrete description of any subspace of $R^n$.

This fact will also provide the crucial link between "linear transformations" and matrices that we first observed in Section 9.5 and which we will discuss in detail in Lessons 30 and 31. In Lessons 28 and 33 we will also demonstrate how an appropriate choice of basis provides the key step in many circumstances, such as solving overdetermined systems of equations and determining the long-term behavior of dynamical systems (like our economic model in Lesson 3).

Now that we have developed the appropriate theory, we may also provide simple explanations for some of the facts we have yet to prove, such as:

– why the dimension of a vector space (i.e., number of vectors in a basis) is a well-defined number,

– why the rank of a matrix is also well-defined, and

– why the reduced row echelon form of a matrix is unique.

This again demonstrates that mathematical theory is not created for its own sake, but to provide a language and a logical framework for explaining observed phenomena.

## 25.1  Bases, Coordinates, and the Isomorphism Principle

In Lesson 22, we saw how a given coordinate system (i.e., basis) allows one to associate the coordinates of the endpoint of a geometric vector (in standard position) with the coefficients of a linear combination of the corresponding basis vectors. For example, once we choose a particular basis $B = \{\vec{\imath}, \vec{\jmath}\}$, the vector in Fig. 25.1 may be uniquely specified as a linear combination of basis vectors:

$$\vec{v} = 3\vec{\imath} + 2\vec{\jmath}, \tag{25.1}$$

and visualized as in Fig. 25.2.

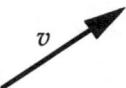

Figure 25.1: A Typical Two-Dimensional Geometric Vector

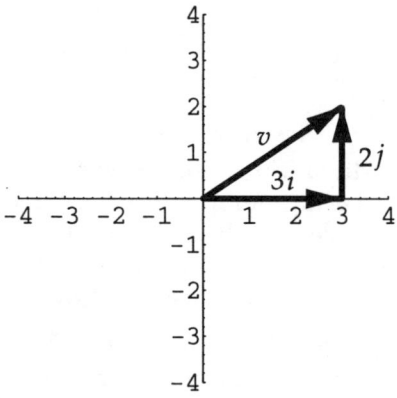

Figure 25.2: Visualizing a Linear Combination

The defining properties of a basis guarantee that we have a 1–1 correspondence between vectors and their coordinates with respect to the given basis. Specifically, every vector has a unique set of coordinates and each set of coordinates corresponds to a unique vector. We write "$\vec{v} \leftrightarrow (3, 2)$" to indicate that $\vec{v}$ corresponds to the coordinates 3 and 2, which is to say that Eq. 25.1 holds.

We may establish a similar correspondence in the vector spaces of column vectors $R^n$ and polynomials $P_n$, once we determine a choice of basis. Both $R^n$ and $P_n$ come with a natural choice of basis. In fact, the spanning sets from Lesson 23 are also independent and hence are bases.

## Exercise 25.1

Prove that each of those spanning sets are independent by appealing directly to Definition 22.5. Specifically:

(a) Show that $S = \left\{ \begin{bmatrix} 1 \\ 0 \\ 0 \end{bmatrix}, \begin{bmatrix} 0 \\ 1 \\ 0 \end{bmatrix}, \begin{bmatrix} 0 \\ 0 \\ 1 \end{bmatrix} \right\}$ is an independent subset of $R^3$, that is, three-dimensional Euclidean space.

(b) Show that $T = \{1, t, t^2\}$ is an independent subset of $P_3$, namely, the vector space of quadratic polynomials.

Although we cannot graph polynomials and column vectors directly, we may still choose a basis for any subspace of these vector spaces. Just as with geometric vectors, a basis establishes a 1–1 correspondence between any given vector and the coefficients expressing that vector as a linear combination of basis vectors. Thus, we often use the geometric language of coordinate systems and coordinates, in place of the algebraic language of bases and coefficients, even when the vectors involved have no natural geometric interpretation.

Moreover, by combining these correspondences, one may associate a vector from one vector space in a sensible manner with a vector in an entirely different space. For example, we may associate a geometric vector in $E^3$ with a set of three coordinates, in the usual way. Using a basis for another vector space of the same dimension, say $P_3$, we may then associate those coordinates with a quadratic polynomial. In this way, we may "translate" between any two vector spaces of the same dimension.

As a specific example, notice that the vector $\vec{v}$ of Fig. 25.1 may be associated with the coordinates $(3, 2)$, with respect to the basis $\{\vec{i}, \vec{j}\}$. These same coordinates correspond to the column vector

$$\begin{bmatrix} 3 \\ 2 \end{bmatrix}$$

with respect to the basis:

$$S = \left\{ \begin{bmatrix} 1 \\ 0 \end{bmatrix}, \begin{bmatrix} 0 \\ 1 \end{bmatrix} \right\}$$

because Eq. 25.1 is satisfied, if we replace $\vec{i}$ and $\vec{j}$ by the corresponding column vectors. Similarly, the same coordinates also correspond to the polynomial

$$P(t) = 3 + 2t$$

with respect to the basis:

$$T = \{1, t\}$$

Because this correspondence preserves the defining vector space operations of addition and scalar multiplication, in a certain sense these three vector spaces may be considered as the same! Formally, we say that all vector spaces of the same dimension are "isomorphic" ("iso-" = "same," "morph-" = "change," so they "change the same way") and the correspondence between them is called an "isomorphism." This Principle of Isomorphism is fundamental to linear algebra. One should have begun to sense this close relationship between these three vector spaces in Ex. 21.6 and 22.5. The following exercise illustrates more precisely what we mean by saying this correspondence preserves addition and scalar multiplication.

## Exercise 25.2

Using the standard bases from Ex. 25.1, determine the column vectors corresponding to $P(t) = 2 - t$ and $Q(t) = 1 + 4t$. Compare these with the column vectors corresponding to $P(t) + Q(t)$ and $3P(t)$. From this example, explain what it means to say that this correspondence preserves addition and scalar multiplication.

Using this correspondence, we may solve a linear algebra problem, which may have been posed in a particular vector space, within any isomorphic vector space. We have already seen this principle in action in Lesson 23. We saw that we could rephrase any question regarding polynomials in terms of column vectors and obtain correct results, when translated back into statements about polynomials. Specifically, to determine a linear combination of $P(t) = 2+3t$ and $Q(t) = -1+4t$ that would yield $R(t) = 5-t$, we may translate this into an equivalent question involving $u = \begin{bmatrix} 2 \\ 3 \end{bmatrix}$, $v = \begin{bmatrix} -1 \\ 4 \end{bmatrix}$, and $w = \begin{bmatrix} 5 \\ -1 \end{bmatrix}$. Once the problem has been solved in the vector space of column vectors, it is a simple matter to translate the answer back into the language of polynomials.

In general, if we are careful to phrase a given problem entirely in the language of linear algebra, that is, only using the linear operations of addition and scalar multiplication, then we may carry out the solution in the most convenient vector space available. That is because all computations in the chosen vector space will be perfectly mirrored in the original vector space of the problem. To use a science fiction analogy, all vector spaces are like parallel universes, where everything that happens in one occurs in all the others in the same way, but only the names for things are changed. Put another way, this correspondence between vector spaces is simply another type of change of viewpoint.

Once we choose a basis $B$ for a given finite-dimensional vector space $V$, this "isomorphism" property follows directly. Logically, we may provide an informal proof of this property in three steps:

1. $B$ establishes a 1–1 correspondence between $V$ and the set of all possible coordinates.

2. Because coordinates with respect to a basis are unique, the properties of a vector space require addition and scalar multiplication to be carried out "coordinatewise."

3. Therefore, the vector space operations in every vector space must reflect the corresponding operations on the set of coordinates, and hence the correspondence between vectors will preserve addition and scalar multiplication.

The next exercise should clarify what is meant in step 2 of this informal proof.

## Exercise 25.3

Let $\{v, w\}$ be a basis and assume that $s \leftrightarrow (3, 2)$ and $t \leftrightarrow (-1, 4)$ (that is, $s = 3v + 2w$, etc.) then:

(a) Write $s$ and $t$ in terms of $v$ and $w$.

(b) Compute $s + t$ and $5s$.

(c) Determine the coordinates of $s + t$ and $5s$, that is, express them as linear combinations of the basis vectors $v$ and $w$.

(d) Describe in general what happens to the coordinates when you add or multiply by a scalar.

Because of the simplicity, concreteness, and direct reflection of coordinates in *column* vectors, we usually convert any linear algebra problem into the equivalent problem phrased in $R^n$. Using column vectors we may perform explicit computations, by computer, to solve linear algebra questions that arise in a variety of different settings. Although the arrow notation of Ex. 25.3 does suggest a 1–1 correspondence, it does not indicate its dependence on the basis $B$. In Lesson 31 we will want to investigate precisely how this correspondence depends on a choice of basis. Thus, we introduce a new, more helpful notation for this correspondence.

**Definition 25.1** If $B = \{b_1, \ldots, b_n\}$ is a basis of an $n$-dimensional vector space $V$, we denote the coordinates $a_i$ of $v \in V$ with respect to $B$, written as a column vector, by

$$C_B(v) = \begin{bmatrix} a_1 \\ \vdots \\ a_n \end{bmatrix} \in R^n, \text{ so that:}$$

$$v = a_1 b_1 + \cdots + a_n b_n \tag{25.2}$$

For example, instead of writing $v \leftrightarrow (1, 2, 3)$, we will now write $C_B(v) = \begin{bmatrix} 1 \\ 2 \\ 3 \end{bmatrix}$.

Notice how we use functional notation to emphasize that this is a functional correspondence. This correspondence provides another important example of a linear transformation. Notice also how we indicate the dependence of the coordinate function on the choice of basis by employing a subscript in the function name. Finally, we should mention that, although a basis is a set of vectors that mathematically has no natural ordering of the elements, when we discuss coordinates, we must take the order of the vectors into account. Thus, our coordinate isomorphism depends on an *ordered* basis.

## 25.2  Bases and Inverses

Since independent and spanning sets are closely related to left- and right-inverses, the concept of a basis is directly connected with the notion of a two-sided inverse. Specifically, by combining Theorems 23.1 and 23.2, we obtain the following theorem.

**Theorem 25.1** *If $A = \begin{bmatrix} v_1 & \cdots & v_q \end{bmatrix}$ is an $p \times q$ matrix and $B = [b_j]$ is a $q \times 1$ column vector, then the following are equivalent:*

(a)  *$AX = B$ has a unique solution for every $B$,*

(b)  *$A$ is bijective (injective and surjective), or we say that it represents a 1–1 correspondence,*

(c)  *$A$ is invertible,*

(d)  *$rank(A) = p = q$,*

(e)  *$Kernel(A) = \{0\}$ and $Image(A) = R^p$, and*

(f)  *$\{v_1, \cdots, v_q\}$ is a basis.*

This theorem provides an easy way of determining if a set $S$ of $n$-dimensional column vectors forms a basis for $R^n$. Namely, place the vectors as columns of a matrix $A$ and determine whether or not $A$ is invertible.

### Exercise 25.4

- Determine if:

$$B = \left\{ \begin{bmatrix} 1 \\ 2 \\ 1 \end{bmatrix}, \begin{bmatrix} 1 \\ 0 \\ -1 \end{bmatrix}, \begin{bmatrix} 1 \\ 1 \\ 1 \end{bmatrix} \right\}$$

forms a basis for $R^3$. Determine the coordinates for $v = \begin{bmatrix} 2 \\ 3 \\ -1 \end{bmatrix}$ in terms of $B$. *Hint*: We must express $v$ as a linear combination of vectors in $B$. What is $C_B(v)$? Verify your answer by appealing to Definition 25.2.

### Exercise 25.5

- Argue that $B = \{1 + 2t + t^2, 1 - t^2, 1 + t + t^2\}$ forms a basis for $P_3$. If $v = 2 + 3t - t^2$, compute $C_B(v)$. *Hint*: Use the Isomorphism Principle. Verify your answer directly in $P_3$.

### Exercise 25.6

- Using the basis $B$ from Ex. 25.5, compute $C_B(w)$ for $w = 1 - t - 2t^2$. Make sure to verify your answer.

## 25.3  Bases Give Concrete Representations for Subspaces

In Section 24.3.1 we constructed a basis $B = \{v, w\}$ for the image of:

$$M = \begin{bmatrix} -2 & 1 & 1 \\ 1 & -2 & 1 \\ 1 & 1 & -2 \end{bmatrix}$$

Specifically, $B$ consisted of the first two columns of $M$, $v = \begin{bmatrix} -2 \\ 1 \\ 1 \end{bmatrix}$ and $w = \begin{bmatrix} 1 \\ -2 \\ 1 \end{bmatrix}$.

This implies that $Image(M)$ is a two-dimensional subspace of three-dimensional Euclidean space.

- We may visualize $S = Image(M)$ by constructing the corresponding coordinate system. One may construct a picture of this by evaluating the BasisPicture[] command in the accompanying Notebook. Notice how the origin is actually in the center of the picture. We have also provided an animation to view this picture from all sides.

Using this basis, we obtain a convenient representation of any vector $u$ in the subspace $S$ in terms of its two coordinates with respect to $B$. For example, the vector $u = 2v - w$ has coordinates $(2, -1)$ with respect to $B$, so we may visualize this in Fig. 25.3.

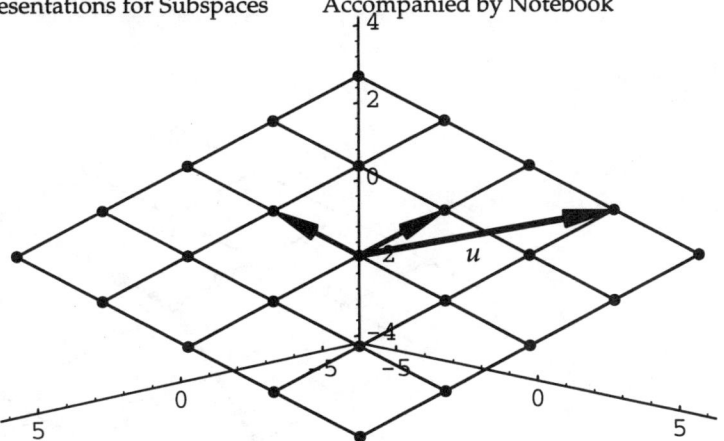

Figure 25.3: Visualizing $u = 2v - w$

Following the same sort of reasoning as in Section 20.3, we may easily infer the general solution to the system:

$$MX = u \qquad (25.3)$$

Since the coordinates for $u$ with respect to the columns of $M$ are given by the vector $\begin{bmatrix} 2 & -1 & 0 \end{bmatrix}^T$ (where we have taken the coordinate for the last column to be 0, since it is unecessary), we can conclude that $X = \begin{bmatrix} 1 \\ -2 \\ 1 \end{bmatrix}$ is one particular solution to Eq. 25.3. We observed in Section 20.3 that the general solution to a linear equation always consists of a copy of the kernel of the system that has been translated to go through any particular solution. We have already computed a basis for $Kernel(M)$ in Ex. 20.6, namely, the set $\left\{ \begin{bmatrix} 1 \\ 1 \\ 1 \end{bmatrix} \right\}$, so that the entire kernel is all multiples $c \begin{bmatrix} 1 \\ 1 \\ 1 \end{bmatrix}$ of this vector. Putting these two facts together, we can describe the general solution of Eq. 25.3 parametrically by the formula:

$$X = \begin{bmatrix} 1 \\ -2 \\ 1 \end{bmatrix} + c \begin{bmatrix} 1 \\ 1 \\ 1 \end{bmatrix} \qquad (25.4)$$

## Exercise 25.7

- Use Gaussian elimination to solve Eq. 25.3 directly and compare your answer with Eq. 25.4.

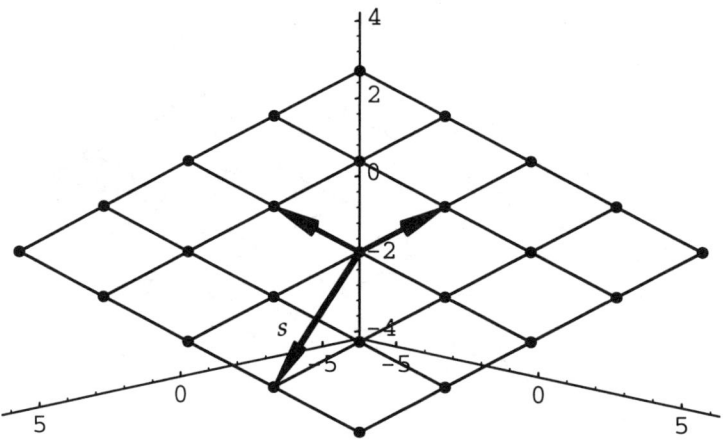

Figure 25.4: Visualizing $s$ in $Image(M)$

### Exercise 25.8

- From Fig. 25.4, determine the coordinates of $s$ with respect to $B$. Use this to determine the general solution to $MX = s$. Compare your solution with that obtained directly by Gaussian elimination.

*Important*: Since all vector spaces are isomorphic, one may visualize *any* subspace in this manner — even if the vectors involved are polynomials, and the geometric intuition obtained from this picture does not mislead.

## 25.4   Four Theoretical Results

Although bases are very useful from a practical standpoint, they are also a valuable theoretical tool. For example, bases allow us to explain why the dimension of a vector space is a well-defined number. In Lesson 22, we originally defined the dimension of a vector space as the number of vectors in any basis for that space. This would be a very poor definition, unless one could be confident that every basis for a given space contains the same number of vectors. This fact follows easily from an interesting little result that we will label as Lemma 25.3, which is commonly known as the "Steinitz Exchange lemma." Intuitively, this lemma says we may "swap" $n$ vectors between two given linearly independent sets $S$ and $T$.

This lemma allows us to illustrate an important type of proof, known as "proof by induction." This style of proof is often employed when the statement to be proven depends on a positive integer $n$. Logically, such a proof consists of a sequence of similar steps. In this case, if we may show that we may "swap" a *single* vector, then by repeating this argument $n$ times, we obtain Lemma 25.3. Thus, we first prove:

**Lemma 25.2** *Given a linearly independent subset $T = \{w_1, \ldots, w_m\}$ and a single, nonzero vector $v$, if we replace any vector in $T$ with $v$, we obtain another linearly independent set $U$.*

**Proof:**

There are two natural possibilities to consider.

*Case 1: $v$ is not in the span of $T$.*

In this case, $T \cup \{v\}$ is then a linearly independent set. In particular, if we then throw out some vector, not equal to $v$, the result $U$ will continue to be linearly independent.

*Case 2: $v$ is in the span of $T$.*

By assumption $v$ is some linear combination of vectors in $T$:

$$v = a_1 w_1 + \cdots + a_m w_m \tag{25.5}$$

where we can assume "without loss of generality" that $a_1 \neq 0$. (If one of the coefficients is not 0, we may simply relabel the vectors in $T$ so that the coefficient appears first in Eq. 25.5.) It remains to show that $U = \{v, w_2, \ldots, w_m\}$ is linearly independent. We prove this by contradiction. That is, we assume that $U$ is *not* linearly independent and derive a contradiction. In this event, we would have a nontrivial equation

$$0 = b_1 v + \cdots + b_m w_m \tag{25.6}$$

(i.e., not all the coefficients are 0). Plugging in Eq. 25.5 yields:

$$
\begin{aligned}
0 = b_1 v + \cdots + b_m w_m &= b_1 (a_1 w_1 + \cdots + a_m w_m) + \cdots + b_m w_m \\
&= (b_1 a_1) w_1 + \cdots + (a_m + b_m) w_m
\end{aligned}
$$

Since $T$ is linearly independent, all of these coefficients must equal 0. In particular, since $a_1$ is not 0, $b_1$ must be 0. Moreover, the first term in Eq. 25.5 drops out leaving a nontrivial relation among the vectors of $T$ (specifically, among $w_2, \ldots, w_m$). This contradicts our assumption of linear independence, so the assumption that $U$ is *not* linearly independent must be *false*. In other words, $U$ must be linearly independent.

Since these two cases cover every possibility and we have proven the lemma true in each case, it must be true in general. Q.E.D.

Now we may prove the Steinitz Exchange lemma by induction.

**Lemma 25.3 (Steinitz Exchange)** *Given two linearly independent sets $S = \{v_1, \ldots, v_n\}$ and $T = \{w_1, \ldots, w_m\}$ where $n \leq m$, one may replace any collection of $n$ vectors in $T$ with the vectors of $S$ to obtain another linearly independent set $U$.*

**Proof:**

Notice that this lemma depends on the positive integer $n$, which is the number of vectors in $S$. Logically, this means that we have *infinitely* many different statements to prove, namely, for each possible value of $n$. A proof by induction is a technique for verifying each of these infinite number of cases by providing two separate arguments. Intuitively, we sometimes compare this process to climbing a ladder.

*The Anchor Step*

First, we prove the lemma assuming that $n = 1$. This is exactly Lemma 25.2, which we have already proven. This means that we can "climb onto" the first rung of the ladder.

*The Inductive Step*

This step is rather subtle. In this step, we *assume* that the lemma is true when $n = i$ (where $i$ is a *particular*, yet *unspecified* value), an assumption known as the "inductive hypothesis." To complete the inductive step, we must demonstrate the lemma when $n = i + 1$; that is, we must show that we may exchange $i + 1$ vectors. Intuitively, this corresponds to climbing from one "ladder rung" to the next.

In this context, our inductive hypothesis states that we have independent sets $S = \{v_1, \ldots, v_{i+1}\}$ and $T = \{w_1, \ldots, w_m\}$, where $i + 1 \leq m$. Since $S'$ contains $i$ vectors, we may apply our inductive assumption to the sets $S' = \{v_1, \ldots, v_i\}$ and $T$ to obtain an independent set $U'$. We may then apply Lemma 25.2 to $v_{i+1}$ and $U'$ to replace one of the vectors in $U'$ that was originally in $T$, to obtain a linearly independent set $U$. Thus, we have shown that we may exchange $i + 1$ vectors, thus, completing the inductive step.

Together these two steps show that we may exchange any number of vectors. Specifically, the anchor step shows that we may exchange the first vector. Then by applying the inductive step repeatedly, with $i = 1, \ldots, n - 1$, one may show that it is possible to exchange $n$ vectors. Intuitively, if we can climb on the ladder and climb from any rung to the next, then we can climb to *any* given rung of the ladder. Q.E.D.

Using Lemma 25.3, it is easy to prove:

**Lemma 25.4** *If $A$ is a set of vectors in a vector space $V$ with $n$ elements, $B$ is another set with $m$ elements and $n < m$, then:*

(a) *if $A$ is a spanning set, $B$ must be linearly dependent;*

(b) *if $B$ is linearly independent, then $A$ cannot be spanning.*

*Note*: This lemma makes *no* assumption about $A$ being contained in $B$ or vice versa. It only depends on their relative sizes. In more intuitive terms, this lemma says that bases are maximal linearly independent sets and minimal spanning sets. That is, bases are "between" the linear independent and spanning sets of a vector space. We will make this statement more precise after we prove the lemma.

**Proof:**

We first prove Lemma 25.4a by contradiction, that is, we assume that $B$ is linearly *independent*. We may drop vectors from $A$ until we obtain a basis, call it $S$, containing $k$ elements, with $k \leq n$. By the Steinitz Exchange lemma, letting $B$ play the role of $T$, we may replace $k$ of the vectors in $B$ by the vectors in $S$ to obtain a linearly independent set $U$. Notice that $U$ must strictly contain $S$, since $k \leq n < m$, so that it contains more vectors. In particular, there is at least one vector $v$ in $U$ that is not in $S$.

This leads to the desired contradiction. Since $S$ is a basis, it is spanning and we may write $v$ in terms of the vectors of $S$. This equation would exhibit a dependence relation within $U$, which contradicts the fact that $U$ is independent. Thus, $B$ must have been linearly dependent.

Statement Lemma 25.4b is simply the "contrapositive" of Lemma 25.4a. That is, while Lemma 25.4a is of the form "$P$ implies $Q$," Lemma 25.4b is of the form "Not $Q$ implies not $Q$." Although it is a well-known fact from mathematical logic that a statement and its contrapositive are logically equivalent, we may provide an explicit proof of Lemma 25.4b by contradiction. Assume that $B$ is linearly independent and $A$ is spanning. Since $A$ is spanning, Lemma 25.4a would imply that $B$ is linearly dependent, which contradicts our assumption. Q.E.D.

We may finally prove that the number of vectors in a basis is well-defined. We also obtain some other useful results for producing a basis. In precise mathematical terms, we have the following corollary.

**Corollary 25.5** *If $V$ is a vector space with a basis $B$ containing $n$ vectors and $A$ is a set of vectors in $V$ with $m$ vectors, then:*

*(a) If $A$ is a basis, then $n = m$.*

*(b) Conversely, if $A$ is independent or spanning and $n = m$, then $A$ is a basis.*

**Proof:**

We prove Corollary 25.5a by contradiction. If $n \neq m$, then either $n < m$ or $m < n$. Both cases lead to similar contradictions.

*Case 1: $n < m$*

Since $A$ is spanning, Lemma 25.4a implies that $B$ is linearly dependent. This contradicts the fact that $B$ is a basis. Thus, we cannot have $n < m$.

*Case 2: $m < n$*

This case is left as an exercise.

Since we cannot have $n < m$ or $m < n$, then we must have $n = m$.

To prove Corollary 25.5b, there are two cases:

*Case 1: A is independent*

In Lesson 24, we saw that we may add vectors to an independent set, such as $A$, to obtain a basis $A'$. Applying Corollary 25.5a to $A'$, we may conclude that $A'$ must contain $n$ vectors. In particular, $A'$ is no bigger than $A$, so that $A = A'$ and $A$ must have been a basis originally.

*Case 2: A is spanning*

This case is also left as an exercise.

Q.E.D.

Together, Lemma 25.4 and Corollary 25.5 imply that, in general, anything "greater than or equal to" a basis is (usually) spanning, while anything "strictly greater than" a basis is linearly dependent. Conversely, anything "less than or equal to" a basis is (usually) linearly independent, while anything "strictly less than" a basis cannot be spanning. *Aside*: One should compare this situation with the corresponding results for left- and right-inverses. By Theorems 23.1 and 23.2, one can see that these results are, in fact, equivalent.

For example, Ex. 23.11 illustrated the fact that if one takes a set $S$ of independent vectors that is not spanning, a randomly chosen vector $v$ will *probably* be independent of $S$, and $S' = S \cup \{v\}$ will be a larger independent set. This is because the difference in size between two subspaces of different dimensions is *very* great (as, for example, a line and a plane). In particular, if we randomly choose a sequence of vectors, we may continue to choose new vectors until we obtain a spanning set, and the set will *probably* remain independent.

Conversely, Ex. 24.8 showed that if one randomly chooses a set $S$ which contains more than enough vectors for a basis, it always possible to drop vectors from $S$ to obtain a linearly independent set $S'$, which *should* be a basis. The only way this procedure will fail to produce a basis is if $S'$ is *too* small. This means that at some point while choosing the vectors of $S$ we chose one or more *dependent* vectors *before* we obtain a spanning set. From our observations in the previous paragraph, this is highly unlikely. Thus, this procedure should *usually* produce a basis.

## Exercise 25.9

Complete the proof of Corollary 25.5a by showing that $m < n$ also leads to a contradiction.

## Exercise 25.10

Finish the proof of Corollary 25.5b by showing that if $A$ is spanning with $n$ vectors, then it must be independent. *Hint*: Think about *throwing away* vectors.

Now that we know that the dimension of a vector space is well-defined, we may prove that the rank of a matrix is also well-defined. We have seen experimentally that the rank of a matrix $A$ is the same as $dim(Image(A))$. Since we have proven that the dimension of a subspace is a well-defined mathematical quantity, the rank must be well-defined, as well.

**Corollary 25.6** *The dimension of the image of $A$ equals the number of leading 1's in any row echelon form of $A$. In other words, $rank(A) = dim(Image(A))$.*

**Proof:**

If $U$ is a row echelon form of $A$, it is clear that the columns containing the leading 1's form a basis for $Column(U)$. The concept of a basis is defined solely in term of dependence relations, which are preserved under Gaussian elimination. In particular, the same columns must form a basis for $Column(A) = Image(A)$. In particular, the number of leading 1's equals the number of vectors in a basis for $Image(A)$, which equals its dimension. Q.E.D.

Since we may now be sure that $rank(A)$ is a well-defined number, we may prove the observations from Ex. 11.7 and 20.12 about rank.

**Corollary 25.7** *If $A$ is $n \times m$, then*

(a) $rank(A) + dim(Kernel(A)) = m$

(b) $rank(A) = rank(A^T)$

**Proof:**

By Corollary 25.6, $rank(A) =$ the number of leading 1's in any row echelon form of $A$. In Section 24.2, we saw by construction that $dimKernel(A) =$ the number of free variables in any row echelon form of $A$. Since every column either contains a leading 1 or is free, Corollary 25.7a follows.

To prove Corollary 25.7b, by appealing to Corollary 25.6, it suffices to show that $Image(A)$ and $Image(A^T)$ have the same dimension. Rephrasing this in terms of the row and column spaces of $A$, we must show that $dim(Column(A)) = dim(Row(A))$. If we let $k$ equal the number of leading 1's in the reduced row echelon form of $A$, by Corollary 25.6, it only remains to show that $k = dim(Row(A))$. If we now let $U$ denote the reduced row echelon form of $A$, we will demonstrate that the $k$ nonzero rows of $U$ form a basis for $Row(A)$.

Just as in the case of $Kernel(A)$, the pattern of 1's and 0's in $U$ guarantees that the nonzero rows of $U$ are independent. Thus, we only must verify that $Row(U) = Row(A)$, that is, that the nonzero rows of $U$ form a spanning set for $Row(A)$. Since Gaussian elimination consists of a sequence of elementary row operations, it suffices to show that each type of row operation does not change the span of the rows. Assume that the nonzero rows of $A$ are $S = \{v_1, \ldots, v_k\}$ and the rows after one row operation are $T = \{w_1, \ldots, w_k\}$. Then there are three cases to consider:

*Case I: Switch two rows:*

This does not change the *set* of rows at all, so that $S = T$ and their spans are the same.

*Case II: Multiply a row by a nonzero number:*

The two sets are essentially the same, except $w_i = cv_i$ for some $i$. We may obtain $w_i$ from $v_i$ and vice versa by scalar multiplication. Thus, $span(S) = span(T)$.

*Case III: Add a multiple of one row to another:*

This time the only difference between $S$ and $T$ is that $w_i = cv_j + v_i$ for some $i \neq j$. This means that we may obtain $w_i$ from $v_i$ and $v_j$, so that $T \subset span(S)$ and $span(T) \subset span(S)$. Conversely, since by this point in Gaussian elimination $v_j = w_j$, we have the equation $v_i = -cw_j + w_i$. In particular, we may obtain $v_i$ from $w_i$ and $w_j$, so that $S \subset span(T)$ and $span(S) \subset span(T)$. Together these two containments imply that $span(S) = span(T)$.

Q.E.D.

As a final application of the theory of bases, we may demonstrate that the reduced row echelon form of a matrix $A$ is unique. Because so many calculations in linear algebra depend on this reduce row echelon form, it is useful to know that that this form is unique. *Note*: This also provides another proof that $rank(A)$ is well-defined.

**Corollary 25.8** *There is only one reduced row echelon form of a matrix $A$.*

**Proof:**

By our observations in Lesson 17, we know that Gaussian elimination and back-addition may be summarized by multiplication by a single invertible matrix $M$ and we may write $A = MU$ where $U$ is a reduced row echelon matrix. To demonstrate this Corollary, we assume that we have two different reduced row echelon decompositions $M_1 U_1 = A = M_2 U_2$ and verify that $U_1 = U_2$.

Since $M_1$ is invertible, we may multiply by $M^{-1}$ to obtain:

$$U_1 = (M_1)^{-1} M_2 U_2 \tag{25.7}$$

Although we cannot guarantee that $M_1 = M_2$, if we let $B = (M_1)^{-1} M_2$, we may show that the effect of $B$ on $U_2$ is negligible, that is, $BU_2 = U_2$. This will complete the proof, since, together with Eq. 25.7, we would have $U_1 = BU_2 = U_2$.

The columns of $U_1$ containing the leading 1's are clearly the leftmost set of linearly independent columns. The corresponding original columns of $A$ will have the same property and so will the corresponding columns of $U_2$. These must also be the columns containing the leading 1's in $U_2$; that is, the leading 1's are in the same positions in $U_1$ and $U_2$. By the Principle of Partitioned Matrices, Eq. 25.7 says that the columns of $U_1$ are simply $B$ times the corresponding columns of $U_2$. Since the columns containing the leading 1's are identical in $U_1$ and $U_2$, $B$ has no effect on these columns. Since all of the other columns of $U_2$ are linear combinations of the leading columns, $B$ has no effect on the remaining columns of $U_2$ either, implying that $BU_2 = U_2$. Q.E.D.

## 25.5  Summary

In this lesson, one should have learned:

- how bases allow us to describe vector spaces in a very concrete way,

- why the dimension of a vector space and the rank of a matrix are well-defined numbers,

- why our observations about rank in Lessons 9 and 14–16 are correct, and

- why the reduced row echelon form of a matrix is unique.

Notice that we have finally have tied together all our loose ends in the study of *exact* solutions to systems of equations. We may describe a system of equations as a matrix equation $AX = B$. We may also describe $A$ as a linear transformation and the process of computing the general solution for $X$ as determining the pre-image of $B$ in the range of this transformation. The only vectors that will have corresponding solutions will be those in $Image(A)$, which may be concretely specified by a basis. Likewise, we may describe the pre-image of $B$ as a translation of the $Kernel(A)$ by any single solution, which in turn may be concretely specified by a different basis. We may utilize left- and right-inverses to compute such solutions, and we have finally placed all of our observations on inverses on a solid, logical foundation. In the next lesson, we will extend these observations to more general linear equations by providing precise formal definitions for the notions of "linear transformation" and "linear subspace."

# Subspaces and Linear Transformations

We have seen how a basis $B$ allows us to represent a geometric vector by its coordinates. In Lesson 22, we did this for vectors in $E^2$, while in Lesson 25 we showed how this technique may even be applied to a plane $S$ embedded in $E^3$. Although these may seem to be rather different examples, because all two-dimensional vector spaces are "isomorphic," logically they are very similar. To this point, we have focused on the vector spaces $E^n$, $R^n$, and $P_n$. In this lesson, we begin to emphasize that planes such as $S$ are vector spaces, as well.

Logically, we only distinguish a vector space such as $S$ from our previous examples by referring to it as a "subspace," in that it is a *subset* of another vector space. Taken together, $E^n$, $R^n$, $P_n$, and their subspaces comprise most of the examples of vector spaces that arise in practice. Thus, we should supply a precise, formal definition of a "linear subspace" and learn to construct simple proofs involving subspaces. Because every subspace may be described as the kernel or image of some linear transformation, we will also discuss the formal definition of a "linear transformation." Specifically, we will:

- provide the formal definition of a linear subspace,

- introduce the formal notion of a linear transformation,

- examine a number of specific examples,

- point out how these two concepts have already appeared in the text, and

- practice reading and doing some more simple proofs.

In Lesson 30, we will see that every linear transformation may be described by a matrix. Using the techniques of Lesson 24, we may then provide a coordinate system (i.e., a basis) for the image or kernel of any transformation. In this way, we may obtain a concrete description of any abstract vector space that we may encounter.

## 26.1   The Definition of a Linear Subspace

Having supplied a rigorous definition of a vector space in Definition 21.1, we may now rigorously define the term "linear subspace."

**Definition 26.1** A linear *subspace*, $U$, of a given vector space, $V$, is a subset of $V$ that is also a vector space under the vector operations of $V$.

A commonly used alternative definition is:

**Definition 26.2** A linear *subspace*, $U$, of a given vector space, $V$, is a nonempty subset of $V$ that is closed under the vector operations of $V$.

By "closed" we mean that if one chooses any two vectors $u$ and $v$ in $U$, then any linear combination of $u$ and $v$ will also be in $U$.

One may prove that Definitions 26.1 and 26.2 are logically equivalent, so either one may be used. Definition 26.1 is convenient from a theoretical point of view, but Definition 26.2 is the easiest to use in practice. Definition 26.1 clearly implies Definition 26.2. Specifically, if $U$ is a vector space, then its addition and scalar multiplication operations are functions with the indicated domain and range:

- addition, $+ : U \times U \rightarrow U$ and

- scalar multiplication, $\cdot : R \times U \rightarrow U$.

This implies that if the inputs of each function are in $U$, then the outputs are also in $U$, so that $U$ is closed under these linear operations. Conversely, if $U$ is closed under addition and scalar multiplication of some larger vector space $V$, then these operations $V$ may be restricted to produce operations on $U$. The vector space axioms are automatically satisfied for these restricted operations since they yield the same values as the operations for $V$, which are known to satisfy the necessary axioms.

We illustrate this definition by considering various subsets of vector spaces that may or may not be linear subspaces. Each of the following subsections will focus on a separate example and formally verify or disprove Definition 26.2. These examples are intended to serve as templates to follow, when one attempts the exercises at the end of the entire section.

## 26.1.1 Example 1

Consider the subset of $R^2$ given by $S = \left\{ \begin{bmatrix} x \\ y \end{bmatrix} \in R^2 \,\middle|\, 2x = y \right\}$. Intuitively, since $S$ is a straight line through the origin, we expect this set to be a linear subspace of $R^2$. To provide a formal verification, we appeal to Definition 26.2.

*S is nonempty:*

Consider $v = \begin{bmatrix} x \\ y \end{bmatrix} = \begin{bmatrix} 1 \\ 2 \end{bmatrix}$. Clearly $2x = 1(1) = 2 = y$, so $v$ satisfies the defining condition of $S$, and so $v \in S$. In particular, $S$ is not empty.

*Closed under +:*

Consider two general vectors $v = \begin{bmatrix} x_1 \\ y_1 \end{bmatrix}$ and $w = \begin{bmatrix} x_2 \\ y_2 \end{bmatrix}$ in $S$. This means that each must satisfy the defining condition of $S$:

$$2x_1 = y_1 \tag{26.1}$$
$$2x_2 = y_2 \tag{26.2}$$

We must show that $v + w = \begin{bmatrix} x_1 \\ y_1 \end{bmatrix} + \begin{bmatrix} x_2 \\ y_2 \end{bmatrix} = \begin{bmatrix} x_1 + x_2 \\ y_1 + y_2 \end{bmatrix}$ is also in $S$, which means that $v + w$ must also satisfy the defining equation for $S$. Adding Eq. 26.1 and Eq. 26.2 yields $2x_1 + 2x_2 = y_1 + y_2$ or $2(x_1 + x_2) = y_1 + y_2$. Since this is the defining equation for $v + w$, we may conclude that $v + w \in S$. Since $v$ and $w$ were arbitrary, we have shown that the sum of any two vectors in $S$ is also in $S$, so that $S$ is closed under addition.

*Closed under · :*

This is quite similar to the previous case. Consider a general vector $v = \begin{bmatrix} x \\ y \end{bmatrix}$ in $S$ and an arbitrary real number $r \in R$. This means that $2x = y$. We must show that $rv = r\begin{bmatrix} x \\ y \end{bmatrix} = \begin{bmatrix} rx \\ ry \end{bmatrix}$ is also in $S$. Multiplying $2x = y$ by $r$ gives $r2x = ry$ or $2(rx) = ry$. Since this is the defining equation for $rv$, we may conclude that $rv \in S$. Since $v$ and $r$ were arbitrary, we have shown that any scalar product of a vector in $S$ is also in $S$, that is, $S$ is closed under scalar multiplication.

Since we have verified all of Definition 26.2, this set $S$ must be a subspace of $R^2$.

## 26.1.2 Example 2

Now consider the set $T = \left\{ \begin{bmatrix} x \\ y \end{bmatrix} \in R^2 \, \middle| \, x^2 = y \right\}$. Intuitively, this is a parabola, which should not be a (linear) subspace. To provide a formal verification (or more accurately a falsification) of this statement, we must show only that Definition 26.2 breaks down at some *one* point. For example,

*Closed under · :*

Consider the specific vector $v = \begin{bmatrix} x \\ y \end{bmatrix} = \begin{bmatrix} 1 \\ 1 \end{bmatrix}$. This is in $T$ because $x^2 = 1^2 = 1 = y$. However, $2v = \begin{bmatrix} 2 \\ 2 \end{bmatrix}$ is not in $T$, since $2^2 = 4 \neq 2$ and the coordinates of $2v$ do not satisfy the defining equation of $T$. Q.E.D.

## Exercise 26.1

Determine two vectors in $T$ that may be used to demonstrate that $T$ is not closed under addition.

At this point, we must stop and logically compare these two examples. When we wish to show that a set *is* a subspace, we must provide a *general* proof that it is closed under linear combinations of *any* vectors and scalars. The reason is that Definition 26.2 implicitly involves the phrase "for every." Specifically, the term "closed" requires certain things to be true for *any* choice of vectors.

On the other hand, to show that a set is *not* a subspace, we are only required to produce *one* linear combination of vectors from the set whose result is not in the set itself. This is because the logical opposite of "for every" (i.e., "always") is "there exists" (i.e., "sometimes"). Practically, this means that to show that Definition 26.2 fails and that a given set is *not* a subspace, we may simply produce *one* specific case where it fails. We call such a negative case a "counterexample." *Note*: Although we were able to show that Definition 26.2 does not hold for this example in two different ways, so that $T$ is quite "far" from being a subspace, we should emphasize that we were only required to determine *one* counterexample to the definition.

## 26.1.3  Example 3

Both of our previous examples have been subspaces of $R^n$. Now we consider a more abstract example within $P_n$. Let $S = \{\, p(t) \in P_3 \mid p(1) = 0 \,\}$ in the space of polynomials of degree less than 3. Notice that if $p(t) = a + bt + ct^2$, then the defining condition for $S$ may be written as $p(1) = a + b + c = 0$. That is, the set $S$ is defined by a linear equation, similar to the example from Section 26.1.1. Thus, we expect this to be a subspace, and that the proof will follow a similar form as before.

*S is nonempty:*

Consider $p(t) = -2 + t + t^2$. Clearly $p(1) = -2 + 1 + 1^2 = 0$, so $p(t)$ satsifies the defining condition of $S$ and $S$ is not empty.

*Closed under $+$:*

Consider two general polynomials $p_1(t) = a_1 + b_1 t + c_1 t^2$ and $p_2(t) = a_2 + b_2 t + c_2 t^2$ in $S$. Call the sum $p_1(t) + p_2(t) = p(t)$. Then $p(1) = p_1(1) + p_2(1) = 0 + 0 = 0$, by the defining condition of $S$ applied to $p_1(t)$ and $p_2(t)$. In particular, $p(t)$ also satisfies the defining equation for $S$, which shows that $S$ is closed under addition.

*Closed under $\cdot$:*

As before, consider a general polynomial $p(t) = a + bt + ct^2$ in $S$ and an arbitrary real number $r \in R$. This means that $p(1) = 0$, and we must show that $rp$ is also in $S$. By the defining equation for $rp$, we have $rp(1) = r0 = 0$, so $rp \in S$ and $S$ is closed under scalar multiplication.

## 26.1.4  Example 4

Notice that the examples from Sections 26.1.1 – 26.1.3 are all defined as solutions to certain equations (i.e., $2x = y$, $x^2 = y$, and $a + b + c = 0$, respectively). Another general type of example is defined in terms of linear combinations. For example, consider the set $S = \{\, p(t) \in P_3 \mid p(t) = a + bt^2 \text{ for some } a \text{ and } b \in R \,\}$. This is a subspace of the space of polynomials of degree less than 3.

Although the general outline for this proof is similar to the previous examples, because the defining condition for $S$ contains the phrase "for some," there are some slight differences.

*S is nonempty:*

Consider $p(t) = 1 + t^2$. This is of the form $p(t) = a + bt^2$ where $a = 1$ and $b = 1$, so $p(t)$ satsifies the defining condition of $S$ and $S$ is not empty.

*Closed under +:*

Consider two general polynomials $p_1(t) = a_1 + b_1 t^2$ and $p_2(t) = a_2 + b_2 t^2$ in $S$. Call the sum $p_1(t) + p_2(t) = p(t)$. Then:

$$
\begin{aligned}
p(t) &= p_1(t) + p_2(t) = \left(a_1 + b_1 t^2\right) + \left(a_2 + b_2 t^2\right) \\
&= (a_1 + a_2) + \left(b_1 t^2 + b_2 t^2\right) \\
&= (a_1 + a_2) + (b_1 + b_2)\, t^2
\end{aligned}
$$

This last equation is of the form $p(t) = a + bt^2$ where $a = a_1 + a_2$ and $b = b_1 + b_2$. In particular, $p(t)$ satisfies the defining equation for $S$, which shows that $S$ is closed under addition.

*Closed under $\cdot$ :*

We leave this case as an exercise.

## Exercise 26.2

Complete the proof that Example 26.1.4 is a subspace by showing that $S$ is closed under scalar multiplication. That is, if $p_1(t) = a_1 + b_1 t^2$ and $r \in R$, show that $p(t) = rp_1(t)$ is of the form $a + bt^2$. Specifically, provide formulas for $a$ and $b$, in terms of $a_1$, $b_1$ and $r$, and verify that $rp_1(t) = a + bt^2$.

## Exercise 26.3

Determine which of the following sets are subspaces. In each case, either supply a formal verification of Definition 26.2 or produce a specific counterexample to the definition. *Hint*: Try to verify closure using specific vectors; if $S$ seems to be closed, decide whether $S$ is defined more easily as solutions to equations or as linear combinations and then mimic the proof from the corresponding example.

(a) $S = \left\{\, v \in R^2 \mid v = a \begin{bmatrix} 1 \\ 0 \end{bmatrix} + b \begin{bmatrix} 0 \\ 1 \end{bmatrix} \text{ for some } \textit{integers } a \text{ and } b \,\right\}$

(b) $S = \left\{\, \begin{bmatrix} x \\ y \end{bmatrix} \in R^2 \,\middle|\, x \geq 0 \text{ and } y \geq 0 \,\right\}$

(c) $S = \{\, p(t) \in P_3 \mid p(t) = 1 + at + bt^2 \text{ for some } a, b \in R \,\}$

(d) $S = \left\{ \, p(t) \in P_3 \mid \int_0^2 p(t)dt = 0 \, \right\}$

(e) $S = \{ \, f(x) = a\sin(x) + b\cos(x) \mid \text{ for some } a, b \in R \, \}$

(f) $S = \left\{ \, \begin{bmatrix} a & b \\ c & d \end{bmatrix} \, \middle| \, \text{ for some a, b, c, d } \in R \, \right\}$

Notice that Ex. 26.3e and Ex. 26.3f involve subsets of two new vector spaces — that of continuous, real-valued functions, $C\,(R)$, and $2 \times 2$ real matrices, $M_{2,2}\,(R)$, respectively. As usual, do not spend an inordinate amount of time on this section. If the proofs seem to take too long to complete, go on and return to them later.

## 26.2  Definition of a Linear Transformation

The notion of a "linear transformation" is one of the most fundamental in linear algebra. Most problems encountered in linear algebra may be phrased in terms of linear transformations. It is an abstract notion like that of a vector space. As with the concepts of vector space and inner product, an abstract approach allows us to tie together many seemingly different examples. For example, it will allow us to view all of our examples of subspaces in terms of kernels and images (and even those will be seen to be related). Also, it will allow us to change our frame of reference (i.e., our basis) on any given problem.

In Lesson 9 we discussed how matrices represent geometric linear transformations. In Section 9.5, we observed that the reverse is *also* true, namely that every linear transformation may be represented as a matrix. In this section, we provide a general definition of a "linear transformation" between *any* pair of vector spaces (not just $E^n$). In Lesson 30, we will show that this correspondence between matrices and linear transformations applies very generally. Out of all the different perspectives on matrices we have employed, this transformational perspective is the most powerful, in that it is the only one that explains *why* matrix multiplication is defined as it is. We will see, in Lesson 30, how matrix multiplication corresponds to the action of a linear transformation on an input vector, as well as the composition of two linear transformations.

As we first mentioned in Lesson 9, the term "transformation" is simply another word for "function." A linear transformation is simply a function between two vector spaces which preserves the linear operations of addition and scalar multiplication. For example, in Lesson 25 we introduced the coordinate transformation $C_B$. If $B$ is a basis of a vector space $V$, $C_B$ is a function from $V$ to $R^n$, where the coordinates of $C_B(v)$ are the coordinates of $v$ with respect to $B$. In Ex. 25.3, we verified that this correspondence preserves the vector space operations, so that this is an example of a linear transformation. In general, the formal definition may be expressed as follows:

**Definition 26.3** A function, $L \,:\, V \,\rightarrow\, W$, where $V$ and $W$ are vector spaces, is said to be a *linear transformation* from $V$ to $W$ iff $L$ preserves the vector space operations of addition and scalar multiplication. That is,

(a)  $L(v + w) = L(v) + L(w)$,      for every $v, w \in V$, and

(b) $L(rv) = rL(v)$,     for every scalar $r$ and vector $v \in V$.

## 26.3  Some Easy Examples

For the remainder of this lesson, we will explore Definition 26.3 by examining various concrete examples. Since most of the remaining exercises will be more algebraic in nature, we first consider two "algebraic" examples. The first example comes from calculus. Specifically, we take $L : P_3 \to P_3$, where $L(p(t)) = p'(t)$. For example:

$$L(t^2 + 3t) = (t^2 + 3t)' = 2t + 3 \qquad (26.3)$$

Linearity of this transformation is one of the most fundamental properties from calculus. It is common to express this property in terms of the two familiar rules:

- "One may differentiate term-by-term, which is to say, the derivative of the sum is the sum of the derivatives" (compare Definition 26.3a), and

- "One may ignore constants which multiply any term while one is differentiating" (compare Definition 26.3b).

To see the connection between these familar rules from calculus and Definition 26.3, we repeat the calculation in Eq. 26.3, putting in *every* logical step:

$$(t^2 + 3t)' = (t^2)' + (3t)' = 2t + 3t' = 2t + 3$$

Substituting the definition of $L$, this reads:

$$L(t^2 + 3t) = L(t^2) + L(3t) = 2t + 3L(t) = 2t + 3$$

Now it becomes clear how we have used Definition 26.3a to break up the sum and Definition 26.3b to pull out the constant 3.

### Exercise 26.4

Find a calculus book and photocopy the page where it states that differentiation is linear. Copy the pages that state that integration and taking limits are linear, as well.

As a second example, consider scalar multiplication by 2 in an arbitrary vector space $V$. Formally, this would be a transformation $M : V \to V$, where $M(v) = 2v$. One may easily verify that $M$ preserves both addition and scalar multiplication. For example, given any two vectors $v$ and $w$, we have

$$M(v + w) = 2(v + w) = 2v + 2w = M(v) + M(w)$$

where the two equalities on the end are simply the definition of $M$, and the middle follows by Axiom III(2) of Definition 21.1.

Similarly, for any vector $v$ and scalar $r$, we have:

$$M(rv) = 2(rv) = (2r)v = (r2)v = r(2v) = rM(v)$$

Here the equalities on the two ends come from the definition of $M$, the next two follow from Axiom II(1) of Definition 21.1, and the middle equality is simply commutativity of ordinary multiplication. Remember that each step in a proof is either based on an appeal to a definition or to a known property. In practice, to verify an equation, we often start at the ends [in this case, $M(rv)$ and $rM(v)$] and continue to rewrite both expressions until we obtain a common expression [in this case, $(2r)v$]. However, notice that we present our proof as a single calculation, from left to right.

Observe how the structure of these two proofs are very similar to the proofs from Section 26.1 for subspaces. To prove that a set $S$ is a subspace, we must show that $S$ is *closed* under linear combinations (i.e., addition and scalar multiplication). Likewise, to show that a function $L$ is a linear transformation, we must verify that $L$ *preserves* linear combinations. As with subspaces, to show that a function is not a linear transformation, we are only required to produce a *single* counterexample.

For example, intuitively the function $K\left(\begin{bmatrix} x \\ y \end{bmatrix}\right) = \begin{bmatrix} x^2 \\ y \end{bmatrix}$ should *not* be a linear transformation. We may demonstrate this by observing that $K\left(2\begin{bmatrix} 1 \\ 1 \end{bmatrix}\right) = K\left(\begin{bmatrix} 2 \\ 2 \end{bmatrix}\right)$ $= \begin{bmatrix} 2^2 \\ 2 \end{bmatrix} = \begin{bmatrix} 4 \\ 2 \end{bmatrix} \neq \begin{bmatrix} 2 \\ 2 \end{bmatrix} = 2\begin{bmatrix} 1 \\ 1 \end{bmatrix} = 2K\left(\begin{bmatrix} 1 \\ 1 \end{bmatrix}\right)$. Thus, $K(rv) \neq rK(v)$ for $r = 2$ and $v = \begin{bmatrix} 1 \\ 1 \end{bmatrix}$.

## 26.4   A Very Important Example

Just before Definition 26.3, we claimed that the coordinate transformation $C_B$ from Section 25.1 is a linear transformation. Because of its importance, we will verify this directly. Although we will assume, for simplicity, that our space is two-dimensional, with a basis $B = \{v_1, v_2\}$, the general argument is very similar.

Remember that $C_B$ is defined by the condition:

$$C_B(s) = \begin{bmatrix} \alpha_1 \\ \alpha_2 \end{bmatrix} \qquad \text{iff} \qquad s = \alpha_1 v_1 + \alpha_2 v_2 \tag{26.4}$$

In other words, $C_B(s)$ is the column vector expressing the coordinates of $s$, with respect to $B$. To verify Definition 26.3a, we must show that:

$$C_B(v + w) = C_B(v) + C_B(w), \qquad \text{for every } v, w \in V$$

To prove this, we take two arbitrary vectors, $v$ and $w$, and introduce notation for their coordinates:

$$C_B(v) = \begin{bmatrix} \alpha_1 \\ \alpha_2 \end{bmatrix} \quad \text{and} \quad C_B(w) = \begin{bmatrix} \beta_1 \\ \beta_2 \end{bmatrix} \tag{26.5}$$

Then we compute $C_B(v + w)$ by determining the coordinates of $v + w$. Using Eq. 26.4, we may rewrite Eq. 26.5 as:

$$v = \alpha_1 v_1 + \alpha_2 v_2 \quad \text{and} \quad w = \beta_1 v_1 + \beta_2 v_2 \tag{26.6}$$

Adding these equations yields:

$$\begin{aligned} v + w &= (\alpha_1 v_1 + \alpha_2 v_2) + (\beta_1 v_1 + \beta_2 v_2) = (\alpha_1 v_1 + \beta_1 v_1) + (\alpha_2 v_2 + \beta_2 v_2) \\ &= (\alpha_1 + \beta_1)v_1 + (\alpha_2 + \beta_2)v_2 \end{aligned} \tag{26.7}$$

## Exercise 26.5

From Eq. 26.7 and Eq. 26.4, determine $C_B(v+w)$. Using Eq. 26.6, determine $C_B(v)+C_B(w)$. Does $C_B$ look linear so far? Explain.

## Exercise 26.6

Supply a similar argument to verify Definition 26.3b, specifically that:

$$C_B(rv) = rC_B(v) \qquad \text{for every scalar } r \in R \text{ and vector } v \in V$$

The definition of a basis implies that $C_B$ establishes a 1–1 correspondence between a vector and its coordinates. In this case, we introduce some additional terminology.

**Definition 26.4** A linear transformation $L : V \to W$ that is also a 1–1 correspondence is called an *isomorphism*. In this case, $L$ posseses a well-defined (two-sided) *inverse*, denoted as $L^{-1} : W \to V$, satisfying the cancellation equations:

$$L^{-1}L(v) = v \quad \text{and} \quad LL^{-1}(v) = v$$

For example, $C_B$ possesses an inverse, $C_B^{-1}$, that does the opposite of $C_B$. Namely, it takes the coordinates of a vector to the vector itself. For example, if $B = \{v_1, \ldots, v_n\}$ and $C_B(s) = \begin{bmatrix} \alpha_1 \\ \vdots \\ \alpha_n \end{bmatrix}$, then $C_B^{-1}\left( \begin{bmatrix} \alpha_1 \\ \vdots \\ \alpha_n \end{bmatrix} \right) = s = \alpha_1 v_1 + \cdots + \alpha_n v_n$.

Notice that when $v_i \in R^m$, we may rewrite $C_B^{-1}$ as:

$$C_B^{-1}\left( \begin{bmatrix} \alpha_1 \\ \vdots \\ \alpha_n \end{bmatrix} \right) = \alpha_1 v_1 + \cdots + \alpha_n v_n = \begin{bmatrix} v_1 & \cdots & v_n \end{bmatrix} \begin{bmatrix} \alpha_1 \\ \vdots \\ \alpha_n \end{bmatrix}. \tag{26.8}$$

That is, in this case, $C_B^{-1}$ is multiplication by the matrix $\begin{bmatrix} v_1 & \cdots & v_n \end{bmatrix}$. We will employ the notation **B** to stand for this *matrix*. Because **B** represents the *linear transformation* $C_B^{-1}$, we will also use **B** interchangeably with $C_B^{-1}$, when $V \subset R^k$.

Because our notation for this matrix **B** is so very *similar* to our notation $B$ for the original *set* of vectors, you should be careful to bear in mind the distinction. Although this notation may lead to confusion if you are not careful, it also leads to a some very helpful and easy-to-remember formulas. For example, since $\left(C_B^{-1}\right)^{-1} = C_B$, if $V = R^n$, we obtain the simple rule:

$$C_B = \left(C_B^{-1}\right)^{-1} = \mathbf{B}^{-1}. \tag{26.9}$$

## 26.5   A Geometric Example

We also claimed that the geometric transformations of Lesson 9 were *linear* transformations. We may finally prove this claim by appealing to Definition 26.3. For example, let $L$ be the transformation that reflects geometric vectors in the plane through the $45°$ line. We will impose the standard Cartesian coordinate system, so that we may represent each vector by its corresponding column vector. In Section 9.5, we used linearity of this transformation to represent it by the matrix $A = \begin{bmatrix} 0 & 1 \\ 1 & 0 \end{bmatrix}$.

To *prove* that $L$ is a linear transformation, we first verify Definition 26.3a. We may do so in three logical steps:

1. $L(v + w)$ is obtained by adding $v$ and $w$, by the parallelogram rule, and then reflecting the result. This sequence of steps may be pictured as in Fig. 26.1.

2. On the other, hand $L(v) + L(w)$ is obtained by first reflecting $v$ and $w$ and then adding the results, again by the parallelogram rule. This sequence of steps may be pictured as in Fig. 26.2.

3. If we reflect the entire parallelogram from Fig. 26.1, we will necessarily obtain the parallelogram in Fig. 26.2, since two of the sides are $L(v)$ and $L(w)$. Thus, the reflection of the diagonal, namely $L(v + w)$, must equal the diagonal of the reflected parallelogram, namely $L(v) + L(w)$.

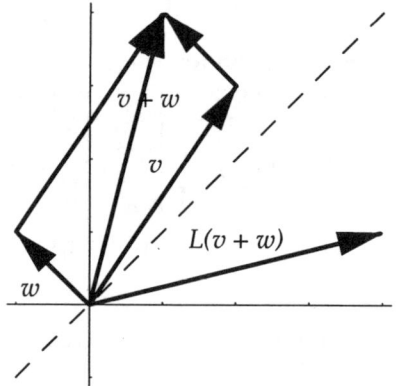

Figure 26.1: Add Then Reflect

Although we drew pictures with specific vectors, it is clear that the argument did not depend on the specific choices. Thus, it is a completely general argument.

This example demonstrates the general geometric character of linear algebra. Because all vector spaces are isomorphic, every linear transformation may be viewed in geometric terms. In particular, since we saw in Section 18.6 that every matrix may be decomposed into geometrically simple matrices, once we demonstrate the general correspondence between matrices and linear transformations, it follows that every linear transformation corresponds to a sequence of simple, "geometric" operations, such as reflections and stretches.

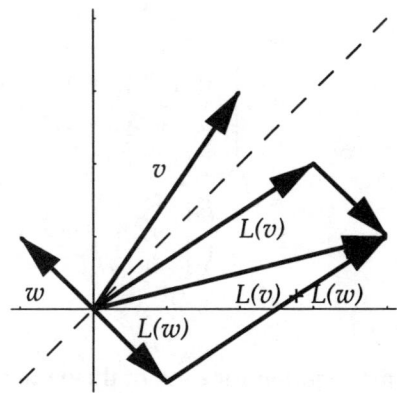

Figure 26.2: Reflect Then Add

## 26.6 The Most Important Example

The most important class of examples are of the form $L : R^m \to R^n$, such as $L : R^3 \to R^2$, where:

$$L\left(\begin{bmatrix} x \\ y \\ z \end{bmatrix}\right) = \begin{bmatrix} x + 2y - z \\ -x + y + 3z \end{bmatrix} \tag{26.10}$$

Notice how the coordinates of the output are linear expressions of the coordinates of the input. Notice also how much this resembles a system of linear equations. We will see that:

**Observation:**

Every linear transformation of *column vectors* is of this form.

Moreover, since every vector space is isomorphic to a vector space of column vectors:

**Observation:**

Every linear transformation $L$ may be written in this form, with respect to particular choices of bases for its domain and range.

We may verify that the $L$ defined by Eq. 26.10 is a linear transformation. Take any two columns vectors $v = \begin{bmatrix} x_1 & y_1 & z_1 \end{bmatrix}^T$ and $w = \begin{bmatrix} x_2 & y_2 & z_2 \end{bmatrix}^T$. Then:

$$\begin{aligned}
L(v + w) &= L\left(\begin{bmatrix} x_1 \\ y_1 \\ z_1 \end{bmatrix} + \begin{bmatrix} x_2 \\ y_2 \\ z_2 \end{bmatrix}\right) = L\left(\begin{bmatrix} x_1 + x_2 \\ y_1 + y_2 \\ z_1 + z_2 \end{bmatrix}\right) \\
&= \begin{bmatrix} (x_1 + x_2) + 2(y_1 + y_2) - (z_1 + z_2) \\ -(x_1 + x_2) + (y_1 + y_2) + 3(z_1 + z_2) \end{bmatrix} \\
&= \begin{bmatrix} x_1 + x_2 + 2y_1 + 2y_2 - z_1 - z_2 \\ -x_1 - x_2 + y_1 + y_2 + 3z_1 + 3z_2 \end{bmatrix}
\end{aligned}$$

$$= \begin{bmatrix} x_1 + 2y_1 - z_1 + x_2 + 2y_2 - z_2 \\ -x_1 + y_1 + 3z_1 - x_2 + y_2 + 3z_2 \end{bmatrix}$$

$$= \begin{bmatrix} x_1 + 2y_1 - z_1 \\ -x_1 + y_1 + 3z_1 \end{bmatrix} + \begin{bmatrix} x_2 + 2y_2 - z_2 \\ -x_2 + y_2 + 3z_2 \end{bmatrix}$$

$$= L\left( \begin{bmatrix} x_1 \\ y_1 \\ z_1 \end{bmatrix} \right) + L\left( \begin{bmatrix} x_2 \\ y_2 \\ z_2 \end{bmatrix} \right) = L(v) + L(w).$$

## Exercise 26.7

Provide the justification for each of the equalities in the previous computation.

## Exercise 26.8

Provide the corresponding argument to show that $L$ preserves scalar multiplication.

## Exercise 26.9

For each of the following transformations $L$, determine its domain and range. Then show that $L$ is linear or provide a counterexample to demonstrate that it is *not* linear.

(a) $L\left( \begin{bmatrix} x \\ y \end{bmatrix} \right) = \begin{bmatrix} x \\ 2y \end{bmatrix}$
(b) $L\left( \begin{bmatrix} x \\ y \end{bmatrix} \right) = \begin{bmatrix} x + y^2 \\ \sqrt{xy} \end{bmatrix}$

(c) $L\left( a + bt + ct^2 \right) = c + bt + at^2$
(d) $L\left( \begin{bmatrix} a & b \\ c & d \end{bmatrix} \right) = a + b + c + d$

We will return to these examples later.

Notice that linear transformations of column vectors look suspiciously like systems of linear equations. In Lesson 20, we observed that any system of linear equations may be written in the form $L(X) = B$. For example, we rewrote Eq. 20.1 as:

$$f(x, y) = (2, 3), \qquad \text{where} \qquad f(x, y) = (4x + 6y, 6x + 9y)$$

Because $f$ is a linear function, we would now denote this as $L : R^2 \to R^2$ and express the input and output as column vectors:

$$L\left( \begin{bmatrix} x \\ y \end{bmatrix} \right) = \begin{bmatrix} 4x + 6y \\ 6x + 9y \end{bmatrix}$$

This is of the form $L(X) = B$, where $X = \begin{bmatrix} x & y \end{bmatrix}^T$ is the unknown input and $B = \begin{bmatrix} 2 & 3 \end{bmatrix}^T$ is the desired output.

We observed in Lesson 20 that this transformational point of view applies to *any* system of equations. Although we could not have said this in Lesson 2, when they were first introduced, a system is said to be linear when the *transformation* that defines it is linear, *not* the other way around. Using this more general transformational point of view, we may now apply the *same* techniques that we have already learned to solve linear equations involving *other* types of linear transformations, such as differentiation, between vector spaces other than $R^n$.

## 26.7   Solving Linear Systems

In Lesson 20, we introduced the notions of the image and kernel of a system of equations or a matrix. These concepts generalize directly to an arbitrary linear transformation.

**Definition 26.5** Given a linear transformation $L : V \to W$, we define:

(a) $Kernel(L) = \{\, v \in V \mid L(v) = 0 \,\}$ and

(b) $Image(L) = \{\, w \in W \mid L(v) = w, \text{for some } v \in V \,\}$

It is clear that the image (kernel) of a system of equations or the associated matrix is the same as the image (kernel) of the corresponding linear transformation of column vectors. Moreover, because many vector spaces are not simply column vectors (such as $P_2$), and many linear transformations are not given by simple additive formulas [such as $L(p(x)) = \int_0^2 p(t)dt$], this definition is more general and ties together many superficially different examples.

We may still apply all of the geometric intuition and exactly the same language that we have developed so far to describe the solutions to any linear equation, that is, to any equation $L(X) = B$, where $L$ is a linear transformation. In particular, we have:

**Theorem 26.1** *If $L$ is a linear transformation, then the equation $L(X) = B$:*

(a) *has at least one solution iff $B \in Image(L)$, and*

(b) *if $v_0$ is a particular solution, then all solutions are of the form $v_0 + n$, where $n$ runs over all vectors $n \in Kernel(L)$. We abbreviate this by saying that the set of all solutions is $v_0 + Kernel(L)$.*

**Proof:**

We say that Theorem 26.1a is a tautology. This means that the two conditions say the same thing, but simply use different language. One should examine the definitions of the terms involved to see that this is the case.

Thus, we move on to Theorem 26.1b. In more mathematical terms this statement says that the set of all solutions $S_1 = \{\, v \mid L(v) = B \,\}$ is equal to the set

$$S_2 = v_0 + Kernel(L) = \{\, v \mid v = v_0 + n \text{ for some } n \in Kernel(L) \,\}$$

To show that two sets are *equal*, it suffices to show that each is *contained* in the other. We start by assuming that $v_0$ is a "fixed" solution (i.e., picked now and assumed to be *constant* throughout the proof), so that $L(v_0) = B$:

*Part 1: $S_1 \subset S_2$:*

If $v$ is any other solution in $S_1$, then $L(v - v_0) = L(v) - L(v_0) = B - B = 0$, so if we let $n = v - v_0, n \in Kernel(L)$ and $v = v_0 + n$. In particular, we have shown that $v$ is in $S_2$.

*Part 2:* $S_2 \subset S_1$:

Conversely, assume that $v$ is in $S_2$. This means that $v = v_0 + n$ for some $n \in Kernel(L)$. Then we have $L(v) = L(v_0 + n) = L(v_0) + L(n) = B + 0 = B$, so that $v$ is also a solution and thus in $S_1$.

We have shown that $S_1 \subset S_2$ and $S_2 \subset S_1$, so that we must have $S_1 = S_2$, which is what we needed to show. Q.E.D.

Geometrically, the operation of adding by a fixed vector like $v_0$ is a translation. In particular, taking $Kernel(L)$ to $v_0 + Kernel(L)$ translates the kernel by some particular solution to the equation $v_0$. In Lesson 20, we observed that this in the case of a particular transformation of two-dimensional column vectors. Now we have proved it for *any* linear transformation. For example, this is why one obtains the term "$+C$" in an indefinite integral. In that case, one is expressing the general solution to a linear equation, where $L$ is the differentiation operator and the constant function $f(x) = 1$ is a basis for $Kernel(L)$.

Every subspace $S$ can be described as either $Image(L)$ or $Kernel(L)$ for an appropriately chosen linear transformation $L$. For instance, consider the example from Section 26.1.4. This set is given as the span of $\{1, t^2\}$, which suggests that it is the image of the transformation given by the formula:

$$L\left(\begin{bmatrix} x \\ y \end{bmatrix}\right) = x(1) + y(t^2)$$

Notice that the domain has the same dimension as the size of the spanning set and the range must be a vector space containing the set, since the image is a subspace of the range. In other words, this formula defines a transformation $L : R^2 \to P_3$. By substituting these particulars into Definition 26.5b, we can see that:

$$
\begin{aligned}
Image(L) &= \{\, w \in W \mid L(v) = w, \text{ for some } v \in V \,\} \\
&= \{\, p \in P_3 \mid L(v) = p, \text{ for some } v \in R^2 \,\} \\
&= \left\{\, p \in P_3 \mid L\left(\begin{bmatrix} a \\ b \end{bmatrix}\right) = p, \text{ for some } \begin{bmatrix} a \\ b \end{bmatrix} \in R^2 \,\right\} \\
&= \{\, p(t) \in P_3 \mid a(1) + b(t^2) = p(t) \text{ for some } a \text{ and } b \in R \,\}
\end{aligned}
$$

so that we have, in fact, represented the desired subspace as an image.

Likewise, the example of Section 26.1.3 is conveniently described as $Kernel(K)$ for an appropriate linear transformation $K$. We know that the kernel of a transformation is a subspace of its domain. Thus, $K$ must have domain $P_3$. By Definition 26.5a, this subspace is described as the solution set to $K(p) = 0$. Since our example is defined by the equation $p(1) = 0$, this suggests that we take $K(p) = p(1)$. Since $p(1) \in R$, this implies that this defines a mapping $K : P_3 \to R^1$. As before, by comparing this with Definition 26.5b:

$$
\begin{aligned}
Image(L) &= \{\, w \in W \mid L(v) = w, \text{ for some } v \in V \,\} = \{\, p \in P_3 \mid L(p) = 0 \,\} \\
&= \{\, p \in P_3 \mid p(1) = 0 \,\}
\end{aligned}
$$

so that we have, in fact, represented the desired subspace as a kernel. Because every subspace possesses a basis, every subspace may be expressed as the image of some linear transformation. Moreover, we will see in Lesson 28 that every subspace may also be expressed as the kernel of some other linear transformation.

### Exercise 26.10

For the example from Section 26.1.1, as well as each of the subspaces from Ex. 26.3, determine a linear transformation for which the given subspace is either an image or a kernel. Make sure to specify the domain, range, and defining formula for each transformation.

## 26.8   Summary

In this lesson, one should have learned:

- the definition of a linear subspace,

- the definition of a linear transformation,

- the definition of the inverse of a linear transformation,

- some of the notation associated with linear transformations,

- a formula for $C_B$ and its inverse,

- how to verify that a given subset is or is not a linear subspace,

- how to verify that a given function is or is not a linear transformation,

- how systems of equations correspond to linear transformations, and

- the geometric structure of the solutions to a linear equation.

In particular, Theorem 26.1 provides a complete picture describing exact solutions to linear systems of equations in terms of the subspaces $Image(L)$ and $Kernel(L)$. Because we may compute bases for each of these subspaces, we have the tools to analyze such systems in great detail.

For the remainder of the text, we begin to focus on two other important topics of linear algebra. In particular, we will discuss a common technique for constructing *approximate* solutions to linear systems, called the Method of Least Squares. We will conclude by considering the heat flow and economic examples of Lesson 3, as "discrete dynamical systems." Specifically, we will discuss eigenvectors and eigenvalues (compare Ex. 18.7).

Specifically, in Lesson 28 we will discuss inner products (or "dot" products) and orthogonality. We will show how inner products allow us to improve on the techniques of Lesson 24 to obtain more useful coordinate systems. We will also gain some insight into the transpose operator $(\cdot)^T$. The concept of "orthogonality" will provide a more visual proof of Corollary 25.7. Moreover, we will suggest a natural geometric technique to construct *approximate* solutions for overdetermined systems.

# Inner Product Spaces

# Inner Products

Accompanied by Notebook

So far, we have only employed the operations of addition and scalar multiplication. Using only vector space operations, we may form linear subspaces (e.g., lines, planes, etc.) and perform parallel translation. Thus, we may describe the general solution set of a linear system as a subspace (i.e., the kernel) that has been translated by any particular solution of the system. Now we introduce an additional operation called an "inner product" of vectors.

An inner product allows us to perform a wider variety of geometric constructions. Using an inner product, we may detect perpendicular lines and planes, and measure lengths and angles between vectors. The standard inner product in $E^2$ and $E^3$, usually called the "dot product" of geometric vectors, is commonly presented in trigonometry and physics texts, but we will generalize this concept to other vector spaces, such as $R^n$ and $P_n$. Specifically, we will:

- provide a formal definition for an inner product on a vector space $V$,

- analyze the standard inner product on Euclidean space,

- demonstrate how an inner product allows us to compute "lengths" and "angles" in an arbitrary vector space $V$, and

- employ an inner product to detect "perpendicularity" of vectors and compute perpendicular projections.

When we employ an inner product in a vector space $V$, mathematically, we refer to $V$ as a "Hilbert space," after the mathematician David Hilbert. Hilbert spaces are used in the study of differential equations, quantum theory, and special relativity. As we continue to study Hilbert spaces, the geometric significance of the transpose operation will become clearer.

## 27.1 Definition of an Inner Product

An inner product of vectors is any operation (on pairs of vectors) that acts like a product. Formally, we define this as follows.

**Definition 27.1** An *inner product* is a symmetric, function $b(v, w)$ of two variables of a vector space, $b : V \times V \to R$, which is a linear transformation in each variable, or in other words, that is *bilinear*. When there will be no confusion, we often write $v \cdot w$ in place of $b(v, w)$. For this text, we will assume that all inner products are *positive definite*, which means that $b(v, v) \geq 0$ for all $v$, with equality only when $v = 0$. Using this notation, the defining conditions on $b$ may be described in more familiar terms:

(a) The condition that $b$ preserves addition in each variable resembles the distributive laws:

$$(v + w) \cdot u = v \cdot u + w \cdot u \qquad \text{and} \qquad u \cdot (v + w) = u \cdot v + u \cdot w$$

(b) The condition that $b$ preserves scalar multiplication in each variable resembles two associative laws:

$$(rw) \cdot u = r(w \cdot u) \qquad \text{and} \qquad u \cdot (rw) = r(u \cdot w)$$

(c) The condition that $b$ is symmetric resembles the commutative law:

$$v \cdot w = w \cdot v$$

(d) The condition that $b$ is positive definite resembles a familiar property for real numbers:

$$v \cdot v \geq 0, \qquad \text{with} \qquad v \cdot v = 0 \Leftrightarrow v = 0$$

Our fundamental example is the standard inner product on $R^n$, which is defined by the multiplication formula:

$$v \cdot w = v^T w. \tag{27.1}$$

*Note*: Although the result is actually a one-dimensional column vector, we commonly blur the distinction between such $1 \times 1$ matrices and ordinary real numbers. Notice also that *Mathematica* overloads the "." symbol, using it to represent the dot product of column vectors, the action of a matrix on a column vector (compare Section 9.2), and the product of two matrices.

One may verify this formula in the associated Notebook. Starting with Eq. 27.1, one begins to observe a connection between the algebraic operation of transposition with the geometric notion of an inner product. This connection is usually explored in more detail in advanced courses on functional analysis and tensor calculus.

For geometic vectors in $E^3$, we usually employ the standard basis, $B = \left\{ \vec{\imath}, \vec{\jmath}, \vec{k} \right\}$ corresponding to the $x$-, $y$-, and $z$-axes, to identify $E^3$ with $R^3$ (via $C_B$). This allows us to define a natural inner product on three-dimensional geometric vectors by Eq. 27.1. For example:

$$\vec{\imath} \cdot \vec{\imath} = 1 \qquad \text{since} \qquad \begin{bmatrix} 1 & 0 & 0 \end{bmatrix} \begin{bmatrix} 1 \\ 0 \\ 0 \end{bmatrix} = 1$$

Likewise:

$$\vec{\imath} \cdot \vec{\jmath} = 0 \qquad \text{since} \qquad \begin{bmatrix} 1 & 0 & 0 \end{bmatrix} \begin{bmatrix} 0 \\ 1 \\ 0 \end{bmatrix} = 0$$

The corresponding formulas for the other inner products of basis vectors should be clear. One may be familiar with these formulas from trigonometry or physics. Using these formulas, one may compute inner products in $E^3$ by direct algebraic manipulation. For example, if $v = \vec{i} + 3\vec{j} - 2\vec{k}$ and $w = 2\vec{i} - 4\vec{j} - 5\vec{k}$, then:

$$
\begin{aligned}
v \cdot w &= \left(\vec{i} + 3\vec{j} - 2\vec{k}\right) \cdot \left(2\vec{i} - 4\vec{j} - 5\vec{k}\right) \\
&= 2\vec{i} \cdot \vec{i} - 4\vec{i} \cdot \vec{j} - 5\vec{i} \cdot \vec{k} + 6\vec{j} \cdot \vec{i} - 12\vec{j} \cdot \vec{j} - 15\vec{j} \cdot \vec{k} - 4\vec{k} \cdot \vec{i} + 8\vec{k} \cdot \vec{j} + 10\vec{k} \cdot \vec{k} \\
&= 2(1) - 4(0) - 5(0) + 6(0) - 12(1) - 15(0) - 4(0) + 8(0) + 10(1) = 0
\end{aligned}
$$

## 27.2   The Geometry of the Inner Product

All important geometric facts about inner products stem from the formula:

$$v \cdot w = |v||w| \cos(\theta) \tag{27.2}$$

where $|v|$ stands for the length of $v$ and $\theta$ is the angle between $v$ and $w$. This may be *proven* in $E^3$, with the inner product introduced in Section 27.1. In an arbitrary Hilbert space, it is not obvious how to define the notions of "length" and "angle." For example, in $P_3$ how should one define the "angle" between two quadratic polynomials? That is why, in general, we employ Eq. 27.2 to *define* the concepts of length and angle. In this way, we may apply our geometric intuition in vector spaces that are not defined geometrically.

For example, to define the length of a vector $|v|$, we may use Eq. 27.2 with $v = w$. In this case, $\theta = 0$, so that Eq. 27.2 becomes:

$$v \cdot v = |v||v| = |v|^2 \qquad \text{or} \qquad |v| = \sqrt{v \cdot v} \tag{27.3}$$

Notice that if $v$ is in $E^3$ with the standard inner product, this is just the usual Euclidean length formula:

$$\left\| \begin{bmatrix} x \\ y \\ z \end{bmatrix} \right\| = \sqrt{x^2 + y^2 + z^2}$$

In particular, Eq. 27.2 is *provable* in Euclidean space when $v = w$. In general, we may prove Eq. 27.2 in $E^n$ using the Law of Cosines. The Law of Cosines says that $C = \sqrt{A^2 + B^2 - 2AB \cos(\theta)}$, for any triangle:

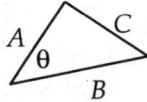

If we have any two vectors $v$ and $w$, the vectors $v$, $w$, and $v - w$ form a triangle:

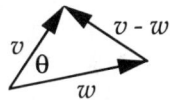

Applying the Law of Cosines with $A = |v|$, $B = |w|$, and $C = |v - w|$ gives:

$$|v - w| = \sqrt{|v|^2 + |w|^2 - 2|v||w| \cos(\theta)} \tag{27.4}$$

A little bit of algebra shows that Eq. 27.4 is equivalent to Eq. 27.2:

**Proof:**

First, square both sides of Eq. 27.4:

$$|v - w|^2 = |v|^2 + |w|^2 - 2|v||w|\cos(\theta)$$

Then apply Eq. 27.3:

$$(v - w) \cdot (v - w) = v \cdot v + w \cdot w - 2|v||w|\cos(\theta)$$

Expanding the left-hand side and and canceling like terms yields

$$v \cdot v - v \cdot w - w \cdot v + w \cdot w = v \cdot v + w \cdot w - 2|v||w|\cos(\theta)$$

so that:

$$v \cdot w + w \cdot v = 2|v||w|\cos(\theta)$$

Finally, Eq. 27.2 follows directly from commutativity of an inner product:

$$v \cdot w + w \cdot v = v \cdot w + v \cdot w = 2v \cdot w$$

Eq. 27.2 may be used to *define* the angle between two vectors $v$ and $w$ by the formula $\theta = \cos^{-1}\left(\frac{v \cdot w}{|v||w|}\right)$. In particular, Eq. 27.2 provides a test to determine when two vectors are perpendicular:

$$v \perp w \quad \text{iff} \quad \theta = 90° \quad \text{iff} \quad v \cdot w = 0 \tag{27.5}$$

As before, this is a *fact* in Euclidean space, and a *definition* in an arbitrary Hilbert space. It is common to employ the term "orthogonal" interchangeably with the word "perpendicular," leading to the following definition.

**Definition 27.2** We say that $v$ and $w$ are *orthogonal* iff $v \cdot w = 0$.

For example, the vectors $v$ and $w$ from Section 27.1 were orthogonal.

## Exercise 27.1

Carefully sketch the vectors $v = \begin{bmatrix} 2 \\ 3 \end{bmatrix}$ and $w = \begin{bmatrix} -6 \\ 4 \end{bmatrix}$. Show that these are perpendicular by actually measuring the angle, and verify that $v \cdot w = 0$. *Note:* You may use *Mathematica* to create the sketch.

## Exercise 27.2

Determine all of the orthogonal pairs of vectors in the set:

$$S = \left\{ \begin{bmatrix} 2 \\ 3 \\ 1 \end{bmatrix}, \begin{bmatrix} 6 \\ 1 \\ -1 \end{bmatrix}, \begin{bmatrix} 1 \\ -2 \\ 4 \end{bmatrix}, \begin{bmatrix} -2 \\ 1 \\ 1 \end{bmatrix} \right\}$$

### Exercise 27.3

As in Ex. 27.1, sketch the vectors $v = \begin{bmatrix} 2 \\ 5 \end{bmatrix}$ and $w = \begin{bmatrix} -3 \\ 7 \end{bmatrix}$. Measure the angle between the two vectors and compare your answer with the value given by Eq. 27.2.

### Exercise 27.4

Determine the angle between $v = \begin{bmatrix} 2 \\ 1 \\ 3 \\ -2 \\ 1 \end{bmatrix}$ and $w = \begin{bmatrix} -1 \\ -2 \\ 3 \\ 2 \\ 1 \end{bmatrix}$.

## 27.3    Inner Products and Projections

Besides lengths and angles, one may employ an inner product to compute perpendicular projections. The necessary formula follows directly from Eq. 27.2 by rewriting it in a slightly different form.

**Theorem 27.1** *Given two vectors $v$ and $w$ in a Hilbert space $V$, if $u$ is the projection of $w$ onto the line through $v$, then:*

$$|u| = \frac{v \cdot w}{|v|} \tag{27.6}$$

$$u = \frac{v \cdot w}{v \cdot v} v \tag{27.7}$$

**Proof:**

Using trigonometry and Fig. 27.1, we may see that $|u| = |w|cos(\theta)$. By Eq. 27.2, this is $\frac{v \cdot w}{|v|}$, proving Eq. 27.6. Since $u$ is a scalar multiple of $v$, we must now determine the correct value for $r$ so that $u = rv$. Using Eq. 27.6, we have $|rv| = |u| = \frac{v \cdot w}{|v|}$. Simplifying, we obtain $|r||v| = \frac{v \cdot w}{|v|}$ or $|r| = \frac{v \cdot w}{|v|^2}$, so that $u = \pm \frac{v \cdot w}{|v|^2} v = \pm \frac{v \cdot w}{v \cdot v} v$. Although one may verify that the positive sign always holds in this equation, we will leave this for now and return to this proof, once we have provided a precise definition for the concept of "perpendicular projection."

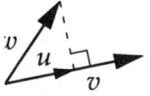

Figure 27.1: Projection of One Vector onto Another

### Exercise 27.5

Compute the projection of $s$ onto $t$ for:

(a) $s = 0$ and $t = \begin{bmatrix} 1 & 2 & 1 & 3 & -1 \end{bmatrix}^T$

(b) $s = \begin{bmatrix} 1 & 2 & 1 & -1 & 3 \end{bmatrix}^T$ and $t = \begin{bmatrix} -3 & 1 & -1 & 1 & 1 \end{bmatrix}^T$

(c) $s = \begin{bmatrix} 2 & -1 & 3 & 0 & 1 \end{bmatrix}^T$ and $t = \begin{bmatrix} 4 & 2 & 1 & 0 & 1 \end{bmatrix}^T$

Although we originally appealed to geometric intuition to derive Eq. 27.7, we formulate a precise, *algebraic* definition of the projection.

**Definition 27.3** The (perpendicular) *projection* of $w$ onto $v$ is that vector $u$ in the span of $v = \{rv \mid r \in R\}$ so that:

$$w - u \text{ is orthogonal to } v, \text{ i.e., } (w - u) \cdot v = 0 \qquad (27.8)$$

## Exercise 27.6

Show where $w - u$ is in Fig. 27.1 and explain how Eq. 27.8 accurately describes this picture.

Although geometric intuition suggests that Eq. 27.8 always has a unique solution, we may provide a formal verification of this fact, which will also complete the proof of Eq. 27.7. By rewriting Eq. 27.8 as

$$w \cdot v = u \cdot v \qquad (27.9)$$

we obtain the following important fact.

**Observation:**

$u$ has the same effect as $w$ when computing inner products with (any multiple of) $v$.

In particular, since $u = rv$, we have that:

$$w \cdot v = v \cdot (rv) = r(v \cdot v) \qquad (27.10)$$

Solving for $r$ yields $r = \frac{w \cdot v}{v \cdot v}$, so that $u = rv = \frac{w \cdot v}{v \cdot v} v$.

Although we are most familiar with the geometry of Euclidean space, many other vector spaces possess important inner products. In fact, many different inner products may be given for the same vector space. An important inner product on $C(R)$ is given by:

$$p \cdot q = \int_{-1}^{1} p(t)q(t)dt \qquad (27.11)$$

This example of a Hilbert space is used in the study of differential equations and when fitting curves to experimental data. A similar example is used for Fourier analysis.

## Exercise 27.7

Verify that Eq. 27.11 defines an inner product by showing that Definition 27.1a – 27.1c hold. *Note*: The proof of Definition 27.1d is more challenging and requires techniques from advanced calculus.

### Exercise 27.8

- Use the inner product from Ex. 27.7 to determine the length of $p(t) = 1 + 2t - t^2$, the angle between $p(t)$ and $q(t) = 3 - t + t^2$, and the projection of $p$ onto $q$.

*Note*: As in Ex. 27.5, by appealing to precise, formal definitions and theorems, we may employ the geometrical language of lengths and angles even when the vectors involved are not easily visualized.

## 27.4   Summary

In this lesson, one should have learned:

- the definition of an inner product on a vector space and of a Hilbert space,
- how inner products lead to formulas for the geometric terms perpendicular, length, angle, and perpendicular projection, and
- how these formulas express our geometric intuition for geometric vectors in $E_3$, thus allowing us to introduce these geometric notions, by analogy, into nongeometric vector spaces, such as $C(R)$ or $P_n$.

In Lesson 28, we will discuss how to extend the notion of orthogonality and orthogonal projection to subspaces. We will also use inner products to create nonstandard coordinate systems with *perpendicular* axes. In linear algebra terms, we will discuss "orthogonal" bases.

# Orthonormal Bases and Projections

## Accompanied by Notebook

In Lesson 27, we formalized the notion of an inner product and derived formulas for familiar geometric constructions solely in terms of the inner product. In particular, we showed how to determine if two vectors are perpendicular and how to construct the perpendicular projection of one vector onto another. In this lesson, we will see how to generalize these notions to subspaces. Namely, we will:

- define what it means for two subspaces $V$ and $W$ to be orthogonal,

- show how two important subspaces, namely, $Kernel(A)$ and $Row(A)$, are orthogonal for any matrix $A$, and

- discuss how to construct the projection of a vector $w$ into a subspace $V$.

Practically, this will allow us to:

- verify that $rank(A) = rank(A^T)$ and $rank(A) + dim(Kernel(A)) = ($ number of columns of $A$) via a compelling *geometric* argument,

- represent any subspace $S$ as $S = Kernel(A)$ for some $A$, that is, giving *equations* for $S$,

- uniquely choose solutions to an underdetermined system, and

- construct the best possible approximate solution to an overdetermined system, via perpendicular projection.

In Lesson 24, we learned to construct bases for particular subspaces. In this lesson, we will begin to discuss how to construct bases consisting of mutually perpendicular, unit-length vectors, which resemble the standard coordinate system for $E_3$. We will show how one may:

- employ an orthonormal basis for a subspace $V$ to compute projections into $V$.

In Lesson 29, we will formalize this procedure, called Gram-Schmidt orthonormalization. In recent years, this process has become even more important than Gaussian elimination. It will lead to a new decomposition formula $A = QR$, which may be used to solve both remaining problems in this text, namely, determining approximate solutions to linear systems and computing eigenvectors and eigenvalues.

# 28.1   Definitions and an Important Example

In this lesson, we will use the word "orthogonal" in a variety of ways. What this term will mean will depend on its context. For example, we may generalize Definition 27.2 to subspaces $V$ and $W$ in an obvious way.

**Definition 28.1** In all three of the following definitions, $w$ is a vector and $V$ and $W$ are subspaces in some vector space.

(a) We say that "$w$ is *orthogonal* to the subspace $V$" iff $w$ is orthogonal to every vector $v \in V$. We denote this by $w \perp V$.

(b) We say that two subspaces "$V$ and $W$ are *orthogonal*" iff every pair of vectors $v \in V$ and $w \in W$ are orthogonal. We denote this by $V \perp W$. *Note*: Since the order does not matter for individual vectors (i.e., $w \perp v$ iff $v \perp w$), it is irrelevant for subspaces, as well.

(c) If $W \subset V$, the *orthogonal complement* of $W$ in $V$ is the set of vectors in $V$ that are orthogonal to $W$, namely $\{v \in V \mid v \perp W\}$. This subset may be shown to be a subspace of $V$, which we denote by $W^{\perp}$.

### Advanced Exercise 28.1

Prove that $W^{\perp}$ is a subspace of $V$.

For Definition 28.1c to make sense, we should verify that $W^{\perp}$ is, in fact, orthogonal to $W$.

**Lemma 28.1** *For any subspace* $W$, $W \perp W^{\perp}$.

**Proof:**

If $w \in W^{\perp}$, $w \perp W$, so that $v \perp w$ for every $v \in W$. Since $v$ was arbitrary, $v \perp w$ for every pair of vectors $v \in V$ and $w \in W$, thus $W^{\perp} \perp W$.

Because both Definition 28.1a and 28.1b are "for every" statements, they tend to be difficult to verify, since they would seem to require an *infinite* number of orthogonality tests between individual vectors. However, one can show that it suffices to verify orthogonality with a spanning set. In other words:

**Lemma 28.2** *If* $V = span\,\{v_1, \ldots, v_n\}$, *then* $w \perp V$ *iff* $w \perp v_i$ *for* $i = 1, \ldots, n$.

**Proof:**

The forward implication is immediate. That is, if we know that $w$ is orthogonal to the span of a set $S = \{v_1, \ldots, v_n\}$ of vectors, then it is orthogonal to each vector in $S$. Explicitly, if $w$ is orthogonal to $V$, then $w$ is orthogonal to every vector in $V$; in particular, $w \perp v_i$ for every vector in the spanning set.

To prove the reverse implication, we must take an arbitrary vector $v \in V$ and verify that $v$ and $w$ are orthogonal, only using the assumption that $w \perp v_i$ for each $i$. This is also straightforward. By the definition of $V$ as a span, we know that $v = a_1 v_1 + \cdots + a_n v_n$ for some choice of coefficients $a_i$. Then $w \cdot v = w \cdot (a_1 v_1 + \cdots + a_n v_n) = a_1 w \cdot v_1 + \cdots + a_n w \cdot v_n$, by the algebraic properties of scalar multiplication. By assumption, each of these terms is 0, so $w \cdot v = 0$, and $w$ and $v$ are orthogonal. Q.E.D.

We have already seen an example of orthogonal subspaces. Using Lemma 28.2, it is easy to prove that the kernel and the row space of any matrix are orthogonal.

**Theorem 28.3** *For any $n \times m$ matrix, $A$, $RowSpace(A) \perp Kernel(A)$. In fact, $Kernel(A) = Row(A)^{\perp}$ in $R^m$.*

**Proof:**

First, we establish some notation. Let $A = \begin{bmatrix} v_1^T \\ \vdots \\ v_n^T \end{bmatrix}$, so that the rows of $A$ are the

(column) vectors $v_i \in R^m$ and, by Definition 24.1, $Row(A) = span\{v_1, \ldots, v_n\}$. Also, let $V = Row(A)$ and $W = Kernel(A)$. By Definition 28.1c and Lemma 28.2:

$$Row(A)^{\perp} = \{w \in R^n \mid w \perp Row(A)\}$$
$$= \{w \in R^n \mid v_1 \perp w, \ldots, v_n \perp w\} \tag{28.1}$$

By Definition 26.5 and the Principle of Partitioned Matrices, for any $w \in Kernel(A)$, we have:

$$0 = Aw = \begin{bmatrix} v_1^T \\ \vdots \\ v_n^T \end{bmatrix} w = \begin{bmatrix} v_1^T w \\ \vdots \\ v_n^T w \end{bmatrix}$$

precisely when $v_1^T w = 0, \ldots, v_n^T w = 0$. In other words, when $v_1 \cdot w = 0, \ldots, v_n \cdot w = 0$, or $v_1 \perp w, \ldots, v_n \perp w$. This implies that:

$$\{w \in R^n \mid v_1 \perp w, \ldots, v_n \perp w\} = \{w \in R^n \mid Aw = 0\} = Kernel(A) \tag{28.2}$$

Combining Eq. 28.1 and Eq. 28.2 yields $Row(A)^{\perp} = Kernel(A)$. Q.E.D.

## Exercise 28.2

Compute a basis for the kernel of $A = \begin{bmatrix} 2 & -1 \\ -4 & 2 \end{bmatrix}$. Compute a basis for $Row(A)$. Sketch both in the plane and verify that they are orthogonal. Describe how they "complement" one another. *Hint*: Think of other circumstances in which this term is used.

Theorem 28.3 has two important uses. It will show us how to choose solutions to underdetermined systems in a unique way. This will in turn help us to construct *approximate* solutions to overdetermined systems.

## 28.2   A Crucial Lemma

In this section we prove an important lemma characterizing orthogonal complements, namely that they are independent and together they span the entire space. This explains why we use the term "complement." Just as complementary angles combine in a nonoverlapping way to form a right angle, complementary subspaces do not intersect (except at the origin) and combine to span the entire space. Formally, the lemma may be stated as follows.

**Lemma 28.4** *Given two subspaces $W, V \subset R^n$, where $V = W^\perp$ in $R^n$:*

(a) $W \cap V = 0$, *so there is no nontrivial "overlap" between $W$ and $V$.*

(b) *If $S = \{w_1, \ldots, w_m\}$ is a basis for $W$ and $T = \{v_1, \ldots, v_k\}$ is a basis for $V$, then $S \cup T = \{w_1, \ldots, w_m, v_1, \ldots, v_k\}$ is a basis for $R^n$. In particular, $m+k = dim(W)+dim(V) = n$.*

(c) *Every $X \in R^n$ may be decomposed uniquely as a sum of vectors $X = w + v$, where $w \in W$ and $v \in V$.*

**Proof:**

To prove Lemma 28.4a, assume that $X \in W \cap V$. On the one hand, $X \in W$, which means that $X$ is orthogonal to everything in $V = W^\perp$. Since $X \in V$, $X$ must be orthogonal to itself, that is, $X \cdot X = 0$, so that $X$ has length 0. Since 0 is the only vector of length 0 in $R^n$, $X = 0$.

To prove Lemma 28.4b, we must show that $S \cup T$ is independent and spanning.

*$S \cup T$ is independent:*

If we had a dependence relation $0 = a_1 w_1 + \cdots + a_m w_m + b_1 v_1 + \cdots + b_k v_k$, then the vector $u = a_1 w_1 + \cdots + a_m w_m = -b_1 v_1 - \cdots - b_k v_k$ is both in $W$ and in $V$. Lemma 28.4 says that this vector must then be 0, so that $0 = u = a_1 w_1 + \cdots + a_m w_m$. Since $S$ is a basis, all of the $a_i$'s must be 0. Similarly, $0 = u = -b_1 v_1 - \cdots - b_k v_k$ so the $b_i$'s must be 0 as well.

*$S \cup T$ spans $R^n$:*

We will prove this by showing that $k = m-n$ (i.e., $V$ is $(m-n)$-dimensional), so that $m = n+k$. Corollary 25.5b will then imply that $S \cup T$ is independent with the right number of vectors, hence a basis (and therefore spanning). To do so, we form the $m \times n$ matrix $A$ with the vectors of $S$ as its rows, that is, $A = \begin{bmatrix} w_1^T \\ \vdots \\ w_m^T \end{bmatrix}$ (compare the proof of Theorem 28.3). By construction, we have $Row(A) = W$. Theorem 28.3 then implies that $V = W^\perp = Row(A)^\perp = Kernel(A)$. By Theorem 25.7b, we also know that $rank(A) = rank(A^T) = dim(Image(A^T)) = dim(Row(A)) = dim(W) = m$. By Theorem 25.7a, we may conclude that $dim(Kernel(A)) = n-m$, so any basis for $Kernel(A) = V$ will contain $n - m$ vectors (Theorem 25.5a). In particular, $k = n - m$.

This proves Lemma 28.4b.

Lemma 28.4c now follows directly from Lemma 28.4b. If $S \cup T = \{w_1, \ldots, w_m,$ $v_1, \ldots, v_k\}$ is a basis for $R^n$, every $X \in R^n$ may be written in one and only one way as $X = a_1 w_1 + \cdots + a_m w_m + b_1 v_1 + \cdots + b_k v_k = w + v$ for $w = a_1 w_1 + \cdots + a_m w_m \in W$ and $v = b_1 v_1 + \cdots + b_k v_k \in V$. Q.E.D.

### Exercise 28.3

For $A = \begin{bmatrix} 2 & 3 & 1 \\ 6 & 1 & -1 \\ 4 & -2 & -2 \end{bmatrix}$, compute a basis for $Row(A)$ and $Kernel(A)$. Use *Mathematica* to verify visually that these are orthogonal subspaces.

*Note*: Together Theorem 28.3 and Lemma 28.4b provide a nice visualization of why:

$$rank(A^T) + dim(Kernel(A)) = \text{number of columns of } A = m$$

Specifically, $rank(A^T) = dim(Image(A^T)) = dim(Row(A))$, which is a subspace of $R^m$ with orthogonal complement $Kernel(A)$.

## 28.3   Representing a Subspace as a Kernel

We have seen that every subspace $V$ of a finite-dimensional vector space has a basis; in particular, $V$ possesses a spanning set and thus may be represented as $Image(A)$ for some transformation $A$. In this section, we show how to determine another transformation $B$, so that $V = Kernel(B)$. Although our demonstration will take place entirely in $R^n$, this technique generalizes directly to arbitrary subspaces. In particular, we will produce a system of equations $BX = 0$, whose solutions are precisely those vectors in $V$. This representation makes it easy to verify whether or not a vector is in $V$, simply by matrix multiplication.

To accomplish this, we require a very natural-looking result, which says that if $V$ is the complement of $W$, then $W$ is the complement of $V$ (and vice versa).

**Theorem 28.5** *Given that $W$ and $V$ are subspaces of $R^n$,*

$$V = W^\perp \text{ in } R^n \qquad \text{iff} \qquad V^\perp = W \text{ in } R^n$$

*or, in other words, $(W^\perp)^\perp = W$ (see Theorem 12.1a).*

**Proof:**

By symmetry of the roles of $V$ and $W$, it suffices to show the implication in either direction. We will assume that $V = W^\perp$ and let $V^\perp = U$. We must show that $U \doteq W$. By Lemma 28.4b, we know that $dim(V) + dim(W) = n = dim(U) + dim(V)$. Thus, $dim(U) = dim(W)$.

Next, we show that $W \subset U$. That is, we must show that any $w \in W$ is also in $U = V^\perp$, i.e., $w \perp V$. Since $V = W^\perp$, this is just $w \perp W^\perp$. Lemma 28.1 guarantees that $W \perp W^\perp$, so that, in particular, $w \perp W^\perp$.

Because $W \subset U$ and they have the same dimension, they must be equal. Specifically, any basis $B$ for $W$ will have the correct number of (independent) vectors to be a basis for $U$. By Corollary 25.5b, $B$ will also span $U$, so that $W = span(B) = U = V^{\perp}$. Q.E.D.

Using Theorems 28.3 and 28.5, we may take any subspace $V$ that is specified by a spanning set, or in other words, that is represented as $V = Image(A)$, and determine a matrix $B$ so that $V = Kernel(B)$. For example, let $V$ be the subspace of $R^3$ spanned by:

$$S = \left\{ \begin{bmatrix} 2 \\ 3 \\ 1 \end{bmatrix}, \begin{bmatrix} 6 \\ 1 \\ -1 \end{bmatrix} \right\}$$

This means that $V = Image(A)$ for $A = \begin{bmatrix} 2 & 6 \\ 3 & 1 \\ 1 & -1 \end{bmatrix}$.

### Exercise 28.4

- Compute a basis $T$ for $Kernel(A^T)$. Form the matrix $B$ with the vectors in $T$ as its rows and compute $Kernel(B)$.

By construction, $Image(A) = V$ and $Kernel(A^T) = span(T) = Row(B)$. We know by Theorem 28.3 that $Kernel(B) = Row(B)^{\perp} = Kernel(A^T)^{\perp} = (Row(A^T)^{\perp})^{\perp}$. By Theorem 28.5, this is $Row(A^T) = Image(A) = V$. Thus, $V = Kernel(B)$.

### Exercise 28.5

- Verify that your basis for $Kernel(B)$ from Ex. 28.4 spans $Image(A)$, by exhibiting the appropriate, explicit dependence relations.

### Exercise 28.6

- Use $B$ to show that $v = \begin{bmatrix} 18 & -5 & -7 \end{bmatrix}^T$ is in $V$.

This procedure works in general. That is, given a description of a subspace $V$ as the image of some matrix $A = \begin{bmatrix} v_1 & \cdots & v_n \end{bmatrix}$ (i.e., given as $V = span\{v_1, \ldots, v_n\}$ via a spanning set), we may compute a matrix $B$ so that $V = Kernel(B)$. Conversely, if we have $V = Kernel(B)$, we may compute a basis for $Kernel(B)$. Since a basis necessarily is a spanning set $\{v_1, \ldots, v_n\}$ for $V$, $V = Image\left( \begin{bmatrix} v_1 & \cdots & v_n \end{bmatrix} \right)$.

## 28.4   Solving Underdetermined Systems of Equations

In Lesson 20, we saw in an informal way the importance of $Image(A)$ and $Kernel(A)$ in solving systems of the form $AX = B$. In Theorem 26.1, we proved these results in general. In this section, we will extend these results, stressing the importance of $Row(A)$ in the solution process. We will show how $Row(A)$ allows us to pick a specific solution to an overdetermined system in a *unique* way.

**Theorem 28.6** *For any $n \times m$ matrix, $A$:*

(a) *$AX = B$ has some solution iff $B \in Image(A)$.*

(b) *If $X_o$ is any solution, then the set of all solutions is:*

$$X_o + Kernel(A) = \{X_o + w \,|\, w \in Kernel(A)\} \,.$$

(c) *If $AX = B$ has any solution, it has a unique solution $X_o \in Row(A)$.*

This makes precise our earlier geometric description of the effect of a matrix $A$ from Section 20.2. We observed that $A$ "collapses" the domain, mapping the result in a 1–1 fashion onto $Image(A)$. This theorem essentially says that $A$ projects the domain onto its row space (by collapsing each translate $X_o + Kernel(A)$ down to a single point) and then mapping the row space in a 1–1 fashion onto $Image(A)$.

**Proof:** (of Theorem 28.6)

Theorems 28.6a and 28.6b follow as a special case of Theorem 26.1, since a matrix $A$ is *also* a linear transformation. We only restate them here for the sake of completeness. To prove Theorem 28.6c, observe that if $X$ is any solution to $AX = B$, by Theorem 28.3 and Lemma 28.4c, we may uniquely decompose $X$ as $X = X_o + w$, with $X_o \in Row(A)$ and $v \in Kernel(A)$. Since $AX_o = A(X-w) = AX - Aw = B - 0 = B$, $X_o = X - w$ is also a solution. In particular, it is the *unique* solution in $Row(A)$. Q.E.D.

We may now supply a *geometric* explanation of Corollary 25.7b. Theorem 28.6c says that $A$ establishes a 1–1 correspondence between $Row(A) = Image(A^T)$ and $Column(A) = Image(A)$. Specifically, if we consider the restriction of the linear transformation given by $A$ to $Row(A)$, we have an isomorphism of vector spaces. In particular, they have the same dimension. Since Theorem 25.6 says that the dimension of each space equals the rank of $A^T$ and $A$, respectively, we obtain another proof of Eq. 16.1.

Theorem 28.6 has another important corollary that may be used to solve overdetermined systems and compute projections.

**Corollary 28.7** *For any $n \times m$ matrix $A$:*

(a) *If $A$ is surjective then $D = AA^T$ is invertible.*

(b) *If $A$ is injective then $D = A^T A$ is invertible.*

**Proof:**

To prove Corollary 28.7a, since $D = AA^T$ is $n \times n$, the Fredholm Alternative (Theorem 16.5) implies that it suffices to show that $D$ has a right-inverse. Likewise, by Theorem 16.2a, it suffices to show that $D$ is surjective, that is, $DX = B$ has a solution for every $B$. Given any $B$, since $A$ is surjective, $AY = B$ has a solution. In fact, by Theorem 28.6c, we know that $AY = B$ has a (unique) solution $Y \in Row(A)$, that is, of the form $Y = A^T X$, so that $B = AY = A(A^T X) = DX$. This means that for every $B$, we have a solution $X$ to $B = DX$. Thus, $D$ is right-invertible and square, hence invertible.

To prove Corollary 28.7b, notice that if $A$ is injective, $m = rank(A) = rank(A^T)$ $= dim(Image(A^T))$. Thus, $Image(A^T) = R^m$, so that $A^T$ is surjective and we may apply Corollary 28.7a to $A^T$ to conclude that $A^T(A^T)^T = A^T A$ is invertible. Q.E.D.

### Exercise 28.7

For $A = \begin{bmatrix} 2 & 6 \\ 3 & 1 \\ 1 & -1 \end{bmatrix}$, verify that $A$ is injective and that $D = A^T A$ is invertible. *Hint:* Use Theorem 16.4.

Although Theorem 28.6c tells us that there exists a unique way of choosing a solution to an underdetermined system $AX = B$, it does not supply a procedure for *computing* this solution. The proof of Corollary 28.7a suggests how one might do this in the special case when $A$ is surjective.

**Corollary 28.8** *If $A$ is surjective, then the unique solution $X_o \in Row(A)$ of $AX = B$ is given by the formula $X_o = A^T(AA^T)^{-1}B$.*

### Proof:

We first observe that since $A$ is surjective, Corollary 28.7a guarantees that $AA^T$ is invertible, so that this formula makes sense. Since $X_o = A^T (AA^T)^{-1} B = A^T Y$, where $Y = (AA^T)^{-1} B$, we may conclude that $X_o \in Image(A^T) = Row(A)$. For this to be the unique solution given by Eq. 28.6c, we only must verify that it is a solution. Since $AX_o = A\left(A^T (AA^T)^{-1} B\right) = (AA^T)(AA^T)^{-1} B = IB = B$, $X_o$ is the solution in the row space of $A$. Q.E.D.

### Exercise 28.8

Verify that $C$ is surjective and use Corollary 28.8 to compute the solution in $Row(A)$ to $CX = B$ where $C = \begin{bmatrix} 2 & 3 & 1 \\ 6 & 1 & -1 \end{bmatrix}$ and $B = \begin{bmatrix} 7 \\ 6 \end{bmatrix}$.

Now that we have a complete description of *underdetermined* systems, it remains to study *overdetermined* systems in more depth. We will solve such systems using the technique of "orthogonal projection."

## 28.5   Computing Projections

Definition 27.3 described the projection of one vector onto another vector. Now we generalize this to the projection of a vector onto a *subspace*.

**Definition 28.2** A vector $u \in V$ is said to be "the (perpendicular) *projection* of $w$ into $V$" iff $w - u$ is orthogonal to every vector in $V$.

At this point, it is not clear that our definition of *the* orthogonal projection makes sense. That is, we have yet to show that, given $w$ and $V$, there are *any* such vectors $u$, much less a *unique* one. In the remainder of this lesson we will show that there is a uniquely defined projection vector and derive three explicit formulas for this projection. We devote so much time to this topic because projection is the standard technique for computing *approximate* solutions to overdetermined systems. Specifically:

1. If $AX = B$ has no exact solution, $B$ is not in $Image(A)$.

2. If we compute the projection $B'$ of $B$ into $Image(A)$, $B'$ will be the nearest point in the image, so it is the best approximation to $B$.

3. We may then solve $AX = B'$ to obtain the best approximate solution.

Since *most* systems that arise in applications are overdetermined, it is important to understand this process in detail.

Given a subspace $V$, we now show one way to compute the projection of $w$ into $V$.

**Theorem 28.9** *Given $V$ with basis $S = \{v_1, \ldots, v_m\}$ and some vector $w$, the projection $u$ of $w$ into $V$ is given by*

$$u = A \left(A^T A\right)^{-1} A^T w \tag{28.3}$$

*where $A = \begin{bmatrix} v_1 & \cdots & v_m \end{bmatrix}$.*

**Proof:**

We must show that this formula for $u$ satisfies the conditions of Definition 28.2. Since $u$ is a multiple of $A$, it is in $Image(A) = V$. Also, we have the following equation:

$$A^T u = A^T \left(A \left(A^T A\right)^{-1} A^T w\right) = \left(A^T A\right) \left(A^T A\right)^{-1} A^T w = I A^T w = A^T w$$

This means that $A^T w = A^T u$, which implies that $0 = A^T w - A^T u = A^T (w - u)$, so that $w - u \in Kernel(A^T)$. By Theorem 28.3, we may conclude that $w - u \perp Row(A^T) = Image(A) = V$, so that $u$ satisfies all of the conditions of Definition 28.2. Q.E.D.

We should stop to observe that Eq. 28.3 is a well-defined formula. Since $S$ is a basis for $V$, $S$ must be linearly independent. Theorem 23.1 implies that $A$ is injective and Corollary 28.7b guarantees that $A^T A$ is invertible, so that Eq. 28.3 is well-defined.

Moreover, we may use the proof of Theorem 28.9 to show that the projection $u$ of Definition 28.2 is uniquely defined. By reversing the argument in the proof, we may show that $w - u \in V$ implies that $A^T w = A^T u$, or in other words, $u$ should be a solution to the equation $A^T X = A^T w$. Since $w$ is *one* solution to this equation, Theorem 28.6c guarantees that $A^T X = A^T w$ has a *unique* solution $u \in Row(A^T) = V$. In particular, the projection of Definition 28.2 is *uniquely* defined.

## Exercise 28.9

Use Theorem 28.9 to compute the projection $u$ of $w = \begin{bmatrix} 9 \\ 2 \\ 4 \end{bmatrix}$ into $V = span(S)$,

where $S = \left\{ \begin{bmatrix} 2 \\ 3 \\ 1 \end{bmatrix}, \begin{bmatrix} 6 \\ 1 \\ -1 \end{bmatrix} \right\}$. Verify that $u$ is in $V$ and that $w - u \perp V$.

### Exercise 28.10

- Use your answer from Ex. 28.9 to construct an approximate solution to $AX = w$ for $A = \begin{bmatrix} 2 & 6 \\ 3 & 1 \\ 1 & -1 \end{bmatrix}$. Because your solution $X$ is only approximate, $AX = w' \neq w$. Determine the error between $AX$ and $w$. Picture this using *Mathematica*.

**Remark:**

We may provide another verification of Theorem 28.9, in the case where $S = \{v\}$. In this case, $A^T A = v \cdot v$ is a scalar, which we may factor out, so that Eq. 28.3 becomes:

$$A \left( A^T A \right)^{-1} A^T w = \begin{bmatrix} v \end{bmatrix} (v \cdot v)^{-1} v^T w = \begin{bmatrix} v \end{bmatrix} (v \cdot v)^{-1} v \cdot w = \frac{(v \cdot w)}{(v \cdot v) \, v}$$

which is the projection by Eq. 27.7.

Although this is a workable formula for computing the projection, in practice, it takes too long and is numerically unstable. That is, the type of approximations that inevitably occur when performing arithmetic on a calculator or computer will cause one's results to become very inaccurate. If we first determine an *orthonormal* basis for $V$, then we may provide an even *simpler* (and numerically stable) formula for this projection. In the next two sections, we define what we mean by an "orthonormal" set of vectors and discuss an important technique for constructing a basis that is orthonormal.

## 28.6    Orthogonal and Orthonormal Sets

When every pair of vectors in a set $S$ are orthogonal, we say that $S$ is an "orthogonal" set of vectors. If each vector in $S$ also has length 1, we say that $S$ is an "ortho*normal*" set (i.e., the vectors are "normalized" or "standardized" to length 1). Formally, we establish the following definitions.

**Definition 28.3** If $S = \{v_1, \ldots, v_n\}$ is a set of vectors, we say that:

(a) $S$ is *orthogonal* iff $v_i \perp v_j$ for all $i \neq j$.

(b) $S$ is *orthonormal* iff $S$ is orthogonal and $v_i \cdot v_i = 1$ for all $i$.

*Note*: Definition 28.3a establishes yet another usage of the term "orthogonal." We will even associate another meaning to this term in Definition 29.1. Since this term is rather over-used in linear algebra, one must be careful to examine the context and grammatical usage of the word to determine what it means in any specific situation.

By the Principle of Partitioned Matrices, these definitions are easy to verify in $R^n$ equipped with the standard inner product.

**Theorem 28.10** *If* $S = \{v_1, \ldots, v_m\} \subset R^n$ *and* $B = \begin{bmatrix} v_1, \cdots, v_m \end{bmatrix}$, *then*

(a) *$S$ is orthogonal iff $B^T B$ is a diagonal matrix.*

(b) *$S$ is orthonormal iff $B^T B$ is an identity matrix.*

**Proof:**

The Principle of Partitioned Matrices implies that:

$$B^T B = \begin{bmatrix} v_1^T \\ \vdots \\ v_m^T \end{bmatrix} \begin{bmatrix} v_1 & \cdots & v_m \end{bmatrix} = \begin{bmatrix} v_1^T v_1 & \cdots & v_1^T v_m \\ \vdots & \vdots & \vdots \\ v_m^T v_1 & \cdots & v_m^T v_m \end{bmatrix}$$

$$= \begin{bmatrix} v_1 \cdot v_1 & \cdots & v_1 \cdot v_m \\ \vdots & \vdots & \vdots \\ v_m \cdot v_1 & \cdots & v_m \cdot v_m \end{bmatrix}.$$

Here we have also used Eq. 27.1 for the standard inner product on $R^n$. From this, we may conclude that $S$ is orthogonal iff all of the off-diagonal inner products are 0. Likewise, $S$ is orthonormal iff the remaining inner products are 1. Thus, both parts of Theorem 28.10 follow immediately. Q.E.D.

We will see that orthogonal and orthonormal sets are very convenient for a number of different calculations. One reason is that orthogonal sets are *automatically* independent.

**Theorem 28.11** *If $S = \{q_1, \ldots, q_k\}$ is an orthogonal set of nonzero vectors, then $S$ is linearly independent.*

**Proof:**

We may prove this by appealing directly to Definition 22.5. If we assume that $0 = a_1 q_1 + \cdots + a_k q_k$, for any $i$, we may form the inner product of both sides with $q_i$, yielding

$$0 \cdot q_i = (a_1 q_1) \cdot q_i + \cdots + (a_k q_k) \cdot q_i$$

or:

$$0 = a_1 (q_1 \cdot q_i) + \cdots + a_k (q_k \cdot q_i)$$

By orthogonality, all of the terms on the right-hand side will equal 0, except the $i$th term, leaving $0 = a_i (q_i \cdot q_i)$. Since $q_i \neq 0$, Definition 27.1d implies that we may divide by $q_i \cdot q_i$ to conclude that $a_i = 0$. Since $i$ was arbitrary, $a_i = 0$ for all $i$, and Definition 22.5 is satisfied. Q.E.D.

# 28.7  Constructing Orthonormal Bases

We have seen that the concept of a basis allows us to construct nonstandard coordinate systems. The main advantage of the standard coordinate system on $E^3$ is that the axes are perpendicular with identical units on each axis. In this section, we will discuss how to construct nonstandard coordinate systems that possess this property.

Specifically, we indicate an inductive procedure for constructing an orthogonal basis $T = \{w_1, \ldots, w_k\}$ from a given spanning set $S = \{v_1, \ldots, v_m\}$. By then dividing each vector of $T$ by its length (i.e., "normalizing" it), we may obtain an "orthonormal basis."

1. Take the first vector $v_1$ in $S$ and call this $w_1$.

2. Take $v_2$ and compute its projection $u$ onto $w_1$. Let $w_2 = v_2 - u$.

3. Take $v_3$ and compute its projection $u$ onto $span\{w_1, w_2\}$. Take the orthogonal component of $v_3$, which is $w_3 = v_3 - u$.

$$\vdots$$

This process is called "Gram-Schmidt orthonormalization." We will devote Lesson 29 to this important algorithm. To prepare for that lesson, we work out a simple example on a set of two vectors.

### Exercise 28.11

- Use this process to construct an orthogonal basis for $V$ from Ex. 28.6. Verify that your two vectors form a basis for $V$.

### Exercise 28.12

- Divide each vector from Ex. 28.11 by its length. Verify that the resulting vectors are unit length and hence form an *orthonormal* basis.

In Lesson 29, we will learn how to carry out this process on larger sets of vectors. We will prove that $T$ will always form a basis for $span(S)$. In fact, we will prove in Theorem 29.1 that $T$ possesses an important spanning property, which characterizes it uniquely. For now, we will discuss some uses of orthonormal bases.

## 28.8   The Projection Matrix

In Theorem 28.9 we have provided one formula for the matrix $P$ that takes any vector $w$ to its projection $u$ into a given subspace $V$. Specifically, Theorem 28.9 says $u = A(A^T A)^{-1} A^T w$, so that this "projection matrix" $P$ for $V$ is given by the formula $P = A\left(A^T A\right)^{-1} A^T$. By employing an orthonormal basis for $V$, we may determine a simpler formula for $P$ that is computationally superior, as well.

**Theorem 28.12** *If $B = \{v_1, \ldots, v_n\}$ is a basis of orthogonal vectors for $V$ and $w$ is any vector, then the projection $u$ of $w$ into $V$ is given by the formula:*

$$u = \frac{v_1 \cdot w}{v_1 \cdot v_1} v_1 + \cdots + \frac{v_n \cdot w}{v_n \cdot v_n} v_n \tag{28.4}$$

*Note*: This formula is the sum of the projections of $w$ onto each vector of $V$.

**Proof:**

Since $u \in V = span(B)$, $u = a_1 v_1 + \cdots + a_n v_n$ for some choice of coefficients $a_i$. Since $w - u$ is orthogonal to $V$, we have $0 = (w - u) \cdot v_i = (w - a_1 v_1 - \cdots - a_n v_n) \cdot v_i = w \cdot v_i - a_1 (v_1 \cdot v_i) - \cdots - a_n (v_n \cdot v_i)$. Since $B$ is orthogonal, $v_j \cdot v_i$ are $0$ for $j \neq i$, so that this simplifies to $0 = w \cdot v_i - a_i v_i \cdot v_i$. Solving for $a_i$ leads to $a_i = \dfrac{v_i \cdot w}{v_i \cdot v_i}$, for all $i$, which implies Eq. 28.4. Q.E.D.

If $B$ is in fact *orthonormal*, Eq. 28.4 simplifies further.

**Corollary 28.13** *If $B = \{v_1, \ldots, v_n\}$ is an orthonormal basis for $V$ and $w$ is any vector, the projection of $w$ into $V$ is given by the formula:*

$$u = (v_1 \cdot w)v_1 + \cdots + (v_n \cdot w)v_n \tag{28.5}$$

*If $Q = \begin{bmatrix} v_1 & \cdots & v_n \end{bmatrix}$ is a matrix with $B$ as columns, then:*

$$u = QQ^T w \tag{28.6}$$

*The matrix $P = QQ^T$ is called the projection matrix for $V$, since it takes any vector $w$ to its projection into $V$.*

### Proof:

Eq. 28.5 follows immediately from Eq. 28.4, since Definition 28.3b guarantees that $v_i \cdot v_i = 1$. Eq. 28.6 follows by the Principle of Partitioned Matrices:

$$u = (v_1 \cdot w)\, v_1 + \cdots + (v_n \cdot w)\, v_n = \begin{bmatrix} v_1 \cdots v_n \end{bmatrix} \begin{bmatrix} v_1 \cdot w \\ \vdots \\ v_n \cdot w \end{bmatrix}$$

$$= \begin{bmatrix} v_1 \cdots v_n \end{bmatrix} \begin{bmatrix} v_1^T w \\ \vdots \\ v_n^T w \end{bmatrix} = \begin{bmatrix} v_1 & \cdots & v_n \end{bmatrix} \begin{bmatrix} v_1^T \\ \vdots \\ v_n^T \end{bmatrix} w = QQ^T$$

Q.E.D.

## Exercise 28.13

Verify that $S = \left\{ \begin{bmatrix} 1 \\ 0 \\ 0 \end{bmatrix}, \begin{bmatrix} 0 \\ 1 \\ 0 \end{bmatrix} \right\}$ forms an orthonormal basis for the $x$-$y$ plane in

three-dimensional space. Use Eq. 28.5 to compute the projection of $\begin{bmatrix} 2 \\ 3 \\ 4 \end{bmatrix}$ into the

$x$-$y$ plane. Compute the projection matrix $P$ for the $x$-$y$ plane. Give an intuitive explanation of your results.

## Exercise 28.14

Use Eq. 28.5 and the orthonormal basis from Ex. 28.12 to compute the projection $u$

of $w = \begin{bmatrix} 9 \\ 2 \\ 4 \end{bmatrix}$ into the subspace $V$ from Ex. 28.11. Compare your answer with the

results of Ex. 28.9.

## Exercise 28.15

Determine the projection matrix for $V$ from Ex. 28.11. *Hint:* Use the results of Ex. 28.12 and Corollary 28.13.

## 28.9   Summary

In this lesson, we have investigated the concept of orthogonal subspaces. One should have learned how:

- $Kernel(A)$ and $Row(A)$ are orthogonal complements,

- to represent $V$ as a $Kernel(B)$, that is, by a system of equations,

- to uniquely choose solutions to underdetermined systems,

- to employ projections to solve overdetermined systems, and

- to compute the projection matrix for a subspace.

Although we have briefly introduced Gram-Schmidt orthonormalization, in Lesson 29 we will examine this process in greater detail. We will learn how to organize our calculations in matrix form, leading to a new type of decomposition of $A$, known as the $QR$ decomposition. As we have mentioned, this will allow us to easily solve an overdetermined system. This decomposition is also key to the most widely used algorithm for computing eigenvectors and eigenvalues.

# Gram-Schmidt Orthonormalization

## Accompanied by Notebook

For any given subspace $V$, we have discussed various techniques to construct a basis $B$ for $V$. In Lesson 24, we showed how to take a spanning set $S$ for $V$ and choose $B$ as an appropriate subset of $S$. If we want our basis to possess special properties, we usually must work a bit harder. For example, in Lesson 28 we saw that *orthogonal* bases are particularly useful. In Section 28.7, we began to sketch a procedure for constructing an orthonormal basis from a spanning set $S$. This procedure is called "Gram-Schmidt orthonormalization," named after its two inventors. Just as Gaussian elimination may be summarized by a $P^T LU$ decomposition, Gram-Schmidt orthonormalization may be described by a matrix decomposition, usually referred to as a $QR$ decomposition.

Performing Gram-Schmidt orthonormalization by hand for a large set of vectors can become rather difficult. What is needed is an procedure that will help us organize the calculation in a clear and systematic way. In this lesson, we will:

- discuss Gram-Schmidt orthonormalization in detail,

- describe an algorithm and matrix format that will enable us to successfully perform Gram-Schmidt orthonormalization by hand, and

- show how to use Gram-Schmidt orthonormalization to compute projections and solve overdetermined systems.

Specifically:

- in Section 29.1, we describe a method of organizing the calculation in a particular matrix format,

- in Section 29.2, we describe how this format directs us through the algorithm,

- in Section 29.3, we compute a specific example in detail, and

- in Section 29.5, we sketch solutions to two application problems.

- You should view the first section in the accompanying Notebook at this point, to discover how to visualize this algorithm. From a geometric standpoint, Gram-Schmidt orthonormalization is a very natural and compelling procedure. By having a picture firmly in mind, one will be better equipped to follow the logical progression directing

**311**

the steps of the algorithm. Other than for this visualization, *Mathematica* is *unnecessary* in this lesson until we reach Ex. 29.2.

## 29.1   General Description of the Algorithm

After introducing some notation, we will describe Gram-Schmidt orthonormalization in detail. The Gram-Schmidt process takes a set of vectors, $S = \{a_1, \ldots, a_n\}$ that spans a subspace $V$ and returns an orthonormal basis for $V$, $S' = \{q_1, \ldots, q_k\}$. *Aside*: Notice that although $S$ has $n$ vectors, $S'$ has only $k$ vectors. That is because, if $S$ is dependent, some vectors may need to be "thrown away" in the process.

We may more precisely characterize this algorithm by explaining how it corresponds to a matrix decomposition. Specifically, if the $a_i$'s form the columns of a matrix $A$, and the $q_i$'s form the columns of $Q$, then Gram-Schmidt orthonormalization will also produce a matrix $R$ so that $A = QR$. In fact, the first $i$ orthonormalized vectors that we compute, $\{q_1, \ldots, q_i\}$, will span the same subspace as the first $i$ *independent* vectors of $S$. It is this additional spanning property of Gram-Schmidt orthonormalization that sets it apart from other possible methods. In terms of a matrix decomposition, this means that $R$ will be in *row echelon form*.

The algorithm that we will discuss may be divided into two stages, which we refer to as "orthogonalization" and "normalization."

1. In the orthogonalization phase, we decompose $A$ into $Q_o R_o$, where $Q_o$ has orthogonal (potentially $0$) columns, and $R_o$ is unit, upper-triangular.

2. In the normalization phase, we convert the columns of $Q_o$ to an *orthonormal* set by eliminating the $0$ columns and dividing the remaining columns by their lengths. We also perform the corresponding operations on $R_o$ necessary to maintain the validity of our decomposition. Specifically, we eliminate the rows corresponding to $0$ columns and multiply the others by the corresponding lengths.

In Section 28.7 we described an inductive procedure for constructing an orthogonal basis $T = \{w_1, \ldots, w_k\}$ from a given spanning set $S = \{v_1, \ldots, v_m\}$:

1. Take the first vector $v_1$ in $S$ and call this $w_1$.

2. Take $v_2$ and compute its projection $u$ onto $w_1$. Let $w_2 = v_2 - u$.

3. Take $v_3$ and compute its projection $u$ onto $span\{w_1, w_2\}$. Take the orthogonal component of $v_3$, which is $w_3 = v_3 - u$.

$$\vdots$$

This accurately describes the orthogonalization stage, although we will use slightly different notation. We will take a set of vectors $\{a_1, \ldots, a_n\}$, expressed as the columns of $A$, and produce a set of *orthogonal* vectors, $\{q_1, \ldots, q_k\}$, expressed as the columns of $Q_o$. *Note*: From the visualization in the accompanying Notebook, one should observe that this process takes each $a_j$ in turn and "straightens it out" to obtain $q_j$, which is that component of $a_j$ perpendicular to its predecessors.

This process may be completely described by the single, inductive formula:

$$q_j = a_j - \left[ \frac{(a_j \cdot q_1)}{(q_1 \cdot q_1)} q_1 + \cdots + \frac{(a_j \cdot q_{j-1})}{(q_{j-1} \cdot q_{j-1})} q_{j-1} \right] \tag{29.1}$$

The expression in square brackets represents the projection $p_j$ of $a_j$ into the span of $\{q_1, \ldots, q_{j-1}\}$, which by construction will be equal to the span of $\{a_1, \ldots, a_{j-1}\}$. One should recognize Eq. 28.4 in this expression. Since $p_j$ is the component of $a_j$ that is parallel to the span of $\{q_1, \ldots, q_{j-1}\}$, $q_j = a_j - p_j$ represents the remaining component of $a_j$, which is then orthogonal to each of the previous $a_i$'s.

Examining Eq. 29.1, we observe that to compute $q_j$, we must first compute:

$$(a_j \cdot q_i), \qquad (q_i \cdot q_i), \qquad \frac{(a_j \cdot q_i)}{(q_i \cdot q_i)}, \qquad \text{and}$$

$$p_j = \left[ \frac{(a_j \cdot q_1)}{(q_1 \cdot q_1)} q_1 + \cdots + \frac{(a_j \cdot q_{j-1})}{(q_{j-1} \cdot q_{j-1})} q_{j-1} \right] \qquad \text{for} \quad 1 \le i < j \le n$$

The main difficulty with the orthogonalization phase of Gram-Schmidt orthonormalization is how to organize all of these computations and then progress steadily to the result. We will display all of this data in five matrices, as in Fig. 29.1. This layout orga-

$$A = \begin{bmatrix} \cdots & a_j & \cdots \end{bmatrix} \qquad R_0^T = \begin{bmatrix} 1 & 0 & \cdots & 0 \\ & \ddots & \ddots & \vdots \\ & & \ddots & 0 \\ \cdots & \frac{(a_j \cdot q_i)}{(q_i \cdot q_i)} & \cdots & 1 \end{bmatrix} \qquad \begin{matrix} \vdots \\ \cdots & (a_j \cdot q_i) & \cdots \\ \vdots \end{matrix}$$

$$L = \begin{bmatrix} \cdots & (q_i \cdot q_i) & \cdots \end{bmatrix}$$

$$P = \begin{bmatrix} \cdots & p_j & \cdots \end{bmatrix} \qquad Q_0 = \begin{bmatrix} \cdots & q_j & \cdots \end{bmatrix}$$

Figure 29.1: Matrix Layout of the Algorithm

nizes the results in a way that will facilitate the necessary calculations, as well as guide us through the process. The orthogonalization phase of our algorithm proceeds left-to-right, *column by column* through $A$. Compare this with Gaussian elimination, which manipulates *rows*. For each column of $A$, we will compute the corresponding columns of $P$, $Q_o$, and $L$. We utilize the space to the right of $R_o^T$ for "scratch-work," which determines the *rows* of $R_o^T$.

We should pause to comment on the names of each matrix. Generally, each name reflects the content of the corresponding matrix, such as, $Q_o$ and $R_o$, which represent the unnormalized factors of the final $QR$ decomposition. However, there are a few anomalies. For example, $P$ contains the *projections* of the columns of $A$. It is *not* the projection matrix of Section 28.8. Although we think of $L$ as containing the *lengths* of the vectors in $Q_o$, it actually contains the lengths *squared*. Notice also how we label the upper-right matrix as $R_o^T$. This is because, for the purposes of computation, it is more convenient to use the *transpose* of $R_o$ at this point of the algorithm, rather than $R_o$ itself.

## 29.2  Verbal Description of Orthogonalization

All the hard work of Gram-Schmidt orthonormalization is done in the orthogonalization phase, so we will discuss it in great detail. The normalization phase is relatively easy, so we will leave that to Section 29.4. In this section, we describe the algorithm in words, emphasizing what each part of the calculation represents geometrically. In Section 29.3, we will step through a specific example in detail.

If you prefer, you may read Section 29.3 first. It is important to be able to understand the logic of the algorithm, as well as the particular computations involved. In the long run, you will remember the algorithm better, if you know what each step represents. Thus, you should read both sections a number of times until it becomes clear how the description in this section corresponds to the steps that are shown in Section 29.3.

Logically, the orthogonalization algorithm consists of repeatedly computing a projection, $p_j$, and an orthogonal complement, $q_j$, for each column, $a_j$, of $A$, working from left to right. This may be described in detail as follows.

1. Compute the projection, $p_j$, of the next column of $A$, $a_j$, into the span of the previous columns of $A$. Each projection takes three steps:

   (a) Compute the inner products $a_j \cdot q_i$ for $i < j$, (i.e., compute the dot product of the current column of $A$ with all the columns of $Q_o$ that one has already determined) and record the results to the right of $R_o^T$.

   (b) Divide each value computed in step 1a by the corresponding entry of $L$ and record the results inside $R_o^T$, in the same row and corresponding column. Logically, this produces the coefficients of the projection, $\dfrac{(a_j \cdot q_i)}{(q_i \cdot q_i)}$ from Eq. 29.1.

   (c) Form the linear combination $\dfrac{(a_j \cdot q_1)}{(q_1 \cdot q_1)} q_1 + \cdots + \dfrac{(a_j \cdot q_{j-1})}{(q_{j-1} \cdot q_{j-1})} q_{j-1}$ to obtain $p_j$. Notice that the coefficients of this linear combination are the values computed in step 1b, while the corresponding vectors are the ones recorded in $Q_o$ directly below. This means that we may compute the linear combination by working our way down through the rows of $Q_o$, multiplying each entry by the corresponding number in $R_o^T$, and recording the totals in the corresponding rows of $P$.

2. Compute the orthogonal component $q_j = a_j - p_j$. Simply subtract the column of $P$ computed in step 1c from the column of $A$ directly above it, and record the result in the corresponding column of $Q_o$. To prepare for the next step of the algorithm, we must also then compute the inner product $q_j \cdot q_j$ and record the result directly above in $L$.

Notice the progression, from $A$ to $Q_o$ to the right of $R_o^T$ to $R_o^T$ to $P$ to $Q$ to $L$. With practice, this "figure-eight" pattern guides one through the algorithm, just as Gaussian elimination proceeds "down-and-right."

## 29.3   Orthogonalization: A Specific Example

By examining a specific example, we may show how the orthogonalization phase looks in practice. For this example, we will take $S = \left\{ \begin{bmatrix} 1 \\ 1 \end{bmatrix}, \begin{bmatrix} 0 \\ 2 \end{bmatrix}, \begin{bmatrix} 1 \\ 0 \end{bmatrix} \right\}$ and construct an orthogonal basis for $V = span(S)$.

*Suggestion*: As you read this section, you should compute your *own* example, say from Ex. 29.3, or by simply changing the numbers in $S$. As one reads each step of this example, one should perform the corresponding step of the algorithm on a separate piece of paper using your *own* example. By proceeding in this more active way, you will learn the algorithm more quickly. This approach requires you to understand each step more fully than by simply reading alone.

*Step 0:*   Form the matrix $A$ with the vectors in $S$ as columns, and set up our matrix format, as in Fig. 29.2. Notice how $P$ and $Q_o$ are the same shape as $A$, while $L$ is a single row and $R_o^T$ is square with the same number of columns. Since $R_o^T$ is always unit lower-triangular, one may immediately fill in the upper-right of the matrix.

$$A = \begin{bmatrix} 1 & 0 & 1 \\ 1 & 2 & 0 \end{bmatrix} \quad R_0^T = \begin{bmatrix} 1 & 0 & 0 \\ & 1 & 0 \\ & & 1 \end{bmatrix}$$

$$L = \begin{bmatrix} & & \end{bmatrix}$$

$$P = \begin{bmatrix} & \end{bmatrix} \quad Q_0 = \begin{bmatrix} & & \end{bmatrix}$$

Figure 29.2: Start of the Algorithm

*Step 1:*   The first step of the algorithm looks a little strange because we are projecting into the span of an empty set of vectors, which is the 0 subspace. The projection $p_1$ is just the 0 vector. Moreover, there are no vectors in $Q_o$ with which to form inner products anyway. This means that we have no coefficients to enter in $R_o^T$. Simply place $p_1 = 0$ in $P$, subtract to obtain $q_1 = a_1$ in $Q_o$, and place $q_1 \cdot q_1$ in $L$, as in Fig. 29.3.

*Step 2:*   The next step illustrates the algorithm a little better. We move on to the next column of $A$ and compute the inner product of this column with each of the computed columns of $Q_o$. Since there is only one so far, this is easy. Place the result to the *right* of $R_o^T$ in the second row, as in Fig. 29.4.

$$A = \begin{bmatrix} 1 & 0 & 1 \\ 1 & 2 & 0 \end{bmatrix} \quad R_0^T = \begin{bmatrix} 1 & 0 & 0 \\ & 1 & 0 \\ & & 1 \end{bmatrix}$$

$$L = \begin{bmatrix} 2 & & \end{bmatrix}$$

$$P = \begin{bmatrix} 0 \\ 0 \end{bmatrix} \quad Q_0 = \begin{bmatrix} 1 \\ 1 \end{bmatrix}$$

Figure 29.3: A Single Vector Is Orthogonal

$$A = \begin{bmatrix} 1 & 0 & 1 \\ 1 & 2 & 0 \end{bmatrix} \quad R_0^T = \begin{bmatrix} 1 & 0 & 0 \\ & 1 & 0 \\ & & 1 \end{bmatrix} 2$$

$$L = \begin{bmatrix} 2 & & \end{bmatrix}$$

$$P = \begin{bmatrix} 0 \\ 0 \end{bmatrix} \quad Q_0 = \begin{bmatrix} 1 \\ 1 \end{bmatrix}$$

Figure 29.4: Inner Product Step

$$A = \begin{bmatrix} 1 & 0 & 1 \\ 1 & 2 & 0 \end{bmatrix} \quad R_0^T = \begin{bmatrix} 1 & 0 & 0 \\ 1 & 1 & 0 \\ & & 1 \end{bmatrix} 2$$

$$L = \begin{bmatrix} 2 & & \end{bmatrix}$$

$$P = \begin{bmatrix} 0 \\ 0 \end{bmatrix} \quad Q_0 = \begin{bmatrix} 1 \\ 1 \end{bmatrix}$$

Figure 29.5: Getting the Coefficients

Divide this result by the corresponding entry in $L$, placing the quotient *in $R_o^T$*, as in Fig. 29.5. This completes the second row of $R_o^T$. Form this linear combination of the columns of $Q_o$ (that is multiply the 1 by the vector below), placing the result in $P$, as in Fig. 29.6. This represents the projection $p_2$ of $a_2$ into $a_1$.

$$A = \begin{bmatrix} 1 & 0 & 1 \\ 1 & 2 & 0 \end{bmatrix} \quad R_0^T = \begin{bmatrix} 1 & 0 & 0 \\ 1 & 1 & 0 \\ & & 1 \end{bmatrix} \, 2$$

$$L = \begin{bmatrix} 2 & & \end{bmatrix}$$

$$P = \begin{bmatrix} 0 & 1 \\ 0 & 1 \end{bmatrix} \quad Q_0 = \begin{bmatrix} 1 \\ 1 \end{bmatrix}$$

Figure 29.6: Computing the Projection

Now subtract this projection from the current column of $A$ (namely $a_2$), placing the resulting column in $Q_o$. This results in Fig. 29.7. This column represents the component $q_2$ of $a_2$ that is orthogonal to $a_1$. We consider $q_2$ as the "orthogonalized" version of $a_2$. To prepare for the next step, compute the inner product of this orthogonalizedvector with itself, placing the result directly above in $L$, as in Fig. 29.8. Notice how each column of $A$ generates a corresponding column of $P, Q_o$, and $L$, and a new row of $R_o^T$.

$$A = \begin{bmatrix} 1 & 0 & 1 \\ 1 & 2 & 0 \end{bmatrix} \quad R_0^T = \begin{bmatrix} 1 & 0 & 0 \\ 1 & 1 & 0 \\ & & 1 \end{bmatrix} \, 2$$

$$L = \begin{bmatrix} 2 & & \end{bmatrix}$$

$$P = \begin{bmatrix} 0 & 1 \\ 0 & 1 \end{bmatrix} \quad Q_0 = \begin{bmatrix} 1 & -1 \\ 1 & 1 \end{bmatrix}$$

Figure 29.7: The First Orthogonalized Vector

*Step 3:*    At this point, we have one remaining vector in $A$ to orthogonalize. This step most effectively illustrates a general step in the algorithm. As before, we move to the next (and final) column of $A$ and form its inner product with the computed columns of $Q_o$, placing the results to the right of $R_o^T$, as in Fig. 29.9. Notice how we move down to the next row of $R_o^T$ to record the results. It is a good idea to work left-to-right through $Q_o$, computing inner products. As one obtains each result one may

$$A = \begin{bmatrix} 1 & 0 & 1 \\ 1 & 2 & 0 \end{bmatrix} \quad R_0^T = \begin{bmatrix} 1 & 0 & 0 \\ 1 & 1 & 0 \\ & & 1 \end{bmatrix} 2$$

$$L = \begin{bmatrix} 2 & 2 & \end{bmatrix}$$

$$P = \begin{bmatrix} 0 & 1 \\ 0 & 1 \end{bmatrix} \quad Q_0 = \begin{bmatrix} 1 & -1 & \\ 1 & 1 & \end{bmatrix}$$

Figure 29.8: Preparing for the Next Step

record it in the same order next to $R_o^T$. In this way, the coefficients will match up with the correct vector.

It is then a simple matter to divide the inner products by the corresponding entries in $L$ (i.e., first by first , second by second , etc.), placing the results in $R_o^T$. This gives Fig. 29.10.

As before, we may form this linear combination of the columns of $Q_o$, placing the result in $P$. Notice how one may work down through $Q_o$ row-by-row, using the co-efficients $\frac{1}{2}$ and $-\frac{1}{2}$ from $R_o^T$ as multipliers. Specifically, we multiply the entries of a row in $Q_o$ by the number in the same column of $R_o^T$, add the results together, and place the sum in the same row of $P$. At this point, our paper should resemble Fig. 29.11.

If we now subtract this projection in $P$ from the current column of $A$, we obtain the component of the last column of $A$ that is perpendicular to the previous *two* columns of $A$. As before, we place this result in $Q_o$, with its length-squared in $L$. In the end, we are left with Fig. 29.12.

### Exercise 29.1

Explain geometrically why we should obtain a zero vector from step 3.

### Exercise 29.2

•     Verify our result by multiplying $R_o$ and $Q_o$. How should you multiply, and what result should you obtain?

## 29.3.1   Some Brief Observations

We interrupt our calculation of this example to prove two important facts concerning the Gram-Schmidt process, originally mentioned in Section 29.1. *Note*: This material is not crucial to the remainder of the text and may be omitted. In particular, at each step the columns of $Q_0$ are orthogonal and span the same subspace as the corresponding columns of $A$. In other words, we have the following theorem, which characterizes the orthogonalization phase of the algorithm.

$$A = \begin{bmatrix} 1 & 0 & 1 \\ 1 & 2 & 0 \end{bmatrix} \quad R_0^T = \begin{bmatrix} 1 & 0 & 0 \\ 1 & 1 & 0 \\ & & 1 \end{bmatrix} \begin{matrix} 2 \\ 1 & -1 \end{matrix}$$

$$L = \begin{bmatrix} 2 & 2 & \end{bmatrix}$$

$$P = \begin{bmatrix} 0 & 1 \\ 0 & 1 \end{bmatrix} \quad Q_0 = \begin{bmatrix} 1 & -1 \\ 1 & 1 \end{bmatrix}$$

Figure 29.9: More Inner Products

$$A = \begin{bmatrix} 1 & 0 & 1 \\ 1 & 2 & 0 \end{bmatrix} \quad R_0^T = \begin{bmatrix} 1 & 0 & 0 \\ 1 & 1 & 0 \\ \frac{1}{2} & -\frac{1}{2} & 1 \end{bmatrix} \begin{matrix} 2 \\ 1 & -1 \end{matrix}$$

$$L = \begin{bmatrix} 2 & 2 & \end{bmatrix}$$

$$P = \begin{bmatrix} 0 & 1 \\ 0 & 1 \end{bmatrix} \quad Q_0 = \begin{bmatrix} 1 & -1 \\ 1 & 1 \end{bmatrix}$$

Figure 29.10: More Coefficients

$$A = \begin{bmatrix} 1 & 0 & 1 \\ 1 & 2 & 0 \end{bmatrix} \quad R_0^T = \begin{bmatrix} 1 & 0 & 0 \\ 1 & 1 & 0 \\ \frac{1}{2} & -\frac{1}{2} & 1 \end{bmatrix} \begin{matrix} 2 \\ 1 & -1 \end{matrix}$$

$$L = \begin{bmatrix} 2 & 2 & \end{bmatrix}$$

$$P = \begin{bmatrix} 0 & 1 & 1 \\ 0 & 1 & 0 \end{bmatrix} \quad Q_0 = \begin{bmatrix} 1 & -1 \\ 1 & 1 \end{bmatrix}$$

Figure 29.11: The Final Projection

$$A = \begin{bmatrix} 1 & 0 & 1 \\ 1 & 2 & 0 \end{bmatrix} \quad R_0^T = \begin{bmatrix} 1 & 0 & 0 \\ 1 & 1 & 0 \\ \frac{1}{2} & -\frac{1}{2} & 1 \end{bmatrix} \begin{matrix} 2 \\ 1 \end{matrix} \quad -1$$

$$L = \begin{bmatrix} 2 & 2 & 0 \end{bmatrix}$$

$$P = \begin{bmatrix} 0 & 1 & 1 \\ 0 & 1 & 0 \end{bmatrix} \quad Q_0 = \begin{bmatrix} 1 & -1 & 0 \\ 1 & 1 & 0 \end{bmatrix}$$

Figure 29.12: The End of Orthogonalization

**Theorem 29.1** *Given a set of vectors $S = \{a_1, \ldots, a_n\}$, if $S' = \{q_1, \ldots, q_n\}$ is defined inductively by Eq. 29.1, then $S'$ is orthogonal. Moreover, if we define $V_j \equiv span\{a_1, \ldots, a_j\}$ and $V_j' \equiv span\{q_1, \ldots, q_j\}$, then we have:*

$$V_j' = V_j \qquad \text{for every } j$$

*In particular, if the zero vectors are dropped from $S'$, one is left with an orthogonal spanning set for $V_n$, which by Theorem 28.11 is an orthogonal basis.*

The spanning property described here distinguishes Gram-Schmidt orthonormalization as unique among all possible methods of constructing orthonormal bases.

**Proof:**

   To show that $S'$ is orthogonal, it suffices to show that each $q_j$ is orthogonal to each of the previous $q_i$ (i.e., for $i < j$). We will use a proof by induction (compare the proof of Lemma 25.3) to verify orthogonality and the spanning property simultaneously. Remember that this technique may be used whenever we wish to prove an infinite collection of statements indexed by a natural number. In this case, we use induction on $j$, which represents the number of currently orthogonalized vectors.

*The Anchor Step: $j = 1$*

   As usual, the first case in an inductive proof is rather easy. Since $q_1 = a_1$, $V_j' = span\{q_1\} = span\{a_1\} = V_j$. Moreover, since any single vector forms an orthogonal set, the orthogonality condition is automatically satisfied.

To prove the inductive step, we show that the $(j - 1)$st statement implies the $j$th statement. Compare this with the proof of Lemma 25.3. Although notationally this may look different, logically both proofs follow the same pattern. Using our ladder analogy, in both cases we show that we may climb from any given rung to the next. In this example, we simply find it convenient to label the given rung as $(j - 1)$ rather than $j$.

*The Inductive Step: $j - 1 \Rightarrow j$*

   We first inductively verify the spanning condition. In other words, we assume that $V_{j-1}' = V_{j-1}$ and we demonstrate that $V_j' = V_j$. Just as in the proofs of Corollary 25.7b and Theorem 26.1b, we do this by verifying that each set is contained in the other.

*Case 1:* $V_j' \subset V_j$

Consider an arbitrary $v' \in V_j' = span\{q_1, \ldots, q_j\}$, so that

$$v' = \alpha_1 q_1 + \cdots + \alpha_{j-1} q_{j-1} + \alpha_j q_j$$

for some coefficients $\alpha_i$. Using Eq. 29.1, we may substitute for $q_j$ to obtain:

$$
\begin{aligned}
v' = {} & \alpha_1 q_1 + \cdots + \alpha_{j-1} q_{j-1} + \\
& \alpha_j \left\{ a_j - \left[ \frac{(a_j \cdot q_1)}{(q_1 \cdot q_1)} q_1 + \cdots + \frac{(a_j \cdot q_{j-1})}{(q_{j-1} \cdot q_{j-1})} q_{j-1} \right] \right\}
\end{aligned}
\tag{29.2}
$$

Notice that this is simply $a_j$ plus a number of terms in $V_{j-1}'$. That is, if we define $q$ by the equation

$$
\begin{aligned}
q = {} & \alpha_1 q_1 + \cdots + \alpha_{j-1} q_{j-1} - \\
& \alpha_j \left[ \frac{(a_j \cdot q_1)}{(q_1 \cdot q_1)} q_1 + \cdots + \frac{(a_j \cdot q_{j-1})}{(q_{j-1} \cdot q_{j-1})} q_{j-1} \right]
\end{aligned}
$$

then we may rewrite Eq. 29.2 as $v' = q + \alpha_j a_j$. Since $q$ is a linear combination of $\{q_1, \ldots, q_{j-1}\}$, it is in $V_{j-1}'$, which by our induction hypothesis equals $V_{j-1}$. Thus, we may express $q$ as $q = \beta_1 a_1 + \cdots + \beta_{j-1} a_{j-1}$ for some coefficients $\beta_i$. In particular, we have:

$$v' = q + \alpha_j a_j = \beta_1 a_1 + \cdots + \beta_{j-1} a_{j-1} + \alpha_j a_j \in V_j$$

Since $v'$ was arbitrary, this shows that every vector in $V_j'$ is in $V_j$, so that $V_j' \subset V_j$.

*Case 2:* $V_j \subset V_j'$

The reverse direction is similar, except this time we start with an arbitrary $v = \beta_1 a_1 + \cdots + \beta_{j-1} a_{j-1} + \beta_j a_j \in V_j$. As before, we appeal to our inductive hypothesis to conclude that:

$$\beta_1 a_1 + \cdots + \beta_{j-1} a_{j-1} = \alpha_1 q_1 + \cdots + \alpha_{j-1} q_{j-1}$$

for some coefficients $\alpha_i$, since:

$$span\{a_1, \ldots, a_{j-1}\} = V_{j-1} = V_{j-1}' = span\{q_1, \ldots, q_{j-1}\}$$

Solving Eq. 29.1 for $a_j$ and substituting yields:

$$
\begin{aligned}
v = {} & \beta_1 a_1 + \cdots + \beta_{j-1} a_{j-1} + \beta_j a_j \\
= {} & \alpha_1 q_1 + \cdots + \alpha_{j-1} q_{j-1} + \beta_j a_j \\
= {} & \alpha_1 q_1 + \cdots + \alpha_{j-1} q_{j-1} \\
& + \beta_j \left\{ q_j + \left[ \frac{(a_j \cdot q_1)}{(q_1 \cdot q_1)} q_1 + \cdots + \frac{(a_j \cdot q_{j-1})}{(q_{j-1} \cdot q_{j-1})} q_{j-1} \right] \right\}
\end{aligned}
$$

which is clearly in $span\{q_1, \ldots, q_j\} = V_j'$. As before, this implies that $V_j \subset V_j'$.

Since we have shown that $V_j \subset V_j'$ and $V_j' \subset V_j$, the two sets must be equal.

We prove orthogonality by appealing to our induction hypothesis again. This hypothesis guarantees that $\{q_1, \ldots, q_{j-1}\}$ is an orthonormal spanning set for $V'_{j-1}$. By dropping any zero vectors, we are left with an orthonormal basis and may appeal to Eq. 28.4 to conclude that:

$$u = \frac{(a_j \cdot q_1)}{(q_1 \cdot q_1)} q_1 + \cdots + \frac{(a_j \cdot q_{j-1})}{(q_{j-1} \cdot q_{j-1})} q_{j-1} \qquad (29.3)$$

is the projection of $a_j$ into $V'_{j-1}$. Note: If a particular $q_i = 0$, we must interpret the coefficient $\frac{(a_j \cdot q_i)}{(q_i \cdot q_i)}$ in Eq. 29.3 as 0. Definition 28.2 then implies that $q_j = a_j - u$ is orthogonal to every vector in $V'_{j-1}$. In particular, $q_j$ is orthogonal to $q_1, \ldots, q_{j-1}$.

Logically, we may repeat the previous argument for $j = 2, 3, \ldots$, etc. to show that our Theorem is true for every possible $j$. Q.E.D.

Looking carefully at the proof of Theorem 29.1, one observes that the orthogonality and spanning properties of the orthogonalization stage of the algorithm allows only one possible definition for them, namely Eq. 29.1. This implies that the only choice involved in Gram-Schmidt orthonormalization occurs in the normalization stage, after we have dropped the zero vectors, when we divide each orthogonal vector by its length to obtain unit-length vectors. At this point, we are free to divide by *any* scalar with the *same absolute value* as the length. Over the real numbers there are only two possibilities, $\pm|v|$, for each vector. If we require *positive* pivots in $R$, then even this choice is eliminated, so that such a decomposition is unique.

**Corollary 29.2** *Any real-valued or even complex-valued matrix $A$ possesses a unique decomposition into $QR$, where $Q$ consists of orthonormal columns and $R$ is in row echelon form with positive pivots.*

Finally, we mention that in certain circumstances the matrix $Q$ arising from this decomposition has another important property. If $A$ is $n \times m$ and $rank(A) = m$, then $Q$ will be a square $m \times m$ matrix with orthonormal columns. By Theorem 28.11 and Corollary 25.5b, the columns of $Q$ form an orthonormal basis for $R^m$. In this case, Theorem 28.10b implies that $Q^T$ is a left-inverse for $Q$. Since $Q$ is square, Corollary 16.5 guarantees that this is a two-sided inverse. That is, we have $Q^T = Q^{-1}$. This prompts the following definition.

**Definition 29.1** A matrix $Q$ is *othogonal* iff $Q^T = Q^{-1}$.

Notice that although this matrix has *orthonormal* columns, we refer to $Q$ as an *orthogonal* matrix. This choice of terminology is unfortunate, but also quite well-established.

Orthogonal matrices are used in a variety of settings. One may notice that the defining condition of an orthogonal matrix $Q^T = Q^{-1}$ is satsified by any permutation matrix (see Definition 17.1). That is, permutations are particular examples of orthogonal matrices. Moreover, one may verify that the rotation and reflection matrices from Section 9.3 are also orthogonal. In general, it may be proven that any orthogonal matrix may be decomposed as a product of reflections, known as Householder matrices. Because they are so easy to define and manipulate, such reflection matrices are often used in practice to construct $QR$-like decompositions for a matrix $A$. For example, this is how the built-in *Mathematica* command QRDecomposition[] is programmed. Such a decomposition, however, will *not* in general have positive pivots, as described by Corollary 29.2.

## 29.4  Normalization: Finishing the Algorithm

Now we return to the example from Section 29.3. To this point, we have only orthog-onalized all of the original vectors in $S$ (i.e., columns of $A$) to obtain a decomposition $A = Q_o R_o$. Now we move on to the normalization phase of the algorithm, which is much easier. It consists of two steps:

1. Eliminate the 0 columns in $Q_o$ and $R_o^T$.

2. Divide (multiply) the columns of $Q_o$ ($R_o^T$) by the square root of the corresponding entry in $L$.

If we return to our matrix format, we may proceed left-to-right through $L$, crossing out the columns with a 0 entry. Then for each nonzero entry of $L$, we multiply the col-umn above and divide the column below by the square root of the entry, which is the length of the corresponding orthogonalized vector. In our example, this would yield the results in Fig. 29.13.

$$
R^T = \begin{bmatrix} \sqrt{2} & 0 \\ \sqrt{2} & \sqrt{2} \\ \frac{\sqrt{2}}{2} & -\frac{\sqrt{2}}{2} \end{bmatrix}
$$

$$
L = \begin{bmatrix} 2 & 2 \end{bmatrix}
$$

$$
Q = \begin{bmatrix} \frac{1}{\sqrt{2}} & -\frac{1}{\sqrt{2}} \\ \frac{1}{\sqrt{2}} & \frac{1}{\sqrt{2}} \end{bmatrix}
$$

Figure 29.13: The Normalized Result

- As with Gaussian elimination, we have provided a new command, called QR[] from the "Bases.m" package, to aid in learning this important algorithm. It is described in more detail in the second section of the accompanying Notebook. It will perform the Gram-Schmidt process on any given matrix $A$ and return its $QR$ decomposition. As usual, one should master this algorithm by hand, only using the computer to *verify* one's own calculations. Eventually, after one is proficient, we will see how we may utilize the QR[] command in the same way that we have used RowReduce[] to solve problems more quickly. For the exercises in this lesson, however, one should carry out the algorithm *by hand* for practice.

Because Corollary 29.2 guarantees that there is only one correct decomposition for any matrix, it is relatively easy to verify the results of this algorithm. As with the Gauss-Eliminate[] command, we have supplied a Verbose option, which forces the QR[] com-mand to print out most of the intermediate calculations that one would obtain while performing our hand algorithm, as well as the final decomposition. *Note*: There is a built-in command called QRDecomposition[], but it does not employ the Gram-Schmidt algorithm. In particular, it will not generally obtain *positive* pivots, and it will not print out any intermediate calculations.

### Exercise 29.3

- Compute the $QR$ decompositions for the following matrices:

$$\text{(a)} \begin{bmatrix} 1 & 4 & 12 \\ 2 & 3 & 4 \\ 1 & 2 & -2 \end{bmatrix} \qquad \text{(b)} \begin{bmatrix} 1 & 1 & 8 & 0 \\ 2 & -3 & 3 & 3 \\ 1 & 0 & 4 & 2 \\ 1 & -2 & 3 & 6 \end{bmatrix}$$

- Make sure to verify your answers via multiplication and/or the QR[] command.

As with the $P^T LU$ decomposition, we can use the MakeNiceMatrix[] command to create new practice exercises. However, because of the special form of this decomposition, we have provided the option DecompositionType->QRFactors to create matrices with nice $QR$ decompositions. *Note*: The default value for DecompositionType is "PL-UFactors".

### *Drill Exercise 29.4*

- Use the command in the accompanying Notebook to create two more of your own exercises and compute their $QR$ decomposition.

## 29.5   Applications of the Algorithm

In Corollary 28.13, we provided the most convenient formula to determine the projection $u$ of a vector $v$ into a subspace $V$, namely $u = QQ^T v$. This formula assumes that the columns of $Q$ form an *orthonormal* basis for $V$. However, $V$ is often presented simply via a spanning set $S$. However, the Gram-Schmidt process is designed to convert such a set $S$ to an orthonormal basis. In fact, if we take $A$ to be the matrix with $S$ as its columns and compute its $QR$ decomposition, the columns of $Q$ will be an orthonormal basis for $V$ (see the discussion in Section 29.1).

### Exercise 29.5

- Compute the projection $u$ of $v = \begin{bmatrix} -2 & -1 & 0 \end{bmatrix}^T$ into the subspace spanned by $S = \left\{ \begin{bmatrix} 1 & 0 & 1 \end{bmatrix}^T, \begin{bmatrix} 0 & 1 & 1 \end{bmatrix}^T \right\}$. Verify that $u - v \perp S$ and that $u \in span(S)$.

### Exercise 29.6

- Compute the projection of $\begin{bmatrix} 1 & 1 & 2 & -1 \end{bmatrix}^T$ into the subspace spanned by:

$$S = \left\{ \begin{bmatrix} 1 & 1 & -1 & -1 \end{bmatrix}^T, \begin{bmatrix} 2 & 1 & 2 & 0 \end{bmatrix}^T \right\}$$

As we mentioned in Section 28.5, projections are most commonly used to solve over-determined systems. There we discussed an approach that would seem to require us to solve two separate problems:

1. Compute the projection $B'$ into $Image(A)$.

2. Solve $AX = B'$.

Fortunately, a little algebra suggests a short cut that avoids the computation of $B'$ entirely! Since the columns of $A$ form a spanning set for $Image(A)$, $A = QR$ will provide the orthonormal basis necessary to compute the projection matrix $P = QQ^T$. *Remember*: This $P$ is *not* the same $P$ as in the algorithm; that was a matrix *consisting of* projections, while this is a matrix that *produces* projections. This implies that $B' = PB = QQ^T B$, so that the equation from step 2 of this approach may be rewritten as $AX = B' = QQ^T B$. Since $A = QR$, this equation becomes $QRX = QQ^T B$ or $Q(RX - Q^T B) = 0$, or $RX - Q^T B \in Kernel(Q)$. Since the columns of $Q$ are orthogonal (and hence independent by Theorem 28.11), $Kernel(Q) = 0$ (by Theorem 23.1), so $RX - Q^T B = 0$, or:

$$RX = Q^T B \qquad (29.4)$$

Thus, to solve an overdetermined system $AX = B$, we may solve Eq. 29.4 to obtain an approximate solution $AX = B' \approx B$. One should compare this situation with that in Section 18.2. Given $Q$ and $R$, Eq. 29.4 may be quickly and simply solved by hand:

1. Transpose and multiply to compute $Q^T B$.

2. Use backsubstitution through $R$ to obtain $X$.

In fact, one may perform a similar analysis with the $Q_o R_o$ decomposition and solve:

$$R_o X = (Q_o^T Q_o)^{-1} Q_o^T B \qquad (29.5)$$

This technique avoids the manipulation of radical expressions. However, one must remember to include $(Q_o^T Q_o)^{-1}$, which is the diagonal matrix (by Theorem 28.10a) consisting of the reciprocals of the lengths-squared of the columns of $Q_o$ (that is, the entries of $L$ from our matrix layout). *Important*: If the columns of $A$ are dependent, one must make sure to eliminate all zero columns from $Q_o$ (and the corresponding rows from $R_o$ before applying Eq. 29.5, or $Q_o^T Q_o$ will not be invertible.

## Exercise 29.7

•    Use Eq. 29.4 or Eq. 29.5 to solve the overdetermined systems:

(a) $\begin{bmatrix} 1 & 1 \\ 2 & 1 \\ -2 & 2 \end{bmatrix} X = \begin{bmatrix} 2 \\ 3 \\ -2 \end{bmatrix}$
    
(b) $\begin{bmatrix} 1 & 1 \\ -1 & -1 \\ 2 & 1 \\ 2 & 0 \end{bmatrix} X = \begin{bmatrix} 1 \\ 1 \\ 2 \\ -1 \end{bmatrix}$.

In Section 19.3.1, we observed that the magnitude of the determinant of a $2 \times 2$ matrix $A$ corresponds to the area of the parallelogram formed by the columns of $A$. Using the geometry of the Gram-Schmidt algorithm, we may sketch a proof of this fact in three dimensions, that will generalize directly to *any* dimension. This depends on a basic geometric principle, known as "Cavalieri's Principle," named after the 17th-century Italian geometer Bonaventura Cavalieri. Simply put, Cavalieri's Principle says that shearing transformations do not change the area of a two-dimensional figure or the volume of a three-dimensional region.

If $A = Q_o R_o = QR$ is a $3 \times 3$ matrix, then there are two cases to consider. If the columns are linearly dependent, then $Column(A)$ is a plane, which has *volume* 0. By Theorem 23.1, $A$ is not invertible, so that $|A| = 0$ and the determinant and the volume are equal.

If the columns of $A$ are independent, then $|A| \neq 0$. Using the AnimaQR[] command, we observe that the rectangular box $P'$ formed from the columns of $Q_o$ is the result of a sequence of shearing transformations from the corresponding figure $P$ (called a "parallelopiped") formed from the columns of $A$. We may also see this algebraically. Since $R_o$ is a diagonal matrix with 1's down the diagonal, it consists entirely of elemetary matrices corresponding to Definition 4.1c, and we observed in Lesson 9 that such matrices correspond to shearing transformations.

The volume of $P'$ equals the product of the lengths of its sides. By construction, these lengths also comprise the diagonal $R$, so $vol(P) = vol(P') = |R|$ (see Ex. 19.14). It remains to show that $|Q| = \pm 1$, since then $|A| = |Q||R| = \pm vol(P)$. By assumption, the columns of $A$ are independent. By Corollary 25.5b, this means that $span(A) = R^3$. In particular, the projection of any vector in $R^3$ into $span(A)$ is *itself*! That is, the projection matrix $QQ^T$ is the identity matrix. This implies that $|QQ^T| = |I| = 1$, so that $|Q|^2 = 1$ and $|Q| = \pm 1$.

## 29.6  Summary

In this lesson, we have discussed a specific matrix layout to facilitate hand-computation of Gram-Schmidt orthonormalization. One should have learned how to:

- perform Gram-Schmidt orthonormalization and how this corresponds to the decompositions $A = Q_o R_o = QR$,

- compute the projection matrix for a subspace $V = span(S)$, and

- solve overdetermined systems, using multiplication and backsubstitution.

Although we proved a number of theoretical facts related to the $QR$ decomposition, regarding its uniqueness and volumes, these facts are not necessary to understand the remainder of the text.

Now that we have a complete understanding of all three types of linear systems, in Lesson 31 we will examine the relationship between matrices and linear transformations. We said before that the abstract approach to vector spaces makes them widely applicable, while the use of bases and coordinates makes them practical. The exact same thing may be said for linear transformations, namely, that an abstract approach allows us to apply the most suitable coordinate system to solve any given problem. Using coordinates, we will be able to represent any linear transformation as a matrix, so that we may apply our techniques for solving linear systems to *any* linear equation. This will also lead naturally to Lesson 33.

# Change of Basis and Eigensystems

# Linear Transformations and Matrices

In Lesson 26, we introduced the notion of a linear transformation. In this lesson we examine the direct connection between linear transformations and matrices. We will see that matrices represent linear transformations, in a particular way. All of the properties of matrix algebra, including the strange definition of multiplication, follow most clearly from this description. In this lesson, we will:

- discuss the composition of linear transformations,

- explain how matrix multiplication represents the action of a linear transformation on a vector as well as composition of linear transformations, and

- demonstrate how to compute the matrix corresponding to a linear transformation, given a choice of bases.

One of our most important examples will be the coordinate transformation, $C_B$, corresponding to a given basis $B$.

## 30.1  The Basic Example

In Section 26.6, we examined the linear transformation:

$$L\left(\begin{bmatrix} x \\ y \\ z \end{bmatrix}\right) = \begin{bmatrix} x + 2y - z \\ -x + y + 3z \end{bmatrix} \tag{30.1}$$

We may rewrite this as

$$L\left(\begin{bmatrix} x \\ y \\ z \end{bmatrix}\right) = \begin{bmatrix} 1 & 2 & -1 \\ -1 & 1 & 3 \end{bmatrix}\begin{bmatrix} x \\ y \\ z \end{bmatrix}$$

so that $L$ corresponds to multiplication by $A = \begin{bmatrix} 1 & 2 & -1 \\ -1 & 1 & 3 \end{bmatrix}$, so that $L(v) = Av$.

We observed a similar phenomena in Section 9.5 for linear, geometric transformations in the plane. Namely, that any such transformation could be represented by multiplication by a particular matrix. In particular, we saw that the image of each of the standard basis vectors, $\begin{bmatrix} 1 \\ 0 \end{bmatrix}$ and $\begin{bmatrix} 0 \\ 1 \end{bmatrix}$, determined the columns of the corresponding matrix. Logically, this is because:

**Observation:**

Once we know the effect of a transformation on a basis, we have a complete description of the linear transformation, since every other vector is *uniquely* determined in a *linear* manner in terms of the basis.

This phenomenon occurs quite generally. That is, using only the properties of the standard basis for $R^n$, we may represent *any* linear transformation $L$ of column vectors by a matrix $A$, where the columns of $A$ are the images of the corresponding standard basis vectors under $L$.

**Definition 30.1** We denote the $n$ columns of the $n \times n$ identity matrix by $e_1, \ldots, e_n$. The set $\{e_1, \ldots, e_n\}$ is known as the *standard basis* for $R^n$.

Using this notation, we may state the above observation more formally.

**Theorem 30.1** *Assume that $A$ is an $n \times m$ matrix and $v$ is an $m \times 1$ column vector.*

(a) *Multiplication of $A$ and $v$ is defined so that $L : R^m \to R^n$ given by $L(v) = Av$ is a linear transformation, that is, multiplication of a matrix with a column vector corresponds to evaluation of a linear transformation between spaces of column vectors.*

(b) *Conversely, given a general linear transformation, $L : R^m \to R^n$, there exists a unique matrix $A$ so that $L(v) = Av$. In this case, we say that $A$ represents $L$. The columns $A$ are simply the outputs of $L$ when applied to the standard basis vectors. That is,*

$$(A)._i = L(e_i) \tag{30.2}$$

**Proof:**

We already have proven Theorem 30.1a in a specific example in Section 26.6. The general proof follows directly from the rules for matrix arithmetic. If $v$ and $w$ are any vectors in $R^m$, $r$ is any scalar and $L$ is defined by multiplication with $A$:

$$L(rv + w) = A(rv + w) = A(rv) + Aw = r(Av) + Aw = rL(v) + L(w)$$

If $r = 1$, this verifies that $L$ preserves addition, while if $w = 0$ this shows that $L$ preserves scalar multiplication. Thus, $L$ is a linear transformation.

To prove Theorem 30.1b, we assume only that $L$ is a linear transformation. Since the $m$ vectors $L(e_i)$ are in $R^n$, we may place them as columns of an $n \times m$ matrix, which we will call $A$. If $v = \begin{bmatrix} a_1 \\ \vdots \\ a_m \end{bmatrix}$ is any vector in $R^m$, we may decompose $v$ as:

$$v = a_1 e_1 + \cdots + a_m e_m$$

(compare Section Ex. 23.5.1). Applying $L$ to both sides of this equation yields:

$$
\begin{aligned}
L(v) &= L(a_1 e_1 + \cdots + a_m e_m) = a_1 L(e_1) + \cdots + a_m L(e_m) \\
&= \begin{bmatrix} L(e_1) & \cdots & L(e_m) \end{bmatrix} \begin{bmatrix} a_1 \\ \vdots \\ a_m \end{bmatrix} = Av
\end{aligned}
\tag{30.3}
$$

Notice that the crucial equality is the second one, where we use the fact that $L$ is linear to break up the sum and pull out the coefficients $a_i$. Likewise, the remaining steps are essentially restatements of the definition of matrix multiplication with a column vector.

Looking back at Section 26.4, one should notice that we have seen the phenomena described by Theorem 30.1 in the case of the coordinate transformations $C_B$. This example is so important, we highlight it as a theorem.

**Theorem 30.2** *If $B = \{v_1, \cdots, v_n\}$ is a basis for a subspace $V \subset R^k$, $C_B^{-1}$ corresponds to multiplication by the matrix $\mathbf{B} = \begin{bmatrix} v_1 & \cdots & v_n \end{bmatrix}$. Moreover, if $V = R^k$ (so that $k = n$), then $\mathbf{B}$ is invertible and $C_B$ corresponds to $\mathbf{B}^{-1}$.*

(Compare Ex. 25.4.) *Remember*: At the end of Section 26.4, we established the convention that if $B$ is a basis of column vectors, then $\mathbf{B}$ stands for the matrix with the vectors in $B$ as columns.

### Exercise 30.1

If $V$ posesses a basis $B = \left\{ \begin{bmatrix} 1 \\ 0 \\ -1 \end{bmatrix}, \begin{bmatrix} 2 \\ 1 \\ 0 \end{bmatrix}, \begin{bmatrix} 1 \\ 3 \\ 1 \end{bmatrix} \right\}$, compute the coordinate matrix $C_B$. Use this to determine the coordinates of $v = \begin{bmatrix} 0 \\ 1 \\ -2 \end{bmatrix}$ with respect to $B$.

Verify your result. Randomly generate a *second* vector $w$, determine its coordinates with respect to $B$, and verify your result.

## 30.2  The Origins of Matrix Multiplication

Equation 30.3 is the single most important formula in linear algebra. It provides the first half of the story behind matrix multiplication: Namely, it demonstrates the connection between multiplication of a matrix with a *column* and *evaluation* of a linear transformation on a vector. To complete the picture, we must discuss what a *general* matrix product represents. We will see that this product is defined so as to correspond with the *composition* of two linear transformations. Composition of transformations is just the same as composition of functions that is discussed in precalculus. The *composite* of two transformations $L$ and $M$ is denoted $L \circ M$, or simply $LM$, and is defined by the equation:

$$
L \circ M(v) = L(M(v))
\tag{30.4}
$$

Intuitively, the composite transformation says to "apply $M$, then apply $L$" to any given input $v$.

## Exercise 30.2

Let $L$ be the transformation of the plane which rotates by $90°$ clockwise, and let $M$ be the transformation that reflects around the $y$-axis. Let $K = LM$ be their composite, that is, $K(v) = L(M(v))$, so that we *first* perform $M$ (reflect) and *then* perform $L$ (rotate). Provide another description of $K$ as a *single* geometric operation. *Hint*: Take a piece of paper, try to perform these two operations on some sample vectors, and deduce a pattern.

Examining a general matrix product, we may prove the following:

**Theorem 30.3** *Assume that $A$ is an $n \times m$ matrix and $B$ is an $m \times p$ matrix. Moreover, assume that $L : R^m \to R^n$ and $M : R^p \to R^m$ are the associated linear transformations $L(v) = Av$ and $M(w) = Bw$.*

(a) *Multiplication of two matrices is defined so that it corresponds to the composition of two linear transformations between spaces of column vectors. That is, the matrix for the composite $K = LM$ is simply the matrix $C = AB$.*

(b) *If a linear transformation $L$ is a bijection, with inverse $L^{-1}$, the inverse function is automatically a linear transformation.*

(c) *If $L$ is an invertible transformation, then $A$ is an invertible matrix and the matrix of $L^{-1}$ is $A^{-1}$.*

**Proof:**

To prove Theorem 30.3a, we simply apply Eq. 30.2 and the definitions of the matrices $A$, $B$, and $C$:

$$K(e_i) = (LM)(e_i) = L(M(e_i)) = L((B)_{.i}) = A((B)_{.i}) = (AB)_{.i} = (C)_{.i}$$

By Theorem 30.1b, this equation implies that $C = AB$ is the matrix that represents $K$. Notice that the key equality $A((B)_{.i}) = (AB)_{.i}$ is precisely our observation regarding the definition of matrix multiplication on p. 82. That is, the product of two matrices may be computed by applying the first matrix to each *column* of the second. This is also a special case of the Principle of Partitioned Matrices.

To prove Theorem 30.3b, we must show that $L^{-1}$ preserves addition and scalar multiplication. This follows from the linearity of $L$ and the defining property of an inverse function:

$$L(w) = v \quad \text{iff} \quad L^{-1}(v) = w$$

If $v_1$ and $v_2$ are two vectors, and we denote $L^{-1}(v_1) = w_1$ and $L^{-1}(w_1) = w_2$, then $L(w_1) = v_1$ and $L(w_2) = v_2$. Since $L$ is a linear transformation, $L(w_1 + w_2) = v_1 + v_2$, so that $L^{-1}(v_1 + v_2) = w_1 + w_2 = L^{-1}(v_1) + L^{-1}(v_2)$. Thus, $L^{-1}$ preserves addition. The proof for scalar multiplication is similar. If $L^{-1}(v) = w$ and $r$ is any scalar, then $L(w) = v$ and $rv = rL(w) = L(rw)$. Thus, $L^{-1}(rv) = rw = rL^{-1}(v)$, and $L^{-1}$ preserves scalar multiplication.

Theorem 30.3c follows directly from Theorem 30.3a. Remember that by definition the inverse of $L$ satisfies the equations $L(L^{-1}(v)) = v = Iv$ and $L^{-1}(L(v)) = v = Iv$.

These equations say that the matrices for the transformations $LL^{-1}$ and $L^{-1}L$ are *both* the identity matrix $I$. If the matrix for $L$ is $A$, and the matrix for $L^{-1}$ is $B$, then by Theorem 30.3a, we have that the matrix for $LL^{-1}$ is $AB$, while the matrix for $L^{-1}L$ is $BA$. Together these facts imply the equations $AB = I$ and $BA = I$. Since these are the defining equations for $A^{-1}$, the matrix for $L^{-1}$ is $B = A^{-1}$. Q.E.D.

Given this result, we may return to an important algebraic property of matrices that we have overlooked to this point. We never have proved associativity of matrix multiplication but simply observed that it seemed to hold in examples. Theorem 30.3 allows us to provide an elegant proof of this fact.

**Theorem 30.4** *Multiplication of matrices is associative. That is, if $A$ is $n \times m$, $B$ is $m \times p$, and $C$ is $p \times q$, then:*

$$A(BC) = (AB)C$$

**Proof:**

Let $K$, $L$, and $M$ be linear transformations corresponding to $A$, $B$, and $C$. By Theorem 30.3, we know that $A(BC)$ corresponds to the transformation $K(LM)$, while $(AB)C$ corresponds to $(KL)M$. By uniqueness of the matrix representation in Theorem 30.1b, it suffices to show that these two transformations are equal.

Since a transformation is characterized completely by its output for any given input, it suffices to show that these two transformations yield the same value on any given input $v$. Repeatedly applying the definition of composition (see Eq. 30.4) gives:

$$(K(LM))(v) = K((LM)(v)) = K(L(M(v))) = (KL)(M(v)) = ((KL)M)(v)$$

Thus, $K(LM) = (KL)M$ and $A(BC) = (AB)C$. Q.E.D.

So far we have shown that the definitions of matrix algebra are *not* arbitrary, but are defined so that matrix algebra *exactly* corresponds to the algebra of linear transformations on column vectors, where:

1. a linear transformation corresponds to a matrix,

2. evaluation of a transformation on a column vector corresponds to multiplication with a column vector, and

3. composition of two transformations corresponds to general matrix multiplication.

In Section 30.3, we generalize these results for linear transformations of column vectors to *all* linear transformations between *any* pair of (finite-dimensional) vector spaces.

## 30.3 Representing Arbitrary Transformations as Matrices

Once we choose a basis for the domain and range of a linear transformation $L$, we obtain isomorphisms of both vector spaces with a corresponding vector space of column vectors. Thus, we may consider any abstract, linear transformation (such as the examples from Section 26.3 of differentiation or scalar multiplication) as a linear transformation *of column vectors*. Thus, by Theorem 30.1b, we may represent this composition by a matrix, as well! A simple picture is helpful to illustrate this composite.

For example, let $L$ be any linear transformation from $V$ to $W$. Using the notation from Lesson 20, we denote this in symbols by:

$$L : V \to W$$

If we select a basis $B = \{v_1, \ldots, v_m\}$ for $V$ and $B' = \{w_1, \ldots, w_n\}$ for $W$, then we have the coordinate isomorphisms,

$$C_B : V \to R^m \quad \text{and} \quad C_{B'} : W \to R^n$$

where $n$ and $m$ are the dimensions of $V$ and $W$, respectively. We have already observed that these are linear transformations. Since these are 1–1 correspondences, there is a well-defined inverse for each. Theorem 30.3b shows that these inverse functions are *also* linear transformations. In Section 26.4, we introduced a short-hand notation for $C_B^{-1}$, namely $\mathbf{B} : R^n \to V$. Notice that $\mathbf{B}$ takes the coordinates of a vector $v$ (with respect to the basis $B$) to $v$ itself, so that it is defined by the equation:

$$\mathbf{B}\left(\begin{bmatrix} \alpha_1 \\ \vdots \\ \alpha_n \end{bmatrix}\right) = \alpha_1 v_1 + \cdots + \alpha_n v_n \qquad (30.5)$$

(compare Eq. 26.8). Using this notation, if we form the composite:

$$C_{B'} L\mathbf{B} : R^m \to V \to W \to R^n$$

we obtain a linear transformation of column vectors. By Theorem 30.1, this composite may be represented by some matrix $A$. We can picture this as in Fig. 30.1.

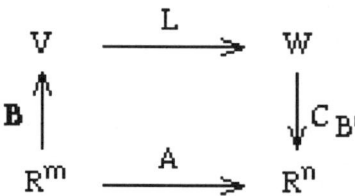

Figure 30.1: A Linear Transformation *vs.* Its Matrix Representation

Such a figure is known as a "commutative" diagram and is meant to represent the equation $C_{B'} L\mathbf{B} = A$. In a commutative diagram, if one may proceed from one point to another by following the arrows in two different ways, those two paths should represent equal compositions of functions. In this case, proceeding up the left side, along the top, and down the right side represents the composite, $C_{B'} L\mathbf{B}$, while the bottom arrow represents $A$. Intuitively, this diagram implies that:

**Observation:**

Evaluation of $L$ on a vector $v$ corresponds to multiplication of the coordinate vector for $v$ with respect to $B$ by the matrix $A$, where one should interpret the resulting column vector as the coordinates of $L(v)$ with respect to $B'$.

This simple diagram describes how one should compute $A$.

**Theorem 30.5** *Consider a linear transformation $L : V \to W$, with bases $B = \{v_1, \ldots, v_m\}$ and $B' = \{w_1, \ldots, w_n\}$ for $V$ and $W$, respectively. If $A$ represents the transformation $C_{B'}LB$, the ith column of $A$ is given by the coordinates of $L(v_i)$ with respect to $B'$. That is:*

$$(A)_{.i} = C_{B'}(L(v_i)) \tag{30.6}$$

*In this case we say that $A$ represents $L$ with respect to the bases $B$ and $B'$, and we indicate this by $L \leftrightarrow A$.*

**Proof:**

By Eq. 30.2, the columns of $A$ are given by:

$$(A)_{.i} = C_{B'}(L(\mathbf{B}(e_i)))$$

To obtain Eq. 30.6, we simply observe that, by Eq. 30.5, $\mathbf{B}(e_i) = 0v_1 + \cdots + 1v_i + \cdots + 0v_m = v_i$. Q.E.D.

In other words, the columns of the matrix $A$ for $L$ (with respect to $B$ and $B'$) may be computed by:

1. evaluating $L$ on each basis vector for the domain, and

2. placing the coordinates of each result $L(v_i)$ with respect to $B'$, in order, as the columns of $A$.

Notice that if $V = R^n$ and $W = R^m$, both equipped with their respective standard bases, then each coordinate transformation is an identity transformation and the matrix of Theorem 30.5 is the same as that of Theorem 30.1.

For example, assume that $V = P_4$ (i.e., the vector space of polynomials of degree 3 or less), $W = P_3$ and $L$ is "differentiation" (i.e., $L(p(t)) = \frac{\partial p(t)}{\partial t}$). In this case, it is natural to employ the standard bases for the domain and range, so that $B = \{1, t, t^2, t^3\}$ and $B' = \{1, t, t^2\}$. To compute the corresponding matrix $A$, by Theorem 30.1b, we must apply the transformation $C_{B'}LB$ to each of the standard basis vectors $e_i$ of the domain $R^4$. Equivalently, Eq. 30.6 says that we should apply $C_{B'}L$ to each basis vector of $B$ and place the results in $A$ as columns. Since $L(1) = 1' = 0$ and $C_{B'}(0) = [\ 0, 0, 0\ ]$, we have that the composite:

$$(C_{B'}LB)(e_1) = C_{B'}(L(\mathbf{B}(e_1))) = C_{B'}(L(1)) = C_{B'}(0) = \begin{bmatrix} 0 \\ 0 \\ 0 \end{bmatrix}$$

Similarly,

$$(C_{B'}LB)(e_2) = C_{B'}(L(\mathbf{B}(e_2))) = C_{B'}(L(t)) = C_{B'}(1) = \begin{bmatrix} 1 \\ 0 \\ 0 \end{bmatrix} \tag{30.7}$$

Notice how we are simply taking the coordinates with respect to $B'$ of the result of $L$ on each of the basis vectors of $B$, just as Theorem 30.5 indicates.

## Exercise 30.3

Justify each step of Eq. 30.7. Compute the remaining columns of $A$ to show that $A =$
$$\begin{bmatrix} 0 & 1 & 0 & 0 \\ 0 & 0 & 2 & 0 \\ 0 & 0 & 0 & 3 \end{bmatrix}.$$

## Exercise 30.4

Let $L : P_3 \rightarrow P_3$ be defined by the formula:
$$L(a + bt + ct^2) = (a + bt + ct^2) + (c + bt + at^2)$$
Compute the matrix for $L$, using the standard basis $B = \{1, t, t^2\}$ on its domain and range. *Note*: $L$ may also be defined by the equation $L(p(t)) = p(t) + t^2 p\left(\frac{1}{t}\right)$.

## Exercise 30.5

Apply Theorem 30.5 to show that the transformation $R : E^2 \rightarrow E^2$ for rotation in the plane by $180°$ corresponds to the matrix $A = \begin{bmatrix} -1 & 0 \\ 0 & -1 \end{bmatrix}$. *Hint*: Use the standard Cartesian coordinate system $\{\vec{i}, \vec{j}\}$ as a basis on the domain and range for $E^2$.

## Exercise 30.6

Assuming only that $V$ is a three-dimensional vector space, show that the matrix for the linear transformation "scalar multiplication by 2," defined by $M(v) = 2v$, is given by the matrix $A = \begin{bmatrix} 2 & 0 & 0 \\ 0 & 2 & 0 \\ 0 & 0 & 2 \end{bmatrix}$. *Hint*: You may assume that $V$ possesses a basis $B = \{v_1, v_2, v_3\}$ and use this for both the domain and range of this transformation.

## 30.4  Summary

In this lesson, one should have learned:

- the definition of the *composition* of two linear transformations,
- that every matrix corresponds to a linear transformation so that multiplication with a column vector corresponds to *evaluation* and multiplication of two general matrices corresponds to *composition*,
- how to compute the matrix $A$ of a linear transformation $L$ with respect to a choice of bases for its domain and range,
- specifically, that the $i$th column of $A$ are the coordinates (with respect to $B'$) of $L(v_i)$, and

– a simple formula for the matrix of $C_B$ for a basis of $R^n$.

In Lesson 31, we will discuss how changing our point of view, namely changing our choice of bases, affects the coordinate representation of a vector and the matrix representation of a transformation. Every problem that we have discussed, namely, solving linear systems and solving for eigenvalues, may be characterized in terms of performing an appropriate change of basis.

# The Effects of Changing Coordinates

## Accompanied by Notebook

In Lessons 26 and 30, we discussed the notion of a linear transformation and how to represent a linear transformation by a matrix, using bases. In this lesson we discuss how the use of different bases affects the corresponding matrix representation. We will discuss coordinate representations of vectors in greater depth, as well.

## 31.1  Coordinate Representations of Vectors

Remember from Section 25.1 that given a basis, $B = \{v_1, \ldots, v_n\}$, we say that $X = \begin{bmatrix} a_1 \\ \cdots \\ a_n \end{bmatrix}$ are the coordinates of $v$ with respect to $B$ precisely when:

$$v = \mathbf{B}X = a_1 v_1 + \cdots + a_n v_n$$

In this way, the column vector $X = C_B(v)$ provides a concrete representation of the vector $v$ in terms of $B$. If $v$ is in $R^n$ (or some other vector space, such as $P_n$, that may be readily converted into a column) and we are given a basis $B$, it is routine to solve for the coordinate vector $C_B(v)$, since it is simply a solution to a system of linear equations. The properties of a basis guarantee that this system always possesses a unique solution. In Theorem 30.2 we showed that if $B$ is a basis for $R^n$, then we may represent $C_B$ as $\mathbf{B}^{-1}$, where $\mathbf{B}$ is the *matrix* with column from the *set* $B$. If $B$ is a basis of some *subspace* $V \subset R^n$, we must be a bit more careful.

### Exercise 31.1

• Determine the coordinates of $v = \begin{bmatrix} 3 & 9 & 2 & 4 \end{bmatrix}^T$ with respect to each of the following bases:

(a) $B_1 = \left\{ \begin{bmatrix} 1 \\ 1 \\ 0 \\ -1 \end{bmatrix}, \begin{bmatrix} 2 \\ -1 \\ 1 \\ 0 \end{bmatrix}, \begin{bmatrix} 1 \\ 2 \\ 1 \\ 2 \end{bmatrix} \right\}$ (b) $B_2 = \left\{ \begin{bmatrix} 5 \\ -1 \\ 2 \\ -1 \end{bmatrix}, \begin{bmatrix} 33 \\ -7 \\ 12 \\ -10 \end{bmatrix}, \begin{bmatrix} 30 \\ -7 \\ 11 \\ -9 \end{bmatrix} \right\}$

We should emphasize that $\begin{bmatrix} 3 \\ 9 \\ 2 \\ 4 \end{bmatrix}$ as well as $C_{B_1}(v)$ and $C_{B_2}(v)$ from Ex. 31.1 all refer to the *same* vector $v$. Each column vector is simply a different description relative to a different choice of basis. Even the original description $v = \begin{bmatrix} 3 & 9 & 2 & 4 \end{bmatrix}^T$ may be viewed as another coordinate representation $C_B(v)$, where $B$ is the *standard* basis $B = \{e_1, \ldots, e_4\}$ for $R^4$.

By employing two different coordinate isomorphisms, $C_{B_1}$ and $C_{B_2}$, we may translate directly from $B_1$-coordinates for $v$ to $B_2$-coordinates. Since the resulting transformation is from $R^n$ to itself, the result is representable by a matrix, which is known as a "transition matrix."

**Definition 31.1** If $B_1$ and $B_2$ are both bases for the same vector space $V$, then the matrix which takes the $B_1$-coordinates for any given vector $v$ to the corresponding $B_2$-coordinates is called the *transition matrix* from $B_1$ to $B_2$.

With a little thought, we may determine some simple techniques for computing such transition matrices.

**Theorem 31.1** *If $V$ is a subspace of $R^n$ and $B_1$ and $B_2$ are bases for $V$, then the transition matrix $M$ from $B_1$-coordinates to $B_2$-coordinates may be computed by the formula*

$$M = L\mathbf{B}_1 \tag{31.1}$$

*where $L$ is any left-inverse of $\mathbf{B}_2$.*

**Proof:**

Consider an arbitrary vector $v \in V$. Denote its coordinates with respect to $B_1$ and $B_2$ as $X$ and $Y$, respectively. Then by our comments at the begining of this section, we have the equations:

$$\mathbf{B}_1 X = v = \mathbf{B}_2 Y \tag{31.2}$$

Since $M$ should take $X$ to $Y$, $M$ is then defined by the equation $MX = Y$. Since $B_2$ is independent, Theorem 23.1 guarantees that $\mathbf{B}_2$ is injective and Theorem 16.2b implies that it has a left-inverse. If $L$ is any left-inverse for $\mathbf{B}_2$, we may apply $L$ to all sides of Eq. 31.2 to obtain:

$$L\mathbf{B}_1 X = Lv = L\mathbf{B}_2 Y = Y = MX \tag{31.3}$$

By choosing $v$ to be the $i$th basis vector in $B_1$, we may assume that $X = e_i$. Thus, Eq. 31.3 implies that $L\mathbf{B}_1 e_i = Me_i$, or in other words, the $i$th columns of both matrices are equal. Since $i$ was arbitrary, this implies that *every* column of $L\mathbf{B}_1$ and $M$ are equal, so that they are the same matrices. Q.E.D.

*Note*: Because composition of functions proceeds from right to left, the order of the terms in this formula may seem "reversed." That is, to convert from $B_1$-coordinates to $B_2$-coordinates, one should use the product of a left-inverse of $\mathbf{B}_2$ and $\mathbf{B}_1$.

**Corollary 31.2** *If* $B = \{v_1, \ldots, v_n\}$ *is a basis of column vectors for* $R^n$ *and* $\mathbf{B} = \begin{bmatrix} v_1 & \cdots & v_n \end{bmatrix}$ *is the corresponding matrix, then the transition matrix* $M$ *from:*

*(a) B-coordinates to the standard basis is given by* $M = \mathbf{B}$; *and*

*(b) the standard basis to B-coordinates is given by* $M = \mathbf{B}^{-1}$.

(See Theorem 30.2.)

**Proof:**

Both cases follow directly from Theorem 31.1. For the first case, if we apply the Theorem 31.1, $\mathbf{B}_1 = \mathbf{B}$ and $\mathbf{B}_2 = \mathbf{I}$, so that $M = L\mathbf{B}_1 = \mathbf{IB} = \mathbf{B}$. In the second, $\mathbf{B}_1 = \mathbf{I}$ and $\mathbf{B}_2 = \mathbf{B}$, so that $M = L\mathbf{B}_1 = \mathbf{B}^{-1}\mathbf{I} = \mathbf{B}^{-1}$. Q.E.D.

## Exercise 31.2

- Using the same $B_1$ and $B_2$ from Ex. 31.1:

(a) Determine the matrix $M$ that converts from $B_1$-coordinates to $B_2$-coordinates.

(b) Verify that your matrix $M$ converts the $B_1$-coordinates of $v$ from Ex. 31.1 to its $B_2$-coordinates.

- *Note*: In the accompanying Notebook, one may learn to use the *Mathematica* commands CoordinateVector[] and TransitionMatrix[]. As usual, one should only employ this commands to check one's work. Most effort should be spent on the *reasoning* behind each calculation and in *explaining* one's results, not simply producing numerical answers.

## Exercise 31.3

- Repeat Ex. 31.2 with the following:

(a) $B_1 = \left\{ \begin{bmatrix} 1 \\ 1 \end{bmatrix}, \begin{bmatrix} 1 \\ 0 \end{bmatrix} \right\}$, $B_2 = \left\{ \begin{bmatrix} 5 \\ 3 \end{bmatrix}, \begin{bmatrix} 2 \\ 1 \end{bmatrix} \right\}$, and $v = \begin{bmatrix} 7 \\ 4 \end{bmatrix}$.

(b) $B_1 = \{1 + t, 1 + t^2, t + t^2\}$, $B_2 = \{1 - t, 1 + t^2, t - t^2\}$, and $v = 5 - 3t + 2t^2$. *Hint:* Use the Isomorphism Principle to relate this situation to one involving column vectors.

We all know that the same idea can be described in many different languages. For example, the words "door," "puerta," "dörr," "Tür," and "porta" all refer to the same concept. In the same way, the same vector may be described in an infinite number of different ways, depending on the basis that we choose. When one converts from one coordinate description to another, one may consider this as "translating" from one language to another. From this point of view, the matrix $M$ from Theorem 31.1 may be viewed as a "translator" from $B_1$-coordinate language to $B_2$-coordinate language. Theorem 31.1 may be remembered intuitively as saying:

**Observation:**

To translate from $B_1$-coordinate language to $B_2$-coordinate language, one should use the matrix $\mathbf{B}_2^{-1}\mathbf{B}_1$.

Keep in mind that $\mathbf{B}_2^{-1}$ must actually stand for a *left*-inverse of $\mathbf{B}_2$ in this context. As we mentioned previously, the order of the matrices in the formula is reversed from the order in the corresponding English sentence, while the inverse is on the first matrix in the formula. If one focuses on this pattern, Eq. 31.1 becomes easy to remember. For any given problem, one should think carefully about which transition matrix $M$ one requires by first expressing in English what translation must be accomplished. Then one may simply apply this rule to determine the appropriate formula for $M$ and plan out the necessary steps to compute the correct left-inverse and product.

## 31.2  Changing Bases for Transformations

Just as the coordinates of a vector depend on the choice of basis for the ambient vector space, the matrix representation of a linear transformation depends on the choice of bases for its domain and range. For example, in Section 30.3 we considered the differentiation transformation, $L : P_4 \to P_3$, where $L\left(p(t)\right) = \frac{\partial p(t)}{\partial t}$. Using the standard bases for the domain and range, $B_1 = \{1, t, t^2, t^3\}$ and $B_2 = \{1, t, t^2\}$, we showed that $L$ corresponds to the matrix:

$$A = \begin{bmatrix} 0 & 1 & 0 & 0 \\ 0 & 0 & 2 & 0 \\ 0 & 0 & 0 & 3 \end{bmatrix}$$

Another natural choice of bases is suggested by calculus. When computing Taylor polynomials, for example, one uses bases such as:

$$B_1' = \{1, \frac{t}{1!}, \frac{t^2}{2!}, \frac{t^3}{3!}\} \quad \text{and} \quad B_2' = \{1, \frac{t}{1!}, \frac{t^2}{2!}\}$$

In the following exercise, we compute the matrix $A'$ of $L$ with respect to these new bases. Later, we will demonstrate a faster method to perform this same computation.

### Exercise 31.4

Compute $A'$ as in Theorem 30.5. For example, to compute the third column of $A'$, one should determine the $B_2'$-coordinates of $L\left(\frac{t^2}{2!}\right)$. Verify that the third column of $A'$ is $\begin{bmatrix} 0 \\ 1 \\ 0 \end{bmatrix}$ and proceed to compute the remaining columns. *Note:* Observe that this was a better choice of bases, in that the corresponding matrix representation is particularly simple.

In Section 31.1, we saw how one may translate directly between different coordinate expressions for a vector via transition matrices. In the same way, one might suppose that we may somehow translate the matrix $A = \begin{bmatrix} 0 & 1 & 0 & 0 \\ 0 & 0 & 2 & 0 \\ 0 & 0 & 0 & 3 \end{bmatrix}$, defined with respect to $B_1$ and $B_2$, directly into the matrix $A'$, defined with respect to $B_1'$ and $B_2'$. To see how this may be accomplished, we again employ commutative diagrams (see Section 30.3).

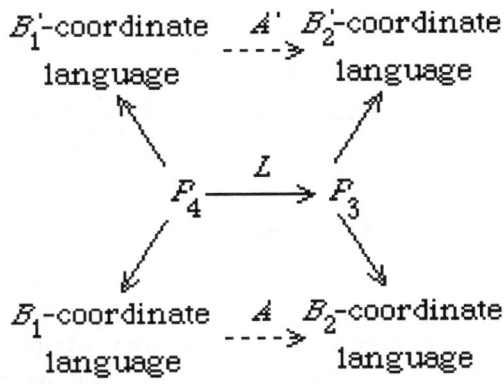

Figure 31.1: Comparing Two Matrix Representations

Intuitively, we picture this situation as in Fig. 31.1. Although all three horizontal arrows represent the same transformation, the middle transformation takes polynomials as input and output, whereas those on the top and bottom manipulate column vectors, which are to be thought of as coordinate descriptions of polynomials with respect to the indicated bases. *Important*: The labels "$B$ coordinate language" stand for the set of *column vectors $R^n$* (of the appropriate dimension) *interpreted* as coordinates with respect the basis $B$.

## Exercise 31.5

For the example above, sketch a copy of Fig. 31.1 replacing each descriptive label by the notation for the set of column vectors of the correct dimension.

Notice that, because there exist transition matrices that convert directly between different coordinate expressions, we may complete this diagram as in Fig. 31.2. This suggests that one could compute $A'$ directly as the composite of:

1. the transition matrix $M_1$ from $B_1'$-coordinates to $B_1$-coordinates,

2. the matrix $A$ representing $L$, which we consider to map $B_1$-coordinate expressions to $B_2$-coordinate expressions, and

3. the transition matrix $M_2$ from $B_2$-coordinates to $B_2'$-coordinates.

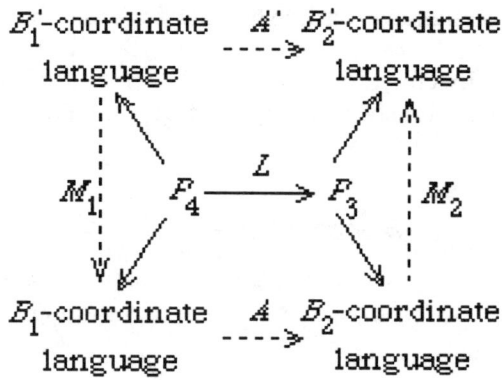

Figure 31.2: Transition Matrices Between Matrix Representations

*Remember*: All of the horizontal arrows in these diagrams represent the *same* mapping of vectors, where only the manner of describing the inputs and outputs is different. This is another example of the Isomorphism Principle that we discussed in Section 25.1. In that section, we saw how calculations in one vector space are mirrored exactly in any other vector space of the same dimension. Now we see that *transformations between* pairs of vector spaces mirror one another as well.

We summarize our observations in the following theorem:

**Theorem 31.3** *Assume that*

- $L : V \to W$,

- $B_1$ *and* $B_1'$ *are bases of* $V$,

- $B_2$ *and* $B_2'$ *are bases of* $W$,

- $A$ *is the matrix representing* $L$ *with respect to* $B_1$ *and* $B_2$,

- $A'$ *represents* $L$ *with respect to* $B_1'$ *and* $B_2'$,

- $M_1$ *is the transition matrix from* $B_1'$*-coordinates to* $B_1$*-coordinates, and*

- $M_2$ *translates from* $B_2$*-coordinates to* $B_2'$*-coordinates.*

*Then $A$ and $A'$ may be directly related by the formula:*

$$A' = M_2 A M_1 \tag{31.4}$$

Theorem 31.3 may seem daunting stated in such formal language. In practice, one does not appeal to this theorem so much as one *rederives* it, as necessary. That is, in any particular example, one should first draw a diagram such as in Figure 31.3 to describe the given situation and then determine the correct formula by simplying following the arrows (and appealing to Eq. 31.1 to determine the necessary transition matrices).

Theorem 31.3 has an important corollary that we will exploit in Lesson 33, when we consider the "eigenvalue problem."

**Corollary 31.4** *Assume that:*

- *$L : V \rightarrow V$ is a linear transformation from V to itself,*
- *B and B' are bases of V,*
- *A is the matrix representing L with respect to B (for the domain and range),*
- *A' represents L with respect to B' (also on both the domain and range), and*
- *M is the transition matrix from B-coordinates to B'-coordinates.*

*Then A and A' may be directly related by the formula:*

$$A' = MAM^{-1} \tag{31.5}$$

*In this case, we say that A and A' are* similar *(or conjugate) matrices.*

**Proof:**

This follows directly from Theorem 31.3, once we relate the notation of the two results. We do so by assuming that all of the notation of Corollary 31.4 has been established and proceeding to define the corresponding notation for Theorem 31.3, so that we may apply it. Specifically, let $W = V$, $L$, and let $B_1 = B_2$ and $B'_1 = B'_2$ of Theorem 31.3 be given by the $V$, $L$, $B$, and $B'$ of Corollary 31.4. Then $M_2$ of Theorem 31.3 is the same as $M$ of Corollary 31.4. Since $M$ translates from from $B$-coordinates to $B'$-coordinates, $M^{-1}$ must translate from $B'$-coordinates to $B$-coordinates. Because this is plays the role of $M_1$ of Theorem 31.3, Eq. 31.5 follows directly from Eq. 31.4.

## Exercise 31.6

• Compute $A'$ for our differentiation example by appealing to Theorem 31.3. Namely:

(a) Find the change-of-basis matrices $M_1$ from $B'_1$-coordinates to $B_1$-coordinates and $M_2$ from $B_2$-coordinates to $B'_2$-coordinates, where:

$$B'_1 = \{1, \tfrac{t}{1!}, \tfrac{t^2}{2!}, \tfrac{t^3}{3!}\}, \quad B_1 = \{1, t, t^2, t^3\},$$

$$B_2 = \{1, t, t^2\}, \quad \text{and} \quad B'_2 = \{1, \tfrac{t}{1!}, \tfrac{t^2}{2!}\}$$

*Hint*: Using the isomorphism principle of Section 25.1, one may relate this to the equivalent problem for column vectors and apply Corollary 31.2.

(b) Compute $A'$ by the equation $A' = M_2 A M_1$.

Compare your result with your answer to Ex. 31.4.

Because we solved the differentiation example in two different ways, the impact of Theorem 31.3 may have been obscured. To practice using Theorems 30.5 and 31.3, we will examine another concrete example. This example will also demonstrate how much time may be saved by employing Theorem 31.3, if one thinks before computing. This time assume that $K : P_3 \rightarrow P_4$, is given by $K(p(t)) = tp(t) + \frac{p(t) - p(0)}{t}$, and apply $B_1 = \{1, 1 + t, 1 + t + t^2\}$ to obtain coordinates for the domain and $B_2 = \{1, 1 - t, 1 + 2t + t^2, 1 - t + 3t^2 - 3t^3\}$ for the range. To obtain a matrix representation for $K$, remember that Theorem 30.5 tells us to:

1. take each of the vectors of $B_1$ in turn,

2. apply $K$, and

3. express the result in terms of $B_2$-coordinates.

For example, $K(1) = t$, which has $B_2$-coordinates $\begin{bmatrix} 1 \\ -1 \\ 0 \\ 0 \end{bmatrix}$, since $t = (1) - (1-t)$. This

then is the first column of our matrix.

## Exercise 31.7

- Compute the remaining columns of the matrix $S$ of $K$ with respect to to $B_1 = \{1, 1 + t, 1 + t + t^2\}$ and $B_2 = \{1, 1 - t, 1 + 2t + t^2, 1 - t + 3t^2 - 3t^3\}$.

If we had thought a bit more before solving Ex. 31.7, we would have noticed that it is much easier to first compute the matrix of $K$ with respect to the *standard* bases for $P_3$ and $P_4$.

## Exercise 31.8

Show that $K$ is represented by $T = \begin{bmatrix} 0 & 1 & 0 \\ 1 & 0 & 1 \\ 0 & 1 & 0 \\ 0 & 0 & 1 \end{bmatrix}$ with respect to the standard bases.

Now we may solve Ex. 31.7, in the same manner as Ex. 31.6, by appealing to Theorem 31.3. Using a diagram similar to Fig. 31.2, we observe that the matrix for $K$ with respect to $B_1$ and $B_2$ is the product of three matrices, where:

- the first converts from $B_1$-coordinates to standard coordinates,

- the second is the matrix from Ex. 31.8 (i.e., taking standard coordinates to standard coordinates), and

- the third converts from standard coordinates to $B_2$-coordinates.

## Exercise 31.9

- Using the bases from this example, compute the transition matrix:

    (a) $M_1$ from $B_1$ to $\{1, t, t^2\}$          (b) $M_2$ from $\{1, t, t^2, t^3\}$ to $B_2$

*Hint*: Use the Isomorphism Principle to relate each part to an equivalent problem on column vectors and apply Corollary 31.2.

## Exercise 31.10

Compare the matrix $S$ from Ex. 31.7 with the product $M_2 T M_1$, using the transition matrices from Ex. 31.9. What does Theorem 31.3 say should be happen? Identify what each of the symbols of the theorem would stand for in this particular case.

The final exercise of this section provides one final example to illustrate these ideas.

### Drill Exercise 31.11

Given the following bases:

$$B_1 = \left\{ \begin{bmatrix} 1 \\ 0 \end{bmatrix}, \begin{bmatrix} 0 \\ 1 \end{bmatrix} \right\} \qquad \text{and} \qquad B_1' = \left\{ \begin{bmatrix} 1 \\ 0 \end{bmatrix}, \begin{bmatrix} 1 \\ 1 \end{bmatrix} \right\},$$

$$B_2 = \left\{ \begin{bmatrix} 1 \\ 0 \\ 0 \end{bmatrix}, \begin{bmatrix} 0 \\ 1 \\ 0 \end{bmatrix}, \begin{bmatrix} 0 \\ 0 \\ 1 \end{bmatrix} \right\} \qquad \text{and} \qquad B_2' = \left\{ \begin{bmatrix} 1 \\ 2 \\ 1 \end{bmatrix}, \begin{bmatrix} 0 \\ 3 \\ -1 \end{bmatrix}, \begin{bmatrix} 0 \\ 0 \\ 1 \end{bmatrix} \right\},$$

compute the matrix for $L\left(\begin{bmatrix} x \\ y \end{bmatrix}\right) = \begin{bmatrix} x - y \\ 2x + y \\ x - 2y \end{bmatrix}$ with respect to each of the various choices of bases:

(a) $B_1$ and $B_2$     (b) $B_1'$ and $B_2$     (c) $B_1$ and $B_2'$     (d) $B_1'$ and $B_2'$

*Hint*: As before, a little thought can save us a lot of effort. In particular, you should observe how each part relates to the previous ones.

## 31.3  Applications

This section illustrates *philosophically* why the freedom to change bases is so important. *Note*: This section is a bit esoteric and may be skipped without consequence. Using the two ideas of linear transformation and change of basis, most linear algebra problems may be solved, at least *in principle*, in four steps:

1. Phrase the given problem as an equation involving a linear transformation.

2. Convert that equation to a *matrix* equation, using *some* choice of basis.

3. Determine "good" bases to represent the equation in a particularly simple form, so that the solution is immediate.

4. Express the resulting solution in the language of the original problem.

In practice, one may skip some of these steps, but the general framework applies to most problems to which the techniques of linear algebra may be applied. For example, by employing bases, we may provide a precise explanation of Corollary 18.4.

For example, if we wish to solve the linear system:

$$\begin{aligned} x + y &= 3 \\ x - y &= 1 \end{aligned} \tag{31.6}$$

we immediately recognize this system as the functional equation:

$$L(X) = \begin{bmatrix} 3 \\ 1 \end{bmatrix}, \text{ where } L\left(\begin{bmatrix} x \\ y \end{bmatrix}\right) = \begin{bmatrix} x+y \\ x-y \end{bmatrix} \tag{31.7}$$

Representing $L$ with respect to the *standard* basis on $R^2$, we may immediately give its matrix as $A = \begin{bmatrix} 1 & 1 \\ 1 & -1 \end{bmatrix}$ and express our problem as the matrix equation $AX = \begin{bmatrix} 3 \\ 1 \end{bmatrix}$.
We saw in Lesson 17 how to use the $P^T LU$ decomposition to determine a "rank decomposition" $A = M_1 J M_2$ (compare Theorem 18.3).

In this specific case, we have the decomposition:

$$A = \begin{bmatrix} 1 & 0 \\ 0 & 1 \end{bmatrix}\begin{bmatrix} 1 & 0 \\ 1 & -2 \end{bmatrix}\begin{bmatrix} 1 & 0 \\ 0 & 1 \end{bmatrix}\begin{bmatrix} 1 & 1 \\ 0 & 1 \end{bmatrix}\begin{bmatrix} 1 & 0 \\ 0 & 1 \end{bmatrix}$$

$$= \begin{bmatrix} 1 & 0 \\ 1 & -2 \end{bmatrix}\begin{bmatrix} 1 & 0 \\ 0 & 1 \end{bmatrix}\begin{bmatrix} 1 & 1 \\ 0 & 1 \end{bmatrix} = M_1 J M_2$$

Taking $\mathbf{B}_2 = M_1$ and $\mathbf{B}_1 = M_2^{-1}$, we have $A = \mathbf{B}_2 J \mathbf{B}_1^{-1}$, or $\mathbf{B}_2^{-1} A \mathbf{B}_1 = J$. In this case,

$$\mathbf{B}_2 = \begin{bmatrix} 1 & 0 \\ 1 & -2 \end{bmatrix}, J = \begin{bmatrix} 1 & 0 \\ 0 & 1 \end{bmatrix}, \text{ and } \mathbf{B}_1 = \begin{bmatrix} 1 & 1 \\ 0 & 1 \end{bmatrix}^{-1}$$

Although we have been considering a specific example, this analysis is quite general. If we consider the matrices $\mathbf{B}$ and $\mathbf{B}'$ as representing change-of-basis matrices, our analysis suggest that:

**Observation:**

Every matrix may be represented by its "rank matrix" $J$, with respect to an appropriate choice of bases on its domain and range.

This means that, up to a good choice of basis (or in other words, a change of variables), all systems of equations behave as if there was only one nonzero coefficient in each equation!

## Exercise 31.12

- Verify that the bases corresponding to $\mathbf{B}_1$ and $\mathbf{B}_2$ are

$$B_1 = \left\{ \begin{bmatrix} 1 \\ 0 \end{bmatrix}, \begin{bmatrix} -1 \\ 1 \end{bmatrix} \right\} \text{ and } B_2 = \left\{ \begin{bmatrix} 1 \\ 1 \end{bmatrix}, \begin{bmatrix} 0 \\ -2 \end{bmatrix} \right\}$$

and compute the matrix for $L$ with respect to $B_1$ and $B_2$ directly as in Theorem 30.5. *Hint*: You should obtain the matrix $J$.

We have shown that the left-hand side of Eq. 31.7 may be simplified to an identity matrix with respect to $B_1$ and $B_2$. To convert the entire problem, we must also convert the right-hand side to its coordinates with respect to $B_2$.

### Exercise 31.13

Verify that the coordinates of $\begin{bmatrix} 3 \\ 1 \end{bmatrix}$ with respect to $B_2$ are also $\begin{bmatrix} 3 \\ 1 \end{bmatrix}$.

With respect to $B_1$ and $B_2$ our original problem becomes rather simple. The solution set corresponds to all vectors $Y$ whose coordinates with respect to $B_1$ satisfy the equation:

$$JY = \begin{bmatrix} 3 \\ 1 \end{bmatrix}$$

Since $J$ is the identity matrix, we know that there is exactly one solution, whose coordinates with respect $B_1$ are simply:

$$Y = \begin{bmatrix} 3 \\ 1 \end{bmatrix}$$

Translating this solution back in terms of the standard basis, we find that the solution to Eq. 31.7 is:

$$X = 3 \begin{bmatrix} 1 \\ 0 \end{bmatrix} + \begin{bmatrix} -1 \\ 1 \end{bmatrix} = \begin{bmatrix} 2 \\ 1 \end{bmatrix}$$

and the solution to Eq. 31.6 is $x = 2$ and $y = 1$.

## 31.4  Summary

In this lesson, one should have learned that:

- any change of basis may be represented by multiplication by an invertible matrix, and conversely, multiplication by an invertible matrix may be considered as a change of basis;

- to convert from a matrix representation with respect to one set of bases to another, simply multiply in front and in back by the appropriate change-of-basis matrices,

- to determine the appropriate change-of-basis matrices to use, we should draw an appropriate commutative diagram by considering coordinate systems as different languages with matrices as translators between the languages, and

- by choosing appropriate bases for the domain and range of a matrix, we may convert any linear equation to that of a rank matrix.

Our final goal in this text is to consider the eigenvalue problem. One way to view this problem is in terms of the following question:

**Question:**

Given a linear transformation $L : V \to V$, can we find a basis $B$ for $V$ so the matrix that represents $L$ with respect to $B$ is *diagonal*?

Alternatively, using Theorem 31.3, this question may be rephrased entirely in terms of matrix algebra:

**Question:**

Given an $n \times n$ matrix $A$, can we find a basis $B$ for $R^n$ so that $\mathbf{B}A\mathbf{B}^{-1}$ is *diagonal*?

We will see that quite often the answer to both questions is "Yes." When this is so, the matrix $\mathbf{B}$ will correspond to "eigenvectors" for $A$ and the diagonal entries of the resulting matrix will correspond to the "eigenvalues" of $A$ (see Ex. 18.7). Before considering this problem in this generality, however, we will first examine the *practical* significance of eigenvectors and eigenvectors in applications. Specifically, we will return to examine the economic and heat flow examples of Lesson 3 in more detail.

# Discrete Dynamical Systems and Eigensystems

*Mathematica* Notebook

## 32.1 Introduction

In this lesson, we will examine the application of linear algebra to the study of "discrete dynamical systems." Such systems include both the heat flow and the economics models of Lesson 3. This will serve to introduce the concept of an "eigensystem," which consists of a list of scalars, called "eigenvalues," and another corresponding list of vectors, called "eigenvectors." We will discover how:

- eigenvalues determine fundamental long-term growth rates,

- large eigenvalues tend to dominate smaller eigenvalues,

- eigenvectors determine fundamental long-term proportions, and

- initial conditions determine the actual distribution between fundamental behaviors.

In order to illustrate the full range of possible dynamic behavior, we must go beyond our realistic, mathematical models and examine a variety of possible models. In this lesson, we will forego formal mathematical definitions and simply rely on *Mathematica* for computations. In Lesson 33, we will discuss the precise definition of an eigensystem, learn how to perform the necessary calculations by hand, and discover how an eigensystem corresponds to our third, and final, decomposition theorem for matrices. In this lesson, we will instead focus on developing a solid, intuitive understanding of eigensystems before discussing them in formal, mathematical terms.

## 32.2 An Economics Example of a Dynamical System

A "dynamical system" is a mathematical model in which the variables are functions of time. For example, in the economics model from Lesson 3, the variables were:

$$x_n \quad = \quad \text{fraction of American-made cars owned in year } n,$$

$$y_n \;=\; \text{fraction of European-made cars owned in year } n, \text{ and}$$

$$z_n \;=\; \text{fraction of Far Eastern-made cars owned in year } n.$$

Each variable depends on $n$ and so is actually a function. Part of a dynamic model is a description of the way variables *change* over time, called its "transition equations". In this example, we used the following transition equations:

$$x_{n+1} - x_n \;=\; -.1x_n + .07y_n + .05z_n$$
$$y_{n+1} - y_n \;=\; .03x_n - .08y_n + .02z_n$$
$$z_{n+1} - z_n \;=\; .07x_n + .01y_n - .07z_n$$

Adding the subtracted terms from the left-hand side onto the right-hand side, we may rewrite these equations as a recursive matrix equation:

$$v_{n+1} = Av_n \tag{32.1}$$

where:

- `Clear[x,y,z,v,A]; v[n_] = {x[n], y[n], z[n]};`
  `A = {{.9,.07,.05},{.03,.92,.02},{.07,.01,.93}};`
  `Print["A = ", MatrixForm[A], "  and  v[n] = ", MatrixForm[v[n]]];`

Originally, we only examined the *equilibrium* values for this system. Using *Mathematica*, however, it is a simple matter to see how the variables of this system change over time to move from their initial values to their equilibrium state. For example, at one point in time the United States possessed a dominant fraction of the automobile market, with, say

- `x[0] = .6; y[0] = .3; z[0] = .1;`

which correspond to:

- `v[0] = {.6, .3, .1};`

Using the transition equation

- `v[n_] := v[n] = A.v[n-1];`

we may easily compute the values for $v[n]$ for $n = 1, \ldots, 10$:

- `Do[Print["v[",i,"] == ", v[i]],{i, 1, 10}]`

*Note*: We must re-index this equation, so that *Mathematica* may complete the calculations without entering an infinite loop. We have also added an extra assignment to allow *Mathematica* to perform the calculations more quickly.

## Exercise 32.1

Verify the first three of these calculations by successively substituting directly into Eq. 32.1 with $n = 0, 1,$ and 2.

As car manufacturers, we would to use this model to *predict* the long-term behavior of the system, that is, what happens to $v[n]$ as $n$ grows infinitely large. For example, we would want to investigate questions such as the following:

– Will any car manufacturer be eliminated from the market?

– Will market shares reach equilibrium values?

– If so, may we estimate how long this will take?

– If not, will the market undergo cyclical market shifts?

## Exercise 32.2

Change the 10 in the previous Do[] command to a larger value and re-evaluate. How many years does it seem to take for this system to essentially reach equilibrium? Compare the equilibrium values for $x$, $y$, and $z$ with those obtained in Lesson 3.

```
•   Clear[x,y,z]; equilibriumsoln = NSolve[
        {0 == -10 x + 7 y + 5 z,
         0 == 3 x - 8 y + 2 z,
         0 == 7 x + 1 y - 7 z,
         1 == x + y + z},
        {x,y,z}]
    v[Infinity] = {x, y, z}/. equilibriumsoln[[1]]
```

Moreover, we would want to discover how the behavior of the market *depends* on the data of the problem, $A$ and $v[0]$. For example:

– To what degree would the United States need to increase its customer loyalty to obtain a 1% increase in equilibrium market share?

– To what degree would the United States need to increase advertising in the European and Far Eastern markets to obtain a 1% increase in equilibrium market share? Can this be accomplished in a cost-effective manner?

Ultimately, the answers to all of these questions depend on the eigensystem of the transition matrix $A$.

In this example, the eigensystem of $A$ is:

```
•   eigensystem = Eigensystem[A]
```

The output of the Eigensystem[] command consists of two parts, a list of eigenvalues and another list of corresponding eigenvectors. In this case, we have:

```
•   TableForm[Join[{{"Eigenvalues","Eigenvectors"}},
                  Transpose[eigensystem]]]
```

*Note*: "Eigen" is German for "unique" or "special"; thus, an "eigensystem" of a matrix $A$ consists of "special" numbers and vectors that determine the behavior of any dynamical system with transition matrix $A$.

As we continue to examine eigensystems, we will continue to discover that the long-term behavior of the system is determined by the largest, or "dominant," eigenvalue, which in this case is 1. Notice, in particular, that the equilibrium is proportional to the corresponding dominant eigenvector:

```
•   dominanteigenvector = eigensystem[[2,1]]
    ratio = v[Infinity][[1]]/dominanteigenvector[[1]]
    v[Infinity] == ratio dominanteigenvector
```

## 32.3  More General Dynamical Systems

In this section, we will explore a general dynamical system

$$v_{n+1} = Av_n \qquad (v_0 \text{ is given}) \tag{32.2}$$

to discover how its long-term behavior depends on $A$ and the initial value $v_0$. We first observe that one may easily derive a formula for the solution $v_n$ in terms of the transition matrix $A$ and the initial condition $v_0$.

Using Eq. 32.2 with $n = 0$, we obtain:

$$v_1 = Av_0$$

Similarly, $n = 1$ yields:

$$v_2 = Av_1 = A(Av_0) = A^2 v_0$$

One may repeat this process indefinitely to obtain a general formula:

$$
\begin{aligned}
v_3 &= Av_2 = A(A^2 v_0) = A^3 v_0 \\
v_4 &= Av_3 = A(A^3 v_0) = A^4 v_0 \\
&\qquad\qquad \vdots \\
v_n &\quad \cdots \qquad\qquad = A^n v_0
\end{aligned}
\tag{32.3}
$$

We may quickly verify this formula in *Mathematica*, as well:

- ```
  Clear[v,v0,A,n]; v[0] = v0; v[n_] := A.v[n-1];
  Do[Print["v[",i,"] = ", v[i]], {i,0,5}];
  ```

Although simple in form, this formula does not tell us much about the qualitative behavior of the solution. We will see that the qualitative behavior of a system depends primarily on the eigenvectors and eigenvalues of the transition matrix $A$, rather than $A$ itself. For simplicity and to illustrate the full range of possible behavior, in this section we will consider a variety of two-variable systems, without limiting ourselves to any particular application setting.

For example, first consider the dynamical system with:

- ```
  A = {{2./3., 1./6.}, {1./3., 5./6.}}; v[0] = {10.,10.}; n = 0;
  ```

This has as its eigensystem:

- ```
  eigensystem = Eigensystem[A]
  ```

Remember that this is to be read as a pair of lists with corresponding entries, so that:

- eigenvalue 1.0 corresponds to $\{-0.447214, -0.8944276\}$

- eigenvalue 0.5 corresponds to $\{-0.707107, 0.707107\}$

We should really refer to this is *an* eigensystem for $A$. Although we will eventually show that eigenvalues of a matrix are uniquely determined, the associated eigenvectors are not. That is because the eigenvectors associated with a given eigenvalue correspond to a choice of basis for a particular subspace. Since bases are not unique, neither are eigenvectors.
  ```

As a pleasant consequence, however, this means that we may replace any eigenvector that *Mathematica* produces by a nonzero multiple. So, for example, we may give an equivalent eigensystem for $A$ as:

- eigenvalue $1.0$ corresponds to $\{1, 2\}$

- eigenvalue $0.5$ corresponds to $\{-1, 1\}$

As we mentioned in Section 32.2, because 1 is the "largest" eigenvalue, it will generally have the greatest effect on solutions. We call it the "dominant" eigenvalue of $A$ and we call the associated eigenvector $\begin{bmatrix} 1 \\ 2 \end{bmatrix}$ a dominant eigenvector.

To see this, examine the values for $v[n]$ as $n$ grows large:

- ```
  Do[Print["v[",n+1,"] = ",v[n+1] = A.v[n]]; n = n+1,{i,1,5}];
  ```

## Exercise 32.3

- Continue to re-evaluate the previous command until a clear pattern emerges. What do you conclude about the behavior of $v_n$ as $n$ grows infinitely large? Describe any connections between this behavior and the eigensystem. *Hint*: Compare this example with that in Section 32.2.

## Exercise 32.4

- Repeat Ex. 32.3, but with the initial condition:

- ```
  v[0] = {20.,5.}; n = 0;
  ```

## Exercise 32.5

Compare and contrast the results of Ex. 32.3 and Ex. 32.4. What seems to be the effect of changing the initial conditions? *Hint*: The difference must be determined by $v_0$, which we changed, while the similarities must be determined by $A$, which we did not change.

## Exercise 32.6

- Now consider the dynamical system given by:

- ```
  Clear[A,v]; n = 0;
  A = {{5./8., 1./8.},{3./8., 7./8.}}; v[0] = {10.,10.};
  Eigensystem[A]
  ```

- ```
  Do[Print["v[",n+1,"] = ",v[n+1] = A.v[n]]; n = n+1,{i,1,5}];
  ```

Compare and contrast this example with those from Ex. 32.3 – 32.4. Infer how the behavior of the system is affected by the Eigensystem[] for $A$. *Hint*: Changes in behavior must stem from changes in $A$ (and its eigensystem) and $v_0$.

### Exercise 32.7

•     Repeat Ex. 32.6 with:

•
```
Clear[A,v]; n = 0;
A = {{1., 1./2.},{1., 3./2.}}; v[0] = {10.,10.};
Eigensystem[A]
```

    *Hint*: While this system does not reach equilibrium, we may still determine its *rate* of growth. How is the rate reflected in the associated eigensystem?

### Exercise 32.8

•     Repeat Ex. 32.6 with:

•
```
Clear[A,v]; n = 0;
A = {{3./10., 1./10.},{1./5., 2./5.}}; v[0] = {10.,10.};
Eigensystem[A]
```

### Exercise 32.9

•     Repeat Ex. 32.6 with:

•
```
Clear[A,v]; n = 0;
A = {{0.,-1./2.},{-1.,-1./2.}}; v[0] = {10.,10.};
Eigensystem[A]
```

## 32.4  Summary

In this lesson, we have observed that a dynamical system may exhibit a variety of types of behavior, depending in a somewhat mysterious way on the initial data and coefficients of the transition equations. The relationship becomes much clearer if we focus on the eigenvalues and eigenvectors of $A$. Specifically:

    – the initial conditions determine the absolute size of the solution,

    – the dominant eigenvalue determines the growth rate of the solution, and

    – the dominant eigenvector controls the long-term relative proportions of the quantities in the solution.

There is a more subtle effect which the initial condition $v_0$ has on the solution, depending on the *non-dominant* eigenvectors and eigenvalues. In order to observe and explain this phenomenon, we must precisely define the terms "eigenvector" and "eigenvalue" and discover how they lead to another decomposition of a matrix $A$, known as the "Jordan decomposition" of $A$. We will discuss this in detail in Lesson 33.

# Eigenvectors and Eigenvalues

## Accompanied by Notebook

In this lesson, we will continue to investigate linear dynamical systems (compare Lesson 32) and discuss in greater detail how their behavior depends on the eigenvectors and eigenvalues of the associated transition matrix. In Section 32.3, we easily derived a correct, general formula for the solution of such a system, but to get a more *useful* formula, the transition matrix must possess a *basis* of eigenvectors. In this case, we may derive a more detailed description of the solution. Although this formula seems complex at first, it will allow us to describe completely the long-term behavior of the system, in terms of the eigenvalues and eigenvectors of the matrix. Ultimately, one may use this formula to create one's *own* systems that exhibit different types of long-term behavior. To this end, we will discuss:

- the formal definitions of "diagonalizable," "eigenvector" and "eigenvalue,"

- the relationship between eigensystems and diagonalizability,

- how to compute an eigensystem for a given matrix,

- how the solution to a discrete-time, linear, dynamical system is determined by its eigensystem and initial conditions, and

- how to produce a system with a variety of behaviors.

## 33.1 Discrete-Time, Linear, Dynamical Systems

In Lesson 32, we introduced the notion of a discrete time, linear, dynamical system. Our example arose from a certain economic model, but such systems arise in many different settings (e.g., Markov processes, such as Brownian motion of a particle in a lattice or population flow between neighboring states). They all take the form:

$$\begin{array}{rclcl} x_{n+1} & = & .2x_n + .3y_n, & x_0 = 6 \\ y_{n+1} & = & .4x_n + .5y_n, & y_0 = 7 \end{array}$$

This is a "discrete-time" model, because $n$ represents time that is measured in fixed, whole units. We say that variables $x$ and $y$ describe the "state" of the system, in that, if one knows the values of $x$ and $y$, then one posesses a complete description of the model at any given point in time. This is a "dynamical" system, in that the values of $x$ and $y$ *change* over time (i.e., they depend on $n$). The values 6 and 7 are called "initial conditions," because they describe the state of the system at the beginning (in the 0th time period) of the model. This model is *linear*, since the transition function is a linear transformation. As we observed in Lesson 32, such a system may be written in matrix terms as:

$$v_{n+1} = Av_n \qquad (v_0 \text{ is given}) \tag{33.1}$$

*Aside*: This system is also said to be "autonomous," because $A$ does not change over time (i.e., does not depend on $n$), and "first order" (since the state of the system depends only on the previous state, and not on the previous two or three, etc.).

We have already shown that Eq. 32.3 gives the general solution to such a system. However, this formula is not very useful as written, since it is very time consuming to compute matrix powers like $A^n$ directly. In fact, it would be faster to compute the solution directly *without* this formula! This formula, however, is the key to a *better* formula. *If* it is possible to find an invertible matrix $B$ and a diagonal matrix $\Lambda$ so that

$$A = B\Lambda B^{-1}$$

then one may compute such matrix powers very quickly, as follows:

$$
\begin{aligned}
A^n = A \quad \underset{n \text{ times}}{\cdots} \quad A &= \left(B\Lambda B^{-1}\right) \quad \cdots \quad \left(B\Lambda B^{-1}\right) \\
&= \left(B\Lambda B^{-1}\right)\left(B\Lambda B^{-1}\right) \quad \cdots \quad \left(B\Lambda B^{-1}\right)\left(B\Lambda B^{-1}\right) \\
&= B\Lambda \left(B^{-1}B\right) \Lambda \left(B^{-1} \quad \cdots \quad B\right) \Lambda \left(B^{-1}B\right) \Lambda B^{-1} \\
&= \underset{n \text{ times}}{B\Lambda \quad \cdots \quad \Lambda B^{-1}} = B\Lambda^n B^{-1}
\end{aligned}
\tag{33.2}
$$

At first glance this looks like we have simply replaced one difficult matrix power by another. Although this is true on the surface, one should observe that since $\Lambda$ is *diagonal*, its matrix powers are easily computed: if $\Lambda = \begin{bmatrix} \lambda_1 & \cdots & 0 \\ \vdots & \ddots & \vdots \\ 0 & \cdots & \lambda_k \end{bmatrix}$, then

$$
\begin{aligned}
\Lambda^2 &= \begin{bmatrix} \lambda_1 & \cdots & 0 \\ \vdots & \ddots & \vdots \\ 0 & \cdots & \lambda_k \end{bmatrix} \begin{bmatrix} \lambda_1 & \cdots & 0 \\ \vdots & \ddots & \vdots \\ 0 & \cdots & \lambda_k \end{bmatrix} = \begin{bmatrix} \lambda_1^2 & \cdots & 0 \\ \vdots & \ddots & \vdots \\ 0 & \cdots & \lambda_k^2 \end{bmatrix}, \\
\Lambda^3 &= \Lambda\Lambda^2 = \begin{bmatrix} \lambda_1 & \cdots & 0 \\ \vdots & \ddots & \vdots \\ 0 & \cdots & \lambda_k \end{bmatrix} \begin{bmatrix} \lambda_1^2 & \cdots & 0 \\ \vdots & \ddots & \vdots \\ 0 & \cdots & \lambda_k^2 \end{bmatrix} = \begin{bmatrix} \lambda_1^3 & \cdots & 0 \\ \vdots & \ddots & \vdots \\ 0 & \cdots & \lambda_k^3 \end{bmatrix}, \\
&\vdots \\
\Lambda^n &= \begin{bmatrix} \lambda_1^n & \cdots & 0 \\ \vdots & \ddots & \vdots \\ 0 & \cdots & \lambda_k^n \end{bmatrix}
\end{aligned}
\tag{33.3}
$$

*Note*: Here we use $k$ for the dimension of $A$ (since we are already using $n$ for the power).

Although this computation seems to require a rather big assumption (i.e., that $A$ may decomposed as $B\Lambda B^{-1}$, sometimes called a "Jordan"), this assumption is justified in a vast majority of useful cases. We will investigate this assumption, usually referred to as "diagonalizability," in great detail in the next section.

## 33.2    Diagonalizability and Eigensystems

The key to Eq. 33.2 is the assumption that $A$ is "diagonalizable."

**Definition 33.1** We say that $A$ is *diagonalizable* iff there exist matrices $\mathbf{B}$ and $\Lambda$ so that

$$A = \mathbf{B}\Lambda\mathbf{B}^{-1} \tag{33.4}$$

where $\mathbf{B}$ is invertible and $\Lambda$ is diagonal. Notice that we may also write this equation as

$$\mathbf{B}A\mathbf{B}^{-1} = \Lambda$$

which, by Corollary 31.4, means that $A$ can be represented by a diagonal matrix with respect to the corresponding basis given by $B$ on its domain and range.

Notice that we have again used our notation from Section 26.4, because we will want to consider the matrix $\mathbf{B}$ as arising from a corresponding *basis* $B$. Specifically, using the Principle of Partitioned Matrices, one may rewrite Eq. 33.4 in a very different looking form. Multiplying Eq. 33.4 on the right by $\mathbf{B}$ gives:

$$A\mathbf{B} = \mathbf{B}\Lambda \tag{33.5}$$

Partitioning into columns, if $\mathbf{B} = \begin{bmatrix} b_1 & \cdots & b_k \end{bmatrix}$, the set $B = \{b_1, \ldots, b_k\}$ must be a basis (by Theorem 25.1, since $\mathbf{B}$ is invertible). If we also let $\Lambda = \begin{bmatrix} \lambda_1 & \cdots & 0 \\ \vdots & \ddots & \vdots \\ 0 & \cdots & \lambda_k \end{bmatrix}$,

Eq. 33.5 becomes:

$$A \begin{bmatrix} b_1 & \cdots & b_k \end{bmatrix} = \begin{bmatrix} b_1 & \cdots & b_k \end{bmatrix} \begin{bmatrix} \lambda_1 & \cdots & 0 \\ \vdots & \ddots & \vdots \\ 0 & \cdots & \lambda_k \end{bmatrix}$$

or

$$\begin{bmatrix} Ab_1 & \cdots & Ab_k \end{bmatrix} = \begin{bmatrix} \lambda_1 b_1 & \cdots & \lambda_k b_k \end{bmatrix}$$

Setting columns equal leads to the matrix equations:

$$Ab_i = \lambda_i b_i \qquad \text{for } i = 1, \ldots, k$$

In other words, we have the following important theorem:

**Theorem 33.1** *A given $k \times k$ matrix $A$ is diagonalizable iff there exist a basis $B = \{b_1, \ldots, b_k\}$ and corresponding numbers $\lambda_i$ so that*

$$Ab_i = \lambda_i b_i \qquad \text{for } i = 1, \ldots, k \tag{33.6}$$

*In this case, the matrix that diagonalizes A is* $\mathbf{B} = \begin{bmatrix} b_1 & \cdots & b_k \end{bmatrix}$, *and the diagonal form of*

$A$ *is* $\Lambda = \begin{bmatrix} \lambda_1 & \cdots & 0 \\ \vdots & \ddots & \vdots \\ 0 & \cdots & \lambda_k \end{bmatrix}$.

Because of this important result, we give vectors satisfying Eq. 33.6 a special name:

**Definition 33.2** If $v$ is a vector and $\lambda$ a scalar so that

$$Av = \lambda v \qquad\qquad (33.7)$$

we say that $v$ is an *eigenvector* for $A$ and $\lambda$ is its associated *eigenvalue*.

Now we can paraphrase Theorem 33.1 as saying:

**Observation:**
$A$ is diagonalizable iff it has a basis of eigenvectors. Any matrix that diagonalizes $A$ is a basis of eigenvectors for $A$, while the diagonal form of $A$ consists of the eigenvalues of $A$ on the diagonal.

Notice that if one knows that $v$ is an eigenvector, one may determine its associated eigenvalue by "plugging into" Eq. 33.7.

## Exercise 33.1

Verify that $v_1 = \begin{bmatrix} 1 \\ 2 \\ -2 \end{bmatrix}$, $v_2 = \begin{bmatrix} 5 \\ -1 \\ 3 \end{bmatrix}$, and $v_3 = \begin{bmatrix} -2 \\ -1 \\ 0 \end{bmatrix}$ are eigenvectors for

$A = \begin{bmatrix} -8 & 20 & 15 \\ 2 & -2 & -3 \\ -6 & 12 & 11 \end{bmatrix}$ by appealing to Definition 33.2. Determine the corresponding eigenvalues.

## Exercise 33.2

Use Theorem 33.1 to compute the matrix $A$ with eigenvectors $b_1 = \begin{bmatrix} 4 \\ -1 \\ 4 \end{bmatrix}$, $b_2 = \begin{bmatrix} -1 \\ 1 \\ -2 \end{bmatrix}$, and $b_3 = \begin{bmatrix} 5 \\ -1 \\ 3 \end{bmatrix}$ with associated eigenvalues 2, 2, and $-3$, respectively.

*Note*: Eigensystems for a matrix are *not* unique (see Ex. 33.1–33.2); this fact is strikingly apparent when we have repeated eigenvalues.

### Exercise 33.3

Use *Mathematica* to determine a basis $B$ of eigenvectors for $A = \begin{bmatrix} 6 & -2 & -2 \\ -2 & 5 & 0 \\ -2 & 0 & 7 \end{bmatrix}$. Use

Theorem 33.1 to predict the matrix of $A$ with respect to $B$. Use the methods of Lesson 31 to compute this matrix directly.

## 33.3  Computing Eigensystems

To this point, we have depended on *Mathematica* to compute an eigensystem for any given matrix $A$. Since it is important pedagogically that one has some experience with computing eigensystems directly, we will discuss the traditional technique for computing eigenvalues and eigenvectors. One should remember, however, that computing eigensystems is computationally one of the most challenging and important algorithms in mathematics today, and in general requires a sophisticated computer program to do it well. In particular, the simple-minded algorithm that we present here is generally *unworkable*, in practice.

Computing an eigensystem is difficult, because it is a *nonlinear* problem. Given an $n \times n$ matrix $A$, the problem is to find a scalar $\lambda$ *and* a vector $v$ so that $Av = \lambda v$. Since the two unknowns $\lambda$ and $v$ in this equation are multiplied together, this is a difficult nonlinear problem in $(n + 1)$ real-valued unknowns (namely, $\lambda$ and the $n$ coordinates of $v$). The traditional approach is to divide this problem into a one-variable nonlinear problem and a corresponding $n$-variable linear one:

1. Determine all possible values for $\lambda$.

2. Determine the set of all eigenvectors corresponding to each eigenvalue from step 1.

Notice that step 2 is a *linear* problem, since $\lambda$ is assigned a known value at this point.

To perform step 1, we rewrite Eq. 33.7 as:

$$Av - \lambda v = 0 \qquad \text{or} \qquad (A - \lambda I)v = 0 \tag{33.8}$$

This implies that if $\lambda$ is an eigenvalue, $Kernel(A - \lambda I)$ is nontrivial, since it contains the nonzero vector $v$. This implies, by Theorem 23.1, that $A - \lambda I$ is not invertible and has determinant 0. In other words, any eigenvalue $\lambda$ satisfies the equation:

$$|A - \lambda I| = 0$$

In this way, we temporarily eliminate $v$ from the problem and obtain a one-variable, nonlinear equation characterizing any eigenvalue $\lambda$.

For example, if $A = \begin{bmatrix} 5 & -1 \\ -1 & 5 \end{bmatrix}$, this equation becomes:

$$\begin{aligned} 0 &= |A - \lambda I| = \left\| \begin{bmatrix} 5 & -1 \\ -1 & 5 \end{bmatrix} - \lambda \begin{bmatrix} 1 & 0 \\ 0 & 1 \end{bmatrix} \right\| = \left\| \begin{bmatrix} 5 - \lambda & -1 \\ -1 & 5 - \lambda \end{bmatrix} \right\| \\ &= (5 - \lambda)(5 - \lambda) - 1 = \lambda^2 - 10\lambda + 24 \end{aligned}$$

This procedure leads to the following definition:

**Definition 33.3** If $A$ is an $n \times n$ matrix, we refer to $|A - \lambda I|$ as the *characteristic polynomial* of $A$. *Note*: In general, this will be a polynomial of degree $n$ in $\lambda$.

Likewise, we have the following theorem:

**Theorem 33.2** *The eigenvalues of a matrix $A$ are precisely the roots of the characteristic polynomial of a matrix $A$.*

## Exercise 33.4

- Determine the roots of $\lambda^2 - 10\lambda + 24$. Compare your answer against the eigenvalues of $A = \begin{bmatrix} 5 & -1 \\ -1 & 5 \end{bmatrix}$, as given by the Eigensystem[] command.

Having completed step 1 on p. 361, we now return to Eq. 33.8 to determine the eigenvectors corresponding to each value for $\lambda$. By our previous observations, the set of eigenvectors associated with a given eigenvalue $\lambda$ is actually a linear *subspace*, namely, $Kernel(A - \lambda I)$. Because $Kernel(A - \lambda I)$ will contain *infinitely many* vectors, we only compute a *basis* for this subspace.

## Exercise 33.5

- For each of the eigenvalues from Ex. 33.4:
    - Compute the matrix $A - \lambda I$.
    - Determine a basis for $Kernel(A - \lambda I)$.
    - Compare your answer with the results of the Eigensystem[] command.

### *Drill Exercise 33.6*

- For each of the following matrices:
    - Compute the characteristic polynomial.
    - Compute the corresponding eigenvalues.
    - Compute a basis of eigenvectors for each eigenvalue.
    - Verify Eq. 33.7 for each eigenvector.
    - Verify Eq. 33.4, taking $B$ to be the union of all bases of eigenvectors that you have computed.

(a) $\begin{bmatrix} -1 & -3 \\ -3 & 7 \end{bmatrix}$ (b) $\begin{bmatrix} 1 & 3 \\ 4 & -3 \end{bmatrix}$ (c) $\begin{bmatrix} 1 & -1 & 0 \\ -1 & 2 & -1 \\ 0 & -1 & 1 \end{bmatrix}$ (d) $\begin{bmatrix} 2 & 0 & -1 \\ 2 & -1 & 0 \\ -12 & 0 & 1 \end{bmatrix}$

## Exercise 33.7

- For each of the following matrices $A$, either determine a matrix $B$ that diagonalizes $A$ (i.e., $\Lambda = B^{-1}AB$ is diagonal) or explain why there is none.

(a) $\begin{bmatrix} 1 & 4 \\ 1 & -2 \end{bmatrix}$ (b) $\begin{bmatrix} 1 & 16 \\ -1 & 9 \end{bmatrix}$

## 33.4   A Few Comments on Diagonalizability and Eigensystems

We pause to make two observations of a more theoretical nature. First, we point out the limitations of the algorithm from Section 33.3. Although there is the quadratic formula for polynomials of degree 2, there are no general formulas of the same type for higher-degree polynomials. This means that determining the roots of a general polynomial equation is quite a difficult problem. One may only solve polynomial equations, in general, by *using a computer*. Moreover, one will only obtain approximations to the actual solutions. This fact compounds the difficulty in computing eigensystems, since if one uses an *approximate* value for $\lambda$ in Eq. 33.7, one will obtain very inaccurate values for the corresponding eigenvectors. We say that Eq. 33.7 is "numerically unstable," because the "condition number" of $A - \lambda I$ is very large. We will not explain this terminology further in this text; this is covered in advanced texts on numerical analysis and numerical linear algebra. The upshot of this is that *Mathematica* does not compute eigensystems using the naive algorithm from Section 33.3. Instead *Mathematica* uses the $QR$ decomposition repeatedly to diagonalize $A$ directly. This procedure, called the "$QR$ algorithm," was invented in the 1960's and is one of the most important algorithms in linear algebra.

Our second observation concerns diagonalizability. One should have noticed from Ex. 33.7 that not all matrices are diagonalizable; that is, we cannot always find a *basis* of eigenvectors. Thankfully, we have the so-called spectral theorem (or principal axis theorem):

**Theorem 33.3** *An $n \times n$ matrix $A$ is diagonalizable by an orthogonal matrix* **B** *(see Definition 29.1 iff $AA^T = A^T A$. Such a matrix is called a normal matrix. Equivalently, there exists an orthonormal basis B for $R^n$ consisting entirely of eigenvectors of $A$.*

Notice that if $A$ is symmetric, so that $A = A^T$, then $A$ must be normal:

$$AA^T = AA = A^T A$$

Many matrices that arise in practice are symmetric, so many such matrices are diagonalizable. Also, if the characteristic polynomial has $k$ distinct roots, then it must have $k$ associated eigenvectors, which will then be independent. Since this is true for most polynomials (*if* we compute using complex numbers), this is true for most matrices, so most matrices are diagonalizable.

### *Drill Exercise 33.8*

For each of the following symmetric matrices:

 – compute a basis of eigenvectors,
 – verify that pairs of eigenvectors associated with different eigenvalues are orthogonal, and
 – find an orthogonal matrix **B** that diagonalizes the given matrix.

(a) $\begin{bmatrix} 1 & 2 \\ 2 & 4 \end{bmatrix}$          (b) $\begin{bmatrix} 2 & 3 \\ 3 & 2 \end{bmatrix}$

## 33.5   Diagonalizability and Dynamical Systems

If we assume that A has a basis of eigenvectors, $B = \{b_1, \ldots, b_k\}$, we may apply Theorem 33.1, along with Eq. 32.3 and Eq. 33.2–33.3, to obtain the most useful formula for the solution of any discrete, dynamical system with transition matrix $A$. This formula will explain the connections that we observed in Lesson 32 between the solutions to a dynamical system and the eigensystem of its transition matrix. Simply apply the Principle of Partitioned Matrices to $\mathbf{B} = \begin{bmatrix} b_1 & \cdots & b_k \end{bmatrix}$:

$$v_n = A^n v_0 = \mathbf{B}\Lambda^n \mathbf{B}^{-1} v_0 = \begin{bmatrix} b_1 & \cdots & b_k \end{bmatrix} \begin{bmatrix} \lambda_1^n & \cdots & 0 \\ \vdots & \ddots & \vdots \\ 0 & \cdots & \lambda_k^n \end{bmatrix} \mathbf{B}^{-1} v_0$$

$$= \begin{bmatrix} b_1 \lambda_1^n & \cdots & b_k \lambda_k^n \end{bmatrix} \left( \mathbf{B}^{-1} v_0 \right) \tag{33.9}$$

By Corollary 31.2b, we recognize the final term $\mathbf{B}^{-1} v_0$ as $C_B(v_0)$, namely the coordinates of $v_0$ with respect to the basis of eigenvectors $B$. If we let $C_B(v_0) = \begin{bmatrix} c_1 \\ \vdots \\ c_k \end{bmatrix}$, then Eq. 33.9 simplifies even further to:

$$v_n = \begin{bmatrix} b_1 \lambda_1^n & \cdots & b_k \lambda_k^n \end{bmatrix} \begin{bmatrix} c_1 \\ \vdots \\ c_k \end{bmatrix} = c_1 \lambda_1^n b_1 + \cdots + c_k \lambda_k^n b_k \tag{33.10}$$

Notice how Eq. 33.10 is a linear combination of the eigenvectors $b_i$ with coefficients $c_i \lambda_i^n$. When $n = 0$, these simplify to just $c_i$, which are the coefficients of $v_0$ with respect to our basis of eigenvectors, as they should. Moreover, as $n$ grows very large, the most significant term in the sum corresponds to the eigenvalue with *largest magnitude*. That is, the entire expression will essentially look like a multiple of the dominant eigenvector, and the solution will essentially be multiplied by the corresponding eigenvalue at each stage. This formula explains the behavior that we observed in Lesson 32:

- the long-term behavior of a dynamical system is determined by the *dominant* eigenvalue;

- the *rate* of growth is determined by the dominant eigenvalue; and

- the relative proportions between the components of the solution are determined by the dominant eigenvector.

In this brief analysis, we have yet to mention the effect of the initial conditions on a solution. We observed in Lesson 32 that the initial conditions affect the absolute size of the solution. We may infer this effect from Eq. 33.10, since the initial conditions determine the $c_i$'s, which multiply the solutions by constant values. Besides this effect, the initial conditions will have a more subtle impact, in general, on the solution, which we will examine in more detail in Section 33.6.

## 33.6  The Effect of Initial Conditions

Another important inference that we may draw from Eq. 33.10 is that the linear combination of the eigenvectors present in the *initial* value vector $v_0 = c_1 b_1 + \cdots + c_k b_k$ determines the linear combination of eigenvectors present in the *entire* solution. In particular, if $c_i = 0$, then the $i$th term will drop out of the general solution. That means that our analysis after equation Eq. 33.10 was *not quite* correct. We said that as $n$ grows very large, the most significant term in the sum corresponds to the eigenvalue with largest magnitude. We should say, "the eigenvalue with largest magnitude, whose eigenvector has *a nonzero component* in $v_0$" (i.e., $c_i \neq 0$). Stated formally, we have the following theorem, which summarizes all our observations on discrete, dynamical systems.

**Theorem 33.4** *If:*

- *$v_n$ represents the solution of a discrete, dynamical system with transition matrix $A$,*

- *$A$ is diagonalizable with a basis of eigenvectors $B = \{b_1, \ldots, b_k\}$ and corresponding eigenvalues $\lambda_i$, $i = 1, \ldots, k$,*

- *the initial vector $v_0$ has coordinates* $\begin{bmatrix} c_1 \\ \vdots \\ c_k \end{bmatrix}$ *with respect to $B$, and*

- *$|\lambda_j|$ is maximum among the set $\{\, |\lambda_i| \mid c_i \neq 0 \,\}$.*

*then, for large values of $n$:*

- *$v_n$ is approximately proportional to $b_j$,*

- *$v_{n+1} \approx \lambda_j v_n$, and*

- *these approximations grow more and more relatively precise as $n$ grows infinitely large.*

We illustrate this theorem by exploring two concrete examples. Almost all of the remaining exercises of this lesson may be answered by making an appropriate appeal to Theorem 33.4. Our first example models population distribution in the Chicago tri-state area (Illinois, Wisconsin, and Indiana). Assume that:

- $\frac{1}{8}$ of Illinoisians move to Wisconsin each year,

- $\frac{1}{16}$ of Illinoisians move to Indiana,

- $\frac{1}{3}$ of Indianans move to Illinois each year,

- $\frac{1}{18}$ of Indianans move to Wisconsin,

- $\frac{1}{5}$ of Wisconsiners move to Illinois each year, and

- $\frac{1}{15}$ of Wisconsiners move to Indiana.

Likewise, let:

$$
\begin{aligned}
x_n &= \text{population of Illinois in the } n\text{th year,} \\
y_n &= \text{population of Indiana in the } n\text{th year, and} \\
z_n &= \text{population of Wisconsin in the } n\text{th year.}
\end{aligned}
$$

If we assume the following initial populations (in millions)

$$x_0 = 20, y_0 = 10, \text{ and } z_0 = 12$$

we obtain the system $v_{n+1} = Av_n$ where:

$$v_n = \begin{bmatrix} x_n \\ y_n \\ z_n \end{bmatrix}, A = \begin{bmatrix} \frac{13}{16} & \frac{1}{3} & \frac{1}{5} \\ \frac{1}{16} & \frac{11}{18} & \frac{1}{15} \\ \frac{1}{8} & \frac{1}{18} & \frac{11}{15} \end{bmatrix}, \text{ and } v_0 = \begin{bmatrix} 20 \\ 10 \\ 12 \end{bmatrix}$$

## Exercise 33.9

Notice that six of the nine entries of $A$ are specified in the statement of the model. Explain how one may determine the remaining three coefficients of $A$ (compare the example of Section 3.4).

*Note*: This is an example of a "Markov process," in that all the entries of $A$ are positive and its columns add to 1. Thus, we may consider each entry of $A$ as the *probability* of an individual of moving from one state to another.

## Exercise 33.10

• Examine the long-term behavior of this system as in Lesson 32 and *explain* your observations in terms of the eigenvectors and eigenvalues of $A$ and Eq. 33.10. *Note*: You will find when using the Eigensystem[] command in *Mathematica* that it will produce totally different answers depending on whether the given matrix contains decimals or exact rational numbers. In general, for the computations in this lesson you will want to use decimals.

## Exercise 33.11

• Determine two other initial conditions which illustrate the other two possible behaviors of this system, corresponding to the nondominant eigenvectors. *Hint*: Express the initial condition as a linear combination of eigenvectors. Are these initial conditions realistically possible? Explain.

A second interesting example comes from a classic problem of the 16th century Italian mathematician Leonardardo of Pisa, also known as Fibonnacci. He modeled a population of pairs (male/female) of rabbits, assuming that they did not become fertile until their second birthday and that each fertile pair gave birth to another pair of rabbits. The assumptions that the rabbits never die and keep reproducing each year leads to the model

$$x_{n+1} - x_n = x_{n-1} \qquad \text{or} \qquad x_{n+1} = x_{n-1} + x_n$$

where $x_n$ is the total rabbit population (pairs) in the $n$th year. This equation says that the increase in rabbits only comes from the rabbits that were born in the previous year (and hence are fertile in the current year). With the initial condition of $x_0 = 1$ pair of rabbits that become fertile in the next year, this gives the famous Fibonacci sequence: $1, 1, 2, 3, 5, 8, 13, \ldots$.

Although this is not a first-order, dynamical system, we may *convert* it to a first-order system by introducing a new variable $y_n$ = number of *fertile* rabbits in the $n$th year. This technique of introducing auxiliary variables is a very general and useful technique that one may use to convert any higher-order system into a first-order system. In this example, we obtain:

$$\begin{aligned} y_{n+1} &= x_n & y_0 &= 0 \\ x_{n+1} &= y_n + x_n & x_0 &= 1 \end{aligned}$$

### Exercise 33.12

Investigate this system by determining $A$ and examining its long-term behavior, as in Ex. 33.10. Supply two different choices of initial conditions that lead to radically different behavior, as in Ex. 33.11.

The last two exercises require one to combine the results of Theorems 33.1 and 33.4.

### Exercise 33.13

Find a system in $x$ and $y$ and two sets of initial conditions for which, in the long run:

- if we start at one set of initial conditions, $x$ is approximately $\frac{2}{3}$ of $y$ and both are approximately tripling each year, and

- if we start at the other initial condition, $x$ is approximately $5y$ and both are going to $0$ by approximately halving each year.

### Exercise 33.14

Find a system in $x$, $y$, and $z$ so that in the long run, either:

- $x$ is approximately $\frac{1}{3}$ of $y$ and $\frac{2}{5}$ of $z$, and all converge to fixed nonzero values, or

- the values converge to $0$, while oscillating around $0$.

## 33.7  Summary

Through a combination of theoretical calculations and computer experimentation, we have seen that any diagonalizable matrix $A$ is determined by its eigenvectors and eigenvalues, which in turn determine the behavior of the associated dynamical system with transition matrix $A$. Although we have only examined discrete time models, the continuous time models (i.e., differential equations) behave in a very similar fashion. Diagonalizable matrices arise in a wide variety of applications. For example:

- In quantum mechanics, all operators are assumed to be symmetric (Hermitian). The "pure" states are the eigenvectors, and their measured values are the associated eigenvalues.

- In studying nonstandard inner products (as in relativity theory), we find that they are in 1–1 correspondence with symmetric matrices.

- In classifying quadratic polynomials, we find that they are all parabolas, hyperbolas, or ellipses, where the major and minor axes correspond to the eigenvectors of the associated symmetric matrix.

Although we cannot explore these topics, we cannot overemphasize the importance of eigensystems in applications.

In this lesson, we have seen:

- how to compute an eigensytem for a given matrix $A$ by hand,

- how to diagonalize $A$ (if possible), and

- how the solution to a discrete-time, linear, dynamical system is related to the eigensystem of its transition matrix and initial conditions, by Eq. 33.10.

In a more advanced text in linear algebra, we would extend our analysis to include the case when $A$ is *not* diagonalizable. It turns out that every matrix is *almost* diagonalizable. To see this, we would need to introduce the notion of a *generalized* eigensystem, and the diagonal matrix of eigenvalues must be replaced, in general, by a "Jordan" matrix. However, we must stop here or this text will never be finished.

Although this text must come to an end, the areas of possible application and further study of linear algebra continue to grow. It is the hope of this author that each reader of this text has begun to glimpse the beauty of this wonderful subject and will continue to explore it for many years to come.

# Appendix of *Mathematica* Commands

## Accompanied by Notebook

In this Appendix, we give a brief description of all of the *Mathematica* commands that are used in this text. We differentiate between built-in commands that are part of *Mathematica*'s kernel, those that are defined in the standard packages that are distributed with *Mathematica*, and those that are defined in specialized packages which accompany this text. This is not intended as a complete reference to *Mathematica*, but only as a helpful guide for those using this text. Each command may come with optional arguments, which are always written in the form *option->value* and placed as the final arguments of the command. One may use the Function Browser under the Help menu, as well, to obtain descriptions of all defined *Mathematica* commands and options.

## Built-In Commands

**?symbol** A special command that will interrogate *Mathematica* as to the values associated with "symbol."

**%n** Equivalent to Out[$n$]. Without any argument, % is equivalent to Out[$-1$], %% is equivalent to Out[$-2$], etc.

**Array[]** Command that takes two inputs, a pure function and a sequence of dimensions, and creates a multi-dimensional array of the given dimensions with entries computed according to the given function.

**AspectRatio** Option for Plot[] and Plot3D[] that specifies the ratio of width to height of the resulting plot. A value of "Automatic" ensures that the scales on all axes will be equal, with the resulting aspect ratio determined by the domain and range of the plot.

**Axes** Option for Show[]. The value "True" forces *Mathematica* to draw in axes, while the value "None" will suppress the axes.

**Blank[]** Used to specify a placeholder in a function definition. Usually written as $x_-$.

**Clear[]** Clears all values associated with all of its arguments.

**Contours** An option to the ContourPlot[] command that specifies the levels of the given function that are plotted.

**ContourPlot[]** Takes three arguments, namely a function of two variables and two lists describing the region containing the plot (compare Plot[] and Plot3D[]).

**ContourShading** An option to the ContourPlot[] command that specifies whether the regions between contours are shaded or not.

**Cos[]** The cosine function, where the argument is assumed to be given in radians.

**D[]** Takes two arguments. The first is an expression and the second is the variable with which to compute the partial derivative of the given expression.

**Degree** A predefined constant equal to $\frac{\pi}{180}$. Used after an angle measured in degrees to automatically convert it to radians.

**DisplayFunction** An option to the Plot[], Plot3D[], ContourPlot[], Show[], etc. It is used to suppress the creation of a picture. Setting this option to IdentityFunction will suppress the picture, while $DisplayFunction will restore the picture.

**Do[]** Takes two arguments, "expression" and "{var, start, end}." Evaluates "expression" a number of times, with"var" first assigned the value "start" and then repeatedly as the value of "var" is incremented by 1 each time, until it reached the value "end." If the second argument contains a fourth entry "step," the value of "var" will be incremented by "step" each time.

**Dot[]** Takes two arguments, "lhs" and "rhs." Usually written as *lhs.rhs*. This computes the standard inner product of two lists. In particular, this will give the product of two matrices, the dot product of two column vectors, and the image of a column vector under the action of a matrix.

**Equal[]** Takes two arguments, "lhs" and "rhs." Usually written as *lhs* == *rhs*, which specifies a logical comparison between the two quantities. If it can be completely simplified, it will result in True or False. Otherwise, it denotes a symbolic equation.

**Eigensystem[]** Returns the list of eigenvalues of a matrix (repeated according to the multiplicity of the corresponding root) and a list of corresponding eigenvectors.

**Expand[]** Expands a given algebraic expression, by distributing products over sums and combining like terms.

**Format[]** A directive that allows one to produce formatted output. In particular, it controls the printed form of any given expression.

**In[]** Takes a single integer argument $n$. If $n$ is positive, In[$n$] is the $n$th expression that *Mathematica* has evaluated during the current session. If $n$ is negative it refers to the expression evaluated the given number of steps in the past.

**ListDensityPlot[]** Creates a rectangular plot of a matrix, where the numerical entries are converted to shades of gray or colors.

**MatrixForm[]** A directive that indicates that a matrix should be printed in the traditional, rectangular manner.

**N[]** Takes one or two arguments. The first is any expression and the second (if given) is the number of decimal digits with which to express the value of the given expression.

**Needs[]** Attempts to load the package associated with each of its arguments.

**NSolve[]** Similar to Solve[], but uses potentially different algorithms and returns its results in decimal form.

**NullSpace[]** Returns a basis for the kernel of a given matrix.

**Out[]** Similar to In[]. Out[$n$] is the computed result of In[$n$]. *Note*: This does not include "side-effects," such as graphs and cells produced by a Print[] command.

**Part[]** Takes two arguments, "list" and "index." Usually written as "list[[index]]." Returns the $i$th element of list. This can be used with more than one index for matrices and tables (i.e., lists of lists).

**Plot[]** Takes two arguments. The first is either a single real-valued function or a list of real-valued functions. The second is a list that specifies the dummy variable of the function and the domain of the plot. Outputs a graphics object representing the plot, as well as displaying the plot as a "side-effect."

**Plot3D[]** Similar to Plot, except it has three arguments, where the second and third are specifications for the $x$- and $y$-axes.

**PlotPoints** Option for ContourPlot[], Plot[], and Plot3D[] that controls how accurate the resulting plot will be; the more points that are used, the more accurate the resulting plot.

**PlotRange** Option for Plot[] and Plot3D[] that specifies the range of each axis that is actually shown in the plot.

**PlotStyle** Option for Plot[] used to supply style directives for two-dimensional plots.

**Plus[]** Takes two arguments, "lhs" and "rhs." Usually written as *lhs + rhs*, which returns the usual sum.

**Power[]** Takes two arguments, "base" and "exponent." Computes the standard power function. Usually written as *base^exponent*.

**Print[]** Prints out all of its argument in a single cell. Does *not* put spaces between the print-out of each argument.

**Random[]** Takes two arguments "type" and "range." The "type" argument is either "Integer," "Real," "Complex," etc., and "range" is a range specifier. Returns a randomly generated number of the given type in the given range.

**Reverse[]** Takes a list and returns the list with its elements in the reverse order.

**RowReduce[]** Performs Gaussian elimination and backaddition on a given matrix.

**ReplaceAll[]** Takes two arguments, "expression" and "rules." The "rules" argument is either a single rule or a list of rules (compare Rule[]). Usually written as *lhs*/.*rhs*. The result will be the value of "expression" after all of the replacement indicated by the rules are performed.

**Rule[]** Takes two arguments, "lhs" and "rhs." Usually written as *lhs*->*rhs*, which indicates that each instance of "lhs" should be replaced by "rhs" (compare ReplaceAll[]).

**Set[]** Takes two arguments, "lhs" and "rhs." Usually written as $lhs = rhs$, which (immediately) assigns the value on the left to the variable on the right.

**SetDelayed[]** Takes two arguments, "lhs" and "rhs." Usually written as $lhs: = rhs$, which defines a relationship between the variable on the left and the expression on the right. The actual value for the left-hand side is recomputed, based on the given relationship, whenever it is needed.

**Show[]** Displays one or more graphics objects, such as those from Plot[], Plot3D[], ContourPlot[], etc. simultaneously and outputs the corresponding, total graphics object.

**Sin[]** The sine function, where the argument is assumed to be given in radians.

**Solve[]** Takes two arguments, namely, a list of equations (compare Equal[]) and a list of variables. Returns a list of all possible solutions in the indicated variables as a list of replacement rules.

**Subtract[]** Takes two arguments, "lhs" and "rhs." Usually written as $lhs - rhs$. This is automatically converted to $lhs + (-1) * rhs$, which equals the usual difference.

**Subscript[]** A directive to produce subscripts in formatted output.

**Subscripted[]** A directive that indicates that the entries of an array should be printed with traditional subscript notation.

**Sum[]** Similar to Do[], except it totals the result of each evaluation.

**Superscript[]** A directive to produce superscripts in formatted output.

**Table[]** Takes at least two arguments. The first is a formula specifying a typical entry of the output. The remaining arguments specify the range of values taken by each dummy variable in the given formula (compare Plot[] and Plot3D[]).

**TableForm[]** Similar to MatrixForm[].

**Times[]** Takes two arguments, "lhs" and "rhs." Usually written as $lhs * rhs$ or *lhs rhs*, and returns the usual product.

**Transpose[]** Computes the transpose of a matrix.

# Standard Packages

## Graphics'Colors Package

**Blue** Equivalent to RGBColor[0,0,1]. Used as a style directive for plots and value for the Color option.

**Green** Equivalent to RGBColor[0,1,0]. Used as a style directive for plots and value for the Color option.

**Red** Equivalent to RGBColor[1,0,0]. Used as a style directive for plots and value for the Color option.

## LinearAlgebra'MatrixManipulation Package

**BlockMatrix[]** A command to compress a matrix of matrices into a single matrix.

**SubMatrix[]** Takes three arguments. The first is a matrix, the second specifies the position of the upper-left corner of the desired submatrix and the third gives the dimensions of the submatrix. Returns the indicated submatrix.

# Supplied Packages

**ClearInOut[]** A command that clears from memory the specified In[] and Out[] variables. If called with no arguments it clears all such variables. Given a single integer argument $n$," it clears up through the $n$th variables. Given a list of integers, it clears those specified variables. This command is defined in "Start.ma" and is not part of any package.

## Programs'Bases Package

**Basis** An option to RandomVectors to specify a basis for the vector space from which to generate the vectors.

**BasisQ[]** Takes one or two arguments, $S$ and $T$, both sets of column vectors. Similar to SpanningQ[], except BasisQ[] also tests for linear independence.

**CoordinateVector[]** Takes two arguments "basis" and "vector," where all vectors are assumed to be column vectors. Returns the coordinates of "vector" with respect to "basis."

**IndependentQ[]** Takes a list of column vectors and determines whether or not they are linearly independent.

**RandomCombination[]** Takes a list and returns a randomly generated linear combination of its elements, where the coefficients are determined by the options ValueType and ValueRange. The last two options are part of the Matrices package.

**RandomVectors[]** Takes an integer $n$ and returns a list of $n$ randomly generated vectors. The entries are determined by the options Basis, ValueType, and ValueRange. The last two options are part of the Matrices package.

**SpanningQ[]** Takes one or two arguments, $S$ and $T$, both sets of column vectors. Determines if $S$ is a spanning set for $T$. If $T$ is omitted, assumes that $T$ is the standard basis of the same dimension as the vectors of $S$.

**TransitionMatrix[]** Takes two arguments "basis1" and "basis2," which are bases of column vectors for the same subspace of $R^n$. Returns the matrix converting a coordinate vector with respect to "basis1" to the corresponding coordinate vector with respect to "basis2."

## Programs'Decomps Package

**AnimaQR[]** Takes a matrix $A$ and creates a sequence of pictures visualizing the Gram-Schmidt algorithm. Can include options to Show[] and Vector[].

**PtLU[]** Takes a matrix $A$ and returns a permutation matrix $P$, a lower-triangular matrix $L$ and a row echelon matrix $U$ so that $A = P^T LU$.

**GaussEliminate[]** Takes a matrix and returns the corresponding row echelon matrix produced by Gaussian elimination. Always chooses to take the first available row interchange.

**Orthogonalize[]** Takes a set of column vectors and returns the corresponding orthogonal set of vectors produced by the Gram-Schmidt algorithm.

**QoRo[]** Takes a matrix $A$ and returns the decomposition $A = Q_o R_o$ produced by the Gram-Schmidt algorithm as a matrix with orthogonal columns $Q_o$ and a unit, upper-triangular matrix $R_o$.

**QR[]** Takes a matrix $A$ returns the decomposition produced by the Gram-Schmidt algorithm as a matrix with orthonormal columns $Q$ and a row echelon matrix $R$ with positive pivots.

**Tolerance** An option for QR[] that specifies the maximum size of the entries of a "zero" vector.

**Verbose** An option for AnimaQR[], QR[], and GaussEliminate[] that specifies whether or not to show the intermediate computations.

## Programs'GVects Package

**Color** Option to MyArrow[] and Vector[] that specifies the color of a vector.

**HeadScale** Option to MyArrow[] and Vector[] that specifies the size of the arrowhead.

**MyArrow[]** Takes a pair of coordinates "start" and "end" in two or three dimensions and returns a geometric object representing an arrow from "start" to "end."

**TailWidth** Option to MyArrow[] and Vector[] that specifies the width of the tail of an arrow.

**Vector[]** Similar to MyArrow[], but only takes a single set of coordinates and returns the corresponding geometric vector in standard position (i.e., starting at the origin). Will also take a list of vectors and attempt to scale the arrowheads appropriately.

## Programs'Matrices Package

**Adjoint[]** Returns the adjoint of a square matrix.

**Decomposed** Option to MakeNiceMatrix[] that determines whether the result is returned as a single matrix or as list of appropriate factors.

**DecompositionType** Option to MakeNiceMatrix[] that determines from which type of decomposition the matrix is to be formed.

**Determined** Value for SolutionType option that corresponds to a unique solution.

**Full** Value for Rank option specifying that the rank should be as large as possible.

**LeftInverse[]** Returns the cannonical left inverse of a given matrix, if possible. Returns Null if matrix has no left inverse.

**MakeLowerTriangular[]** Takes a single argument "dimension." Returns a square, randomly generated, lower-triangular matrix of the given dimension.

**MakeMatrix[]** Takes two arguments "rows" and "columns." Returns a randomly generated matrix of the given dimensions.

**MakeNiceMatrix[]** Takes two arguments "rows" and "columns." Returns a matrix of the given dimensions with a randomly generated decomposition of the given type (compare DecompositionType[]).

**MakePermutationMatrix[]** Takes a single argument "dimension" and returns a randomly generated permutation matrix of the given dimension.

**MakeReducedRowEchelon[]** Takes two arguments "rows" and "columns." Returns a randomly generated reduced row echelon matrix of the given dimensions.

**MakeRowEchelon[]** Takes two arguments "rows" and "columns." Returns a row echelon matrix of the given dimensions with randomly generated entries.

**MakeSquareSystem[]** Takes a single argument $n$. Returns a randomly generated matrix corresponding to a system of $n$ equations in $n$ unknowns.

**MakeSystem[]** Takes two arguments $n$ and $m$. Returns a randomly generated matrix corresponding to a system of $n$ equations in $m$ unknowns.

**MatrixRank[]** Returns the rank of a given matrix.

**NonSiingular** An option to MakeLowerTriangular[] that specifies whether or not the result is invertible.

**OverDetermined** Value for SolutionType option that corresponds to no solution.

**PLUFactors** Value for DecompositionType option specifying that the matrix should be generated from a conveniently chosen $QR$ decomposition.

**QRFactors** Value for DecompositionType option specifying that the matrix should be generated from a conveniently chosen $P^T LU$ decomposition.

**Randomized** Value for "Rank" indicating that the rank should be randomly generated.

**Rank** Option for MakeNiceMatrix[], MakeReducedRowEchelon[], and MakeRowEchelon[]. Determines the rank of the result. As an option for MakeSquareSystem[] or MakeSystem[] it specifies the rank of the left-hand side of the resulting system. The value "Full" specifies the maximum possible rank.

**RightInverse[]** Similar to LeftInverse[], except with a right inverse.

**SolutionType** An option to MakeSquareSystem[] or MakeSystem[] that takes one of the three values "Determined," "OverDetermined," or "UnderDetermined" and that specifies the type of system to generate.

**SwitchRows[]** Takes a matrix $A$ and two integers $i$ and $j$. Returns the matrix obtained by switching rows "$i$" and "$j$" of $A$.

**UnderDetermined** Value for SolutionType option that corresponds to infinitely many solutions.

**Unital** An option to MakeLowerTriangular[] that specifies whether or not the result has all 1's along the diagonal.

**ValueRange** Option for MakeLowerTriangular[], MakeMatrix[], MakeNiceMatrix[], MakeReducedRowEchelon[], MakeRowEchelon[], MakeSquareSystem[], MakeSystem[], RandomCombination[], and RandomVector[]. Determines the size of the entries of the result.

**ValueType** Similar to ValueRange, usually given the value Integer, Real, or Complex. Determines the type of entries of the result.

## Programs'MPicts Package

**BasisPicture[]** Takes two arguments, a set of column vectors in two or three dimensions and a positive integer $n$. Constructs a grid corresponding to a coordinate system built out of the given vectors, where each axis extends $n$ units in each direction. Can include options to Vector[].

**MatrixPicture[]** Takes a $2 \times 2$ matrix $A$ and plots a sequence of colored points in the image of $A$, plotting each pre-image in the corresponding color.

**MatrixPlot[]** Takes a $2 \times 2$ matrix $A$ and plots a sequence of labeled points and their images under the action of $A$.

**MatrixPlotCircle[]** Takes a $2 \times 2$ matrix $A$ and plots a colored unit circle and its image under the action of $A$.

**MatrixPlotSquare[]** Takes a $2 \times 2$ matrix $A$ and plots a colored unit square and its image under the action of $A$.

**ShowBitmap[]** Takes a matrix of 0's and 1's and plots the result as a grid of white and black squares, respectively.

**ShowVectors** Option to BasisPicture[] that indicates whether or not to show the original set of vectors.

**SpanningPicture[]** Takes a set of column vectors in two or three dimensions along with a list of range descriptions and plots all possible linear combinations with coefficents taken over the given ranges. Can include options to Vector[].

## Programs'Plots Package

**CoordinatePlot[]** Takes a "function," $f$, and plots a subset of its domain and the corresponding image points in its range in corresponding colors. If $f$ is a real-valued function of a single variable (i.e., $y = f(x)$), it takes two more arguments, namely, the name of the free variable $x$ and a "range" specification for $y$ in the range. If $f$ is a vector-valued function of two variables (i.e., $(u, v) = f(x, y)$), the next two arguments are "range" specifications for $x$ and $y$ (compare ContourPlot[]). Both versions display the plot as a side-effect and return the plot as output. Can include options to Show[].

**DirectionPlot[]** Takes two arguments "function" and "range," where "function" is a vector-valued function of a single variable and "range" specifies the domain of the plot (compare Plot[]). Constructs a ParametricPlot[] in two or three dimensions The color of the plot varies from Red to Blue to indicate the change in the input parameter. Displays the plot as a side-effect and returns the plot as output. Can include options to Show[].

**PlotJoined** Option to PointPlot[] that specifies whether or not to join the points in each list by line segments.

**PointPlot[]** Takes one or more lists of points in two or three dimensions and plots the results. Displays the plot as a side-effect and returns the plot as output.

## Programs'Vects Package

**InnerProduct[]** Takes two column vectors and returns their inner product, assuming the standard inner product of Euclidean space.

**VectorLength[]**  Takes a single column vector and returns its length, assuming the standard inner product of Euclidean space.

# Index

Page numbers are **bold** when they represent the definition or main source of information about the indexed term. Entries are **bold** when they are the names of *Mathematica* commands.

$*$, *see* **Times** command
$+$, *see* **Plus** command
$-$, *see* **Subtract** command
$.$, *see* **Dot** command
$/.$, *see* **ReplaceAll** command
$:=$, *see* **SetDelayed** command
$=$, *see* **Set** command
$==$, *see* **Equal** command
$C(R)$, **276**
$C_B$, *see* coordinate isomorphism
$E^n$, **201**
$I$, *see* identity matrix
$M_{2,2}(R)$, **276**
$P_n$, **203**
$R^n$, **200**
$[[\ ]]$, *see* **Part** command
B notation, **279**
$\hat{}$, *see* **Power** command
$x_-$, *see* **Blank** command
$->$, *see* **Rule** command
? command, 13, **A-1**

**Adjoint** command, **A-7**
adjoint, **181**
**AnimaQR** command, **A-6**
**Array** command, 166, **A-1**
**AspectRatio** option, 16, **A-1**
associative law
    of matrix addition, 94
    of matrix multiplication, 335
    of scalar multiplication, 207
    of vector addition, 206

augmented matrix, **38**, *see* matrix,
    partitioned
autonomous, **360**
**Axes** option, 211, **A-1**

backaddition, **46**
backsubstitution, **46**
basic, *see* variable, basic
**Basis** option, 233, 236, **A-5**
basis, **225**
    standard, 125, **332**
**BasisPicture** command, 219, **A-8**
**BasisQ** command, 251, **A-5**
bilinear, **290**
**Blank** command, 67, **A-1**
**BlockMatrix** command, 106–107, **A-5**
**Blue** value, **A-4**
bullet, 6

characteristic polynomial, **364**
**Clear** command, 8, **A-1**
**ClearInOut** command, 10, 11, **A-5**
codomain, *see* range
coefficients, **108**, 214
cofactor, **179**
    expansion, **179**
**Color** option, **A-6**
column space, **250**, *see* image of a
    matrix
column vectors, **200**
    addition, 200
    in *Mathematica*, 210
    scalar multiplication, 200

commutative law
    of inner products, 290
    of matrix addition, 96
    of vector addition, 206
complement, *see* orthogonal
        complement
composition, **333**, 334
conjugate, *see* similar
contour plot, **16**
**ContourPlot** command, 16, 20, **A-2**
**Contours** option, 18, **A-2**
**ContourShading** option, 18, **A-2**
contrapositive, **265**
coordinate isomorphism, **259**, 278–279,
        333
coordinate system, 216
    nonstandard, 215–220
**CoordinatePlot** command, 72, 185, **A-9**
coordinates, **219**, 248, 255–259
**CoordinateVector** command, 343, **A-5**
**Cos** command, 70, **A-2**

**D** command, 73, **A-2**
**Decomposed** option, **A-7**
decomposition
    $P^T LU$, **148**, 146–152, 155–163
    $QR$, **322**, 311–325
    Jordan, **361**
    rank, **161**, 349–351
**DecompositionType** option, 324, **A-7**
**Degree** value, 70, **A-2**
dependence relation, **224**, *see* linear
        dependence
dependent, *see* linearly dependent
determinant function, 137, **170**
**Determined** value, 62, **A-7**
diagonalizability, **361**, 361–363,
        365–366
dimension, **226**
**DirectionPlot** command, **A-9**
discrete-time, **359**
**DisplayFunction** option, 18, **A-2**
distributive law
    of inner products, 290
    of matrix multiplication, 97

of scalar multiplication, 96
**Do** command, 232, 354, **A-2**
domain, **68**
    notation for, 187
dominant, **357**
**Dot** command, 67, **A-2**, *see* matrix and
        inner product
dummy variable, **16**
dynamical system, **353**, **360**

**Eigensystem** command, 355, **A-2**
eigensystems, 363–367
eigenvalue, **362**
eigenvector, **362**
elementary
    matrix, **143**, 141–146
    row operation, *see* row operation
elimination
    Gauss-Jordan, *see* backaddition
    Gaussian, *see* Gaussian
        elimination
**Equal** command, 16, **A-2**
equation
    linear, *see* linear equation
    nonlinear, **20**
**Expand** command, 234, **A-2**
expansion
    column, 179
    row, 179

first order, **360**
**Format** command, 166, **A-2**
Fredholm Alternative, 135, 194
free, *see* variable, free
**Full** value, 62, **A-7**

Gauss, Carl Friedrich, 35
**GaussEliminate** command, 53, **A-6**
Gaussian elimination, 35–37, 42
    in *Mathematica*, 51–55
geometric vectors, **200**, 200–203
    addition, 201–202
    in *Mathematica*, 210–211
    scalar multiplication, 203
    subtraction, 202

Gram-Schmidt orthonormalization, 307–308, 311–324
graph, **15**
graph paper, 216–217
**Green** value, **A-4**

**HeadScale** option, 211, **A-6**

identity matrix, **99**
iff, **100**
image
    of a function, **185**
    of a linear transformation, 283
    of a matrix, 189, 191–192, 195, 242, 267, 303
    of a point, **68**, 185
**In** command, 10, **A-2**
inconsistent, *see* system, overdetermined
independent, *see* linearly independent
**IndependentQ** command, 237, **A-5**
inductive hypothesis, **264**
initial conditions, **360**
inner product, **289**
**InnerProduct** command, **A-9**
**Integrate** command, 295
**Inverse** command, 166
inverse
    computing a left-, 127
    computing a right-, 124–125, 127
    formula, 181
    left-, **118**, 192–193
    right-, **118**, 193–194
    transformation, **279**
    two-sided, **114**
invertibility, **116**
    left-, 132–135, 194, 238
    non-, 71, 100, 116, 136
    right-, 132–135, 194, 242
    two-sided, 132–135
isomorphism, **279**
Isomorphism Principle, 255–259

kernel
    of a linear transformation, 283

of a matrix, **189**, 190–192, 195, 238, 267, 299, 301, 303
**LeftInverse** command, **A-7**
linear
    combination, **108**, 212–215
    dependence relation, **224**, 228, 231–234
    equation, **19**
    operation, **19**
linear combination, **214**
linear transformation, *see* transformation, linear
linearly
    dependent, 192, **224**
    dependent on, **224**
    independent, **224**, 235–238
**ListDensityPlot** command, **A-2**
lower-triangular, **128**

**MakeLowerTriangular** command, **A-7**
**MakeMatrix** command, **A-7**
**MakeNiceMatrix** command, 153, **A-7**
**MakePermutationMatrix** command, **A-7**
**MakeReducedRowEchelon** command, **A-7**
**MakeRowEchelon** command, **A-7**
**MakeSquareSystem** command, 55, **A-7**
**MakeSystem** command, 61, **A-7**
**Map** command, 180
Markov process, **368**
matrices in *Mathematica*, 49–50, 87–90
matrix, **38**
    action on a point, *see* **Dot** command, 67
    addition, 77
    equality, 78, 79
    identity, **100**
    inverses, *see* inverse
    multiplication, 80–83
    partition, 38, **103**
    scalar multiplication, 78, 79
    sub-, *see* **SubMatrix** command, **103**

subscripts, 50, **79**
subtraction, 78, 79
transposition, *see* transposition
    operator
**MatrixForm** command, 50, **A-2**
**MatrixPicture** command, 69, 188, **A-8**
**MatrixPlot** command, 68, **A-8**
**MatrixPlotCircle** command, **A-8**
**MatrixPlotSquare** command, 69, **A-8**
**MatrixRank** command, 56, **A-7**
memory, freeing for
    the *Mathematica* Front End, 11
    the *Mathematica* Kernel, 10
minor determinant, 178
**Minors** command, 180
**MyArrow** command, 210, 211, **A-6**

**N** command, 9, 70, **A-3**
**Needs** command, 11–13, **A-3**
**NIntegrate** command, 175
**NonSingular** option, **A-7**
normal, **365**
**NSolve** command, **A-3**
**NullSpace** command, 246, **A-3**
nullspace, **246**

onto, *see* transformation, surjective
**Options** command, 62
orthogonal
    complement, **298**
    matrix, **322**
    set, **306**
    subspaces, **298**, 299
    vector and subspace, **298**
    vectors, **292**
**Orthogonalize** command, **A-6**
orthonormal set, **306**
orthonormalization, *see* Gram-Schmidt
    orthonormalization
**Out** command, 10, **A-3**
**OverDetermined** value, **A-7**

packages, *see* **Needs** command
parametric equation, **189**
**Part** command, 50, **A-3**
**Partition** command, 88

partition, *see* matrix partition
permutation, **146**
pivot, 42
**Plot** command, 6, **A-3**
**Plot3D** command, 18–19, **A-3**
**PlotJoined** option, 191, **A-9**
**PlotPoints** command, 191
**PlotPoints** option, **A-3**
**PlotRange** option, 17, **A-3**
**PlotStyle** option, **A-3**
plotting
    by contours, *see* **ContourPlot**
        command
    intersection points, 17
    multiple
        functions, 16
        three-dimensional graphs, 18
        two-dimensional contours, 18
        two-dimensional graphs, 17
    single functions, 6
**PLUFactors** value, **A-7**
**Plus** command, 9, **A-3**
**PointPlot** command, **A-9**
polynomial functions, 203–204
    addition, 203–204
    in *Mathematica*, 212
    scalar multiplication, 203–204
positive definite, **290**
**Power** command, 9, **A-3**
pre-image, **185**
Principle of Partitioned Matrices,
    106–107
**Print** command, 9, **A-3**
projection
    into a subspace, **304**, 305, 308, 309,
        324
    onto a vector, 293, **294**
projection matrix, **309**, 324
proof
    by cases, 97–98, 101, 116–117,
        132–135, 263, 265–268
    by contradiction, 159
    by induction, **262**, 263–264,
        320–322
    formal, 95–96

informal, 94–95
**PtLU** command, 152, **A-6**

Q.E.D., 95
**QoRo** command, **A-6**
**QR** command, 323, **A-6**
**QRDecomposition** command, 324
**QRFactors** value, **A-8**

**Random** command, 232, **A-3**
**RandomCombination** command, 232,
    **A-5**
**Randomized** value, **A-8**
**RandomVectors** command, 232, **A-5**
range, **68**
    notation for, 187
**Rank** option, 61, **A-8**
rank, *see* **MatrixRank** command, 56,
    90, 131, 134, 136, 194–195, 267
    column, **137**
    decomposition, *see*
        decomposition, rank
    row, **137**
rank matrix, **161**
**Red** value, **A-5**
**Remove** command, 12
**ReplaceAll** command, **A-3**
**Reverse** command, 180, **A-3**
**RightInverse** command, **A-8**
row
    echelon, **41**
    operation, **38**
    operations in *Mathematica*, 51–52
    reduced echelon, **46**
row space, **250**, 299, 303
**RowReduce** command, 53, **A-4**
**Rule** command, **A-4**

scalar multiplication, *see* column,
    geometric, matrix, or
    polynomial
**Set** command, 7, **A-4**
**SetDelayed** command, 8, **A-4**
**Show** command, 18, **A-4**
**ShowBitmap** command, **A-9**
**ShowVectors** option, **A-9**

similar, **347**
**Sin** command, 70, **A-4**
singular, 100, **116**
solution, **15**
    checking with *Mathematica*, 91
    from a left-inverse, 114
    from a right-inverse, 118
    general, 59–61
    set, **15**
solutions
    existence of, 132, 190, 242, 283
    structure of, 189, 283, 303
    uniqueness of, 132, 190, 238
**SolutionType** option, 61, **A-8**
**Solve** command, 30, **A-4**
solving
    parallel systems, 123–125
    series systems, 156–158
span, **191**, **223**, 242
spanning set, 238–242
**SpanningPicture** command, 192, **A-9**
**SpanningQ** command, 241, **A-5**
spans, **191**, *see* spanning set
**Sqrt** command, 16
standard position, **211**
state, **359**
Steinitz Exchange lemma, 263
**SubMatrix** command, 106–107, **A-5**
**Subscript** command, **A-4**
**Subscripted** command, 166, **A-4**
subscripts, *see* matrix subscripts
subscripts in *Mathematica*, 50, *see* **Part**
    command
subspace, **272**
**Subtract** command, 9, **A-4**
**Sum** command, **A-4**
**Superscript** command, **A-4**
**SwitchRows** command, 52, **A-8**
symmetric, **128**
system, **17**
    overdetermined, **60**, 131, 305
    underdetermined, **60**, 131, 302
    uniquely determined, **59**, 131

**Table** command, 232, **A-4**

**TableForm** command, 355, **A-4**
**TailWidth** option, 211, **A-6**
**Times** command, 9, **A-4**
**Tolerance** option, **A-6**
transformation
    bijective, **132**, 135
    injective, **132**, 135, 238, 303
    linear, **276**
    matrix representation, 73–75, **332**,
        337
    nonlinear, 71
    surjective, **132**, 135, 242, 303
transition equations, **354**
transition matrix, **342**, 343, 345
**TransitionMatrix** command, 343, **A-6**
**Transpose** command, 88, **A-4**
transposition
    matrix, **146**
    operator, *see* **Transpose** command,
        100–103
transposition operator, **84**, 136, 267

**UnderDetermined** value, 61, **A-8**
**Unital** option, **A-8**
upper-triangular, **128**

**ValueRange** option, **A-8**
**ValueType** option, **A-8**
variable
    basic, **46**
    free, **46**
**Vector** command, 211, **A-6**
vector space, 199–204, **206**, 256–258
**VectorLength** command, **A-9**
vectors, **206**
    column, 108, *see* column vectors,
        **200**
    geometric, *see* geometric vectors
    in *Mathematica*, 209–212
    polynomial, *see* polynomial
        functions
**Verbose** option, 153, 323, **A-6**
**ViewPoint** option, 191
visualizing
    a general function, 187
    a linear function, 189
    a one-variable function, 184

zero matrix, 99